Plant Gene Research

Basic Knowledge and Application

Edited by

E. S. Dennis, Canberra

B. Hohn, Basel

Th. Hohn, Basel (Managing Editor)

P. J. King, Basel

J. Schell, Köln

D. P. S. Verma, Columbus

Springer-Verlag Wien New York

Temporal and Spatial Regulation of Plant Genes

Edited by D. P. S. Verma and R. B. Goldberg

Springer-Verlag Wien New York

Prof. Dr. Desh Pal S. Verma
Biotechnology Center, Ohio State University, Columbus

Prof. Dr. Robert B. Goldberg
Department of Biology, University of California, Los Angeles

With 55 Figures

Softcover reprint of the hardcover 1st edition 1988

Library of Congress Cataloging-in-Publication Data

Temporal and spatial regulation of plant genes /
edited by Desh Pal S. Verma and
Robert B. Goldberg.
p. cm. -- (Plant gene research. ISSN 0175-2073)
Includes bibliographies and index.

1. Genetic regulation. 2. Plant molecular genetics.
I. Verma, D. P. S. (Desh Pal S.), 1944–
II. Goldberg, Robert B. III. Series.
QK981.4.T46 1988 581.87'328--dc19 88-19965

ISSN 0175-2073
ISBN-13: 978-3-7091-7448-7 e-ISBN-13: 978-3-7091-6950-6
DOI: 10.1007/978-3-7091-6950-6

Preface

First attempts to isolate plant genes were for those genes that are abundantly expressed in a particular plant organ at a specific stage of development. However, many important gene products are produced in a very minute quantity and in specialized cell types. Such genes can now be isolated using a variety of approaches, some of which are described in this volume. The rapid progress during the last decade in regeneration of a number of crop plants and the availability of molecular tools to introduce foreign genes in plants is allowing the engineering of specific traits of agricultural importance. These genes must, however, be regulated in a spatial and temporal manner in order to have desired effects on plant development and productivity.

The habitat of plants necessitate adaptive responses with respect to the environmental changes. Starting from germination of the seed, the plant begins to sense environmental cues such as moisture, light, temperature and the presence of pathogens, and begins to respond to them. Little is known about various signal transduction pathways that lead to biochemical and morphogenetic responses, in particular, transition from vegetative to reproductive phase. With the availability of tools to generate specific mutations *via* transposon tagging, identification and isolation of genes affecting these processes may be facilitated. Transfer of these genes into heterologous environments will allow understanding of the complex processes that control plant development. This volume outlines regulation of expression of some of the genes expressed in specific plant organs and tissue types from seed germination to fruit ripening, and the manner in which these genes respond to some environmental signals. Several chapters are devoted to the reproductive phase of plant life, where application of molecular biology is providing some new insight. Two examples of transgenic plants carrying useful traits, herbicide resistance and protection from viral infection, are presented in this volume.

The observations that the heterologous genes can function in a tissue-specific manner in transgenic plants suggest that the signals responsible for this regulation arc highly conserved. Attempts are being made in a number of laboratories to identify and isolate genes encoding such *trans*-acting factors that regulate plant genes in a spatial and temporal manner. This may help our understanding of the signal transduction mechanisms in plants. The next volume in this series will concentrate upon the chloroplast and other cellular organelles and their communication with the nucleus.

Columbus and Los Angeles,
June 1988

D. P. S. Verma and
R. B. Goldberg

Contents

Chapter 6 **Anther- and Pollen-Expressed Genes**
J. P. Mascarenhas, Albany, N.Y., U.S.A.

Chapter 7 **Self-Incompatibility Genes in Flowering Plants**
E. C. Cornish, J. M. Pettitt, and A. E. Clarke, Melbourne, Victoria, Australia

Chapter 1

Arabidopsis as a Tool for the Identification of Genes Involved in Plant Development

Ruth Finkelstein[1], Mark Estelle[2], Jose Martinez-Zapater[1], and Chris Somerville[1]

[1] MSU-DOE Plant Research Laboratory, Michigan State University, East Lansing, MI 48824; [2] Biology Dept., Indiana University, Bloomington, IN 47405, U. S. A.

Contents

I. Introduction

The sessile lifestyle of plants requires that they be able to alter their growth in order to adapt to environmental changes. Much research has been devoted to determining how plants perceive and respond to environmental cues such as light intensity and quality, photoperiod, gravity, water stress, and temperature. In the case of light-regulated phenomena, various photoreceptors have been identified whose absorption maxima coincide with regions of the action spectra of the observed phenomena. However, the transduction pathways from perception to cellular response have not been elucidated. Similarly, much correlative evidence suggests that phytohormones are involved in transducing environmental signals into biochemical or morphological effects. In this case neither the elements linking the environmental effects with changes in hormone levels or sensitivity nor those mediating the hormonally induced responses have been identified. Furthermore it is not known whether the correlations between environmental or hormonal signals and the observed responses reflect a causal, direct relationship or if they represent parallel or overlapping paths toward the same result.

Despite their developmental plasticity, the ways in which plants respond to their environment are constrained by their genetic makeup. Neither the nature of these intrinsic controls nor the mechanisms by which they interact with environmental signals are known, but we can ask a number of general questions to direct our dissection of these control mechanisms. Do plants have regulatory loci which control the conversion of the apical meristem from vegetative to reproductive growth? Are the effects of flowering-inducing treatments, such as appropriate photoperiod or vernalization, mediated by products of these loci or do they act through alternate pathways, possibly overriding intrinsic controls of the putative regulatory loci? Do regulatory loci, analogous to the homoeotic loci of insects, control determination of primordia to develop into specific organs? A variety of homoeotic-like mutations affecting floral development have been identified [e. g. teopod in maize (Poethig, 1986), agamous in *Arabidopsis* (Koornneef, 1982)] but we do not know if these affect regulatory loci per se or metabolic functions. Plant development is much more flexible than that of animals and it is very possible that the mechanisms of regulation of determination and differentiation are also different. Therefore, we should be careful not to try to force our interpretation of plant development into unjustified parallels with animal models.

A mutational approach to development offers several benefits. Mutants allow us to define the individual components of the system and to analyze the role of each component separately. Furthermore, by marking relevant genes, mutants may provide a way to clone the genes and identify the gene products involved in regulating development. However, to use this approach we must be able to distinguish between mutations affecting genes that regulate development and those required for growth. Because alterations in basic metabolism can have pleiotropic effects on development it can be difficult to deduce the molecular basis of a morphological mutation. For example, polyamine mutants of tobacco have dramatically altered flower morphology, such as ovaries filled with anthers (Malmberg and McIndoo, 1983), which could be interpreted as homoeotic effects except that these mutations have numerous adverse effects on plant growth. One way to avoid this problem is to work forwards from mutations which affect the amount or activity of phytohormones and photoreceptors, known to be involved in various developmental processes. Analysis of these mutants should help to define the role of the affected loci in controlling development. In theory, by isolating a series of mutants affecting a developmental process it should be possible to determine the epistatic relationships and thereby dissect the transduction pathway. However, unlike most biochemical pathways, where it is possible to propose a sequence of chemical events, we have almost no idea what the intermediate steps might be in a developmental pathway. Thus, in order to pursue this approach, we need mutants with differential effects on the final response, allowing us to determine which effect is dominant in the recombinant, (i. e. which is epistatic). This technique has been used successfully to deduce the order of gene action for *Drosophila melanogaster* homoeotic loci (Kaufman and Abbott, 1984).

Ultimately, to fully dissect the components of a developmental process, we need to be able to identify the affected gene, its product and its function. Various techniques which have been used to identify and isolate eukaryotic genes whose products are unknown include cloning by complementation (Henikoff *et al.*, 1981), chromosome walking to a region delineated genetically (Bender *et al.*, 1983), and tagging by insertional mutagenesis with a transposon (Bingham *et al.*, 1981; Federoff *et al.*,1984) or some other foreign DNA (Schnieke *et al.*, 1983). The large genome size and relatively low transformation efficiency of most plants makes cloning by complementation impractical. Similarly the large component of repetitive DNA in most plant genomes complicates chromosome walking in the few plants with adequate genetic maps, such as maize or tomato.

The small crucifer *Arabidopsis thaliana* offers many advantages as a model organism for a molecular analysis of plant development. Its small size and short generation time permit rapid isolation and characterization of mutants. It has a small genome with relatively little repetitive DNA (Leutwiler *et al.*, 1984), making chromosome walking feasible. Finally, numerous developmental mutants of *Arabidopsis* have already been isolated and partially characterized (Koornneef, 1983). In this review we have attempted to summarize the available information concerning mutations in *Arabidopsis* affecting hormone levels and response, tropic responses to environmental stimuli such as light and gravity, and induction of flowering by chilling and photoperiod. The morphological mutants and the embryo lethals have recently been reviewed (Haughn and Somerville, 1987).

II. Phytohormone Mutants

A. Introduction

Analysis of the mechanism of hormone action is complicated by several unknowns: (a) the site(s) of action for a given hormone, (b) the compartmentation of a hormone under various conditions, and (c) which of absolute levels, compartmentation or sensitivity to a hormone is more significant in regulating plant development (Trewavas, 1981). The physiological analysis of the role of phytohormones during growth and development relies on correlations between levels of endogenous or applied hormones and an observed response. These correlations may be tested by genetic analysis,

Table 1. Summary of phytohormone mutants

Gene symbol revised[a]	Gene symbol original[b]	Phenotype	Reference
aux1	aux	2,4-D and IAA resistant altered root development	Maher and Martindale, 1980
axr1	axr-1	2,4-D and IAA resistant decreased apical dominance, root branching, stamen length, and geotropism	Estelle and Somerville, 1986
dwf	Dwf	2,4-D and IAA resistant dwarf, reduced geotropism	Maher and Martindale, 1980
er1	ER1	resistant to applied ethylene	Bleeker, 1987
ga1	ga-1	GA responsive dwarf, some alleles non-germinating	Koornneef and van der Veen, 1980
ga2	ga-2	same as *ga1*	Koornneef and van der Veen, 1980
ga3	ga-3	same as *ga1*	Koornneef and van der Veen, 1980
ga4	ga-4	GA responsive dwarf	Koornneef and van der Veen, 1980
ga5	ga-5	same as *ga4*	Koornneef and van der Veen, 1980
gai	Gai	GA-insensitive dwarf	Koornneef *et al.*, 1985
aba	aba	ABA deficient, non-dormant, wilty	Koornneef *et al.*, 1982
abi1	Abi-1	ABA-resistant, non-dormant, wilty	Koornneef *et al.*, 1984
abi2	abi-2	ABA-resistant, non-dormant, wilty	Koornneef *et al.*, 1984
abi3	abi-3	ABA-resistant, non-dormant	Koornneef *et al.*, 1984

[a] The revised nomenclature follows the guideline suggested by an *ad hoc* committee at the Third International *Arabidopsis* Meeting, Michigan State University, April 1987 (Haughn and Somerville, 1987).

[b] In the original nomenclature dominant mutations were capitalized.

using mutants that are either deficient in, or insensitive to, a given class of phytohormone. The insensitive mutants should be particularly useful in helping us to understand the mechanism of hormone action. Several classes of mutants could give rise to an insensitive phenotype: receptor mutants, mutants with alterations further downstream in the transduction pathway(s), or mutants affecting uptake or metabolism. However, as with the physiological studies, interpretation of mutant analysis of hormone metabolism and action is complicated by interactions between hormones. For instance, because auxin stimulates ethylene synthesis, an auxin-insensitive mutant may also have reduced ethylene responses. It is also possible that multiple parallel or overlapping pathways result in the same effect. Thus, although a hormone-deficient mutant may still be capable of a response thought to be mediated by that hormone, one can conclude only that the hormone is not required for the effect, not that it is not involved. Furthermore, several of the phytohormones may be essential for plant growth and development, making it impossible to isolate null mutations affecting either synthesis or response. It should be possible to circumvent the latter problem by isolating conditional mutants. The phytohormone mutants of *Arabidopsis* are listed, along with summaries of their phenotypes, in Table 1. To date, mutations affecting synthesis or sensitivity have been isolated for all the major plant growth regulators except cytokinin.

B. Auxin

The major naturally occurring auxin, indole-3-acetic acid, was identified in 1924 by Went and has since been shown to influence a variety of growth processes including cell expansion during elongation (Jacobs and Ray, 1976), apical dominance (Phillips, 1975), root initiation (Blakely *et al.*, 1972), and tropic behavior (Shen-Miller, 1973). In addition, auxin is essential for growth of cells or tissues in culture, whether supplied exogenously (Jablonski and Skoog, 1954) or endogenously in tumors (Cheng, 1972). The diversity of these growth processes, and in particular the auxin requirement in tissue culture, suggests that auxin plays a fundamental role in plant growth and development and is probably essential for viability. Unfortunately our knowledge of the role of auxin in plant development does not extend beyond this general statement. The extent to which auxin directly regulates any particular aspect of growth is still controversial (Trewavas, 1981) and the biochemical details of auxin biosynthesis as well as mode of action are largely unknown.

Because auxin appears to be essential for plant growth, a mutation which eliminates either an auxin biosynthetic activity or a function directly required for auxin action will likely be lethal. Although it is theoretically possible to isolate conditional mutants, the identification of this class of mutants may be difficult since it is not clear that exogenous application of auxin will rescue an auxin-deficient phenotype. An alternative approach is to isolate mutants which are resistant to exogenous application of auxin.

In *Arabidopsis,* supraoptimal concentrations of auxin severely inhibit the growth of newly germinated seedlings. Auxin resistant mutants were first isolated by Maher and Martindale (1980) by selecting plants whose roots grew into perlite which had been moistened with medium containing a growth-inhibiting concentration (2 μM) of the artificial auxin 2,4-dichlorophenoxy-acetic acid (2,4-D). In the best characterized mutant line, resistance was shown to be due to a single recessive mutation named *aux1*. The *aux1* mutation confers an approximately 14 fold higher level of resistance to both IAA and 2,4-D. In addition, *aux1* mutants display altered root development: the roots have an increased rate of extension growth and are not geotropic. The majority of the mutant lines isolated were reported to have a phenotype similar to *aux1*. However in the second auxin-resistant line examined in detail, resistance was due to a single dominant mutation named *dwf*. This mutation was apparently lethal when homozygous but in the heterozygous condition confers resistance to approximately 1000-fold higher concentrations of 2,4-D relative to wild type. *Dwf* plants are severely stunted and, like *aux1* plants, display defects in geotropic response.

More recently, another auxin-resistant locus, named *axr1*, has been identified by selecting for resistance to 5 μM 2,4-D (Estelle and Somerville, 1987). Recessive mutations of *axr1* confer resistance to both 2,4-D and IAA. In addition *axr1* mutants exhibit a number of morphological aberrations including an apparent decrease in apical dominance, a reduction in root branching, a reduction in stamen elongation which results in self-sterility in some alleles, and a dramatic decrease in both shoot and root geotropic response.

Although a complementation analysis between the *axr1* mutants and either *aux1* or *dwf* has not been done, on the basis of phenotype it appears that the mutations define at least two and perhaps three genes which are involved in the responses to exogenous auxin. The *aux1* gene appears to function primarily in the roots since a mutant phenotype is evident only in the roots. The other two genes, *axr1* and *dwf*, apparently function in many and perhaps all tissues of the plant since mutations in each have an extremely pleiotropic phenotype. The significance of these patterns is unclear without more information on the biochemical function of the products of these genes. However, it appears that some components of the cellular machinery involved in auxin action may be tissue specific while others may function throughout the plant.

The primary biochemical defect has not been determined for any of the auxin-resistant mutants of *Arabidopsis*. Therefore, we cannot yet link any aspect of a mutant phenotype to a specific defect in auxin biology. However, many of the developmental defects observed in auxin-resistant mutants involve processes thought to be mediated by auxin. One of the most striking examples of this is the loss of normal geotropic response in the auxin-resistant mutants. According to the Cholodny-Went hypothesis (Went and Thimann, 1937) the differential growth of the upper and lower sides of a gravity stimulated root or shoot is due to an asymmetric distribution of auxin. Recent attempts to use a sensitive immunoassay for IAA have failed to detect any asymmetry in auxin distribution, bringing into question the involvement of auxin in geotropism (Merten and Weiler, 1983). However, all of the auxin-resistant mutants identified in *Arabidopsis* display defects in geotropism, indicating a role for auxin in this process. In addition, auxin resistance has been documented in geotropically defective mutants of tomato (Kelly and Bradford, 1986) and barley (Tagliana *et al.*, 1986).

C. Ethylene

Ethylene is a simple olefin which exists as a gas under normal physiological conditions. The growth regulating activity of ethylene was first described in 1901 by Neljubow. Since then, the exogenous application of ethylene has been shown to influence a large number of growth processes ranging from germination to senescence. The pathway for endogenous production of ethylene has been characterized (Yang and Hoffman, 1984) and is found to be most active in ripening fruit, senescing tissue, and in tissues that have been exposed to an environmental stress such as wounding. The precise role of the ethylene produced at these times has not been determined and the molecular mechanisms of ethylene action are unknown.

Several mutants of *Arabidopsis* have recently been isolated which display a reduced sensitivity to exogenous ethylene (Bleeker *et al.*, 1987). These mutants were recovered by germinating seeds from a mutagenized population in the dark in an atmosphere containing 5 ppm ethylene. When wild type seedlings are exposed to these conditions, both hypocotyl and root growth are inhibited and the cotyledons do not unfold. Two ethylene resistant mutants were recovered in this screen. In one of these mutants, ethylene resistance is conferred by a dominant mutation called *er1*. Plants which are either homozygous or heterozygous for the *er1* mutation are completely insensitive to exogenous ethylene. The ethylene responses so far examined include inhibition of seedling root and shoot growth, unfolding of the cotyledons, inhibition of inflorescence growth, and the ethylene promotion of chlorophyll degradation in excised leaves. The second ethylene-resistant mutant has not been characterized. Despite the dramatic and apparently complete loss of ethylene sensitivity, the *er1* mutants are vigorous and display no gross abnormalities in growth, flower development or maturation of fruit. These results suggest that in *Arabidopsis* grown under controlled laboratory conditions, the ability to respond to ethylene is not essential for normal growth and development.

The *er1* mutant and other similar mutants with lesions affecting ethylene biosynthesis and response may be extremely useful in the study of the role of ethylene. Although ethylene does not appear to directly regulate growth of *Arabidopsis* grown under optimum conditions it may play an important role during plant growth in natural conditions where plants are exposed to a number of environmental stresses. In addition, the availability of these mutants should facilitate the identification of the proteins involved in sensing and responding to changes in ethylene concentration.

D. Gibberellins

Gibberellins (GAs) were discovered as a result of their capacity to induce stem elongation of rice plants (Kurosawa, 1926). Initially, gibberellins were isolated from fungal cultures (Yabuta and Sumiki, 1938; Stodola *et al.*, 1955; Brian *et al.*, 1954). In the 1950s GA-like substances were isolated from higher plants by West and Phinney (1956), working with seeds and fruits of several angiosperms, and Radley (1956), working with pea seedlings. Since then gibberellins have been identified in a wide range of species and are thought to be ubiquitous in higher plants. Chemical characterization of gibberellins has identified 72 different forms (Sponsel, 1987). The biological activity of the various forms differs depending on the response observed and the species tested. In addition, the distribution of the different gibberellins varies both temporally and spatially within a given plant, as well as among species.

Physiological studies have suggested that GAs are involved in stimulating cell division at the shoot apices (Sachs *et al.*, 1959), stem elongation in young seedlings and mature plants (Ecklund and Moore, 1968), periods of rapid growth during embryogenesis (Murakimi, 1961), and production of hydrolytic enzymes required for mobilization of seed storage reserves immediately following germination (Jones, 1973), as well as many other developmental processes. In addition to problems in interpretation due to the unknowns of hormone uptake efficiency and compartmentation, the diversity of GA activity and distribution further complicates analysis of GA function.

A genetic approach can avoid some of these problems. Indeed, much of the analysis of GA function has involved comparison of dwarf and normal cultivars of various species. In many cases the dwarf cultivars were found to have reduced levels of GAs and studies with dwarf mutants of rice, pea, and maize (Phinney, 1984) have done much to elucidate the GA biosynthetic pathway and interrelationships among GAs. For example, analysis of the maize dwarf mutants has shown that GA_1 is the only GA active for inducing shoot elongation. Theoretically, with a complete set of mutants producing blocks at all steps of biosynthesis, it should be possible to analyze the role of a specific GA in a particular process.

GA sensitive mutants of *Arabidopsis* were initially isolated as dark green, sterile dwarves whose phenotype could be converted to wild type by repeated application of GA (Koornneef and van der Veen, 1980), similar to GA-deficient mutants in other species. By screening M^2 families, these workers also isolated some mutants which required exogenous GA for germination and, without further GA application, developed into dark green dwarves. The GA sensitive mutants were isolated at a frequency of 6 per 1000 M^2 families. All were monogenic recessives and could be placed into one of 5 complementation groups, 3 of which were comprised of both germinating and non-germinating dwarves (*ga1, ga2, ga3*), while mutants at the other 2 loci (*ga4* and *ga5*) were all germinating dwarves. The fact that non-germination, unlike dwarfism, is not always associated with these mutations implies that germination may require less GA than is necessary for elongation growth. The physiological characterization of mutations at *ga1, ga2* and *ga3* suggests that the products of these loci are involved in GA biosynthesis. Attempts to measure GA levels in these mutants by bioassays of HPLC fractionated extracts showed no detectable GA-like activity (Zeevaart, 1983) and subsequent studies tracing metabolism of radioactive precursors have shown that the blocks are early in the pathway (Zeevaart, 1986). In contrast, HPLC fractionation of extracts from *ga4* and *ga5* revealed an almost wild type profile of GA activity, with only one fraction reduced, suggesting that these loci control steps later in the GA biosynthetic pathway.

More recently, Koornneef *et al.* (1985) isolated another class of dwarf mutant similar to the *ga* mutants in that stem growth, apical dominance, and seed germination were all reduced. Unlike the *ga* mutants, this mutation *(gai)* is dominant and does not affect floral development. The GA profile of this mutant was similar to that of wild type plants and the dwarf phenotype was not reversible by application of GA, suggesting that the defect might be in recognition or transduction of the GA signal. Because even wild type plants show only a small response to applied GA, the *gai* mutation was introduced into a line carrying the *ga1* mutation to produce a doubly mutant line (i. e. *gai ga1*). Because the *ga1* mutant exhibits a pronounced response to applied GA, the double mutant should provide a more sensitive background for bioassays. GA sensitivity was tested in terms of the effects of applied GA on stem elongation, apical dominance and germination. For all responses tested, both the *gai* and the *gai ga1* double mutant were insensitive. The GA responses of a mutant with a reduced number of GA receptors would be expected to saturate at lower GA concentrations than wild type and show only slight differences at low concentrations. Although saturation was not shown for either wild type or *gai* plants, the observation that the *gai* mutant showed a reduced shoot elongation response to a wide range of GA concentrations was interpreted to mean that this mutation did not affect the number of GA receptors.

The non-germinating GA-responsive mutants (e. g. *ga1*), have been used to examine the effects of GA on seed germination (Koornneef *et al.*). Analysis of GA effects on germination is complicated by the fact that time in storage and environmental factors, such as light and cold treatments, also affect germination. Therefore, germination was assayed in either light or dark, with or without previous cold treatment (imbibition for 7 days at 2° C). Recently harvested wild type seed of some races (e. g. Landsberg) is deeply dormant and requires GA (at least 10 μM GA_{4+7}) for germination in darkness. In contrast, concentrations of GA_{4+7} up to 1 mM did not induce germination of the *gai* mutant in darkness. Thus the *gai* mutant is at least 100-fold less sensitive than the wild type for GA-induced dark germination. This mutant is not completely GA-insensitive because applied GA will induce germination of *gai* seeds in the light. However, they still require at least 10-fold more GA than the wild type for germination in light. Both the GA-insensitive and GA-deficient mutants require less GA to induce germination in light than in darkness, leading Karssen and Lacka (1985) to suggest that light results in both increased GA synthesis and sensitivity. Light-induced GA synthesis is also indicated by the lack of light-induced germination when GA synthesis is blocked by tetcyclasis (Karssen, pers. comm.). Because illumination results in decreased dormancy, the effects of cold treatment are most dramatic for seeds maintained in darkness. Although cold treatment alone did not permit germination of wild type seeds in darkness it did result in increased sensitivity to applied GA. In contrast, cold treatment had no effect on GA sensitivity of *gai* seeds. Finally, *gai* seed which has been stored for more than one year has reduced dormancy, as does wild type, and therefore requires less GA for germination. The biochemical basis of the change in GA sensitivity following long-term storage or exposure to light or cold is not known. Indeed, it is not clear whether these treatments have the same primary effects since the *gai* mutant distinguishes between these signals. Thus, the increase in GA sensitivity following storage or exposure to light suggests that there are multiple pathways involved in GA-induction of germination.

E. Abscisic Acid

Abscisic acid was discovered more or less simultaneously by Addicott and coworkers, studying substances which accelerate leaf abscission in cotton, and Wareing and collaborators, working on inhibitors involved in bud dormancy of woody plants. Comparison of the molecular properties of “abscisin II” and “dormin” revealed that they were the same compound (Addicott *et al.*, 1968), henceforth known as abscisic acid (ABA). ABA is generally regarded as a growth inhibitor, because it is associated with dormancy of both seeds (Black, 1983) and buds (Wareing and Phillips, 1983), inhibition of cell elongation, and decreased RNA and protein synthesis in some tissues (Walbot, 1978). In addition, it appears to be involved in many stress responses, such as stomatal closure in response to water stress (Van Steveninck, 1983; Davies and Mansfield, 1983) and induction of freezing tolerance (Chen and Gusta, 1983). It is also correlated with a variety of developmental processes: embryogenesis (Walbot, 1978), transition from submerged to aerial growth of aquatic weeds (Anderson, 1978) and transition from juvenile to adult growth of woody perennials (Rogler and Hackett, 1975).

Much physiological and genetic evidence in a variety of species suggests that ABA plays a role in controlling seed dormancy. For example, viviparous mutants of maize are ABA-deficient or insensitive (Brenner *et al.*, 1977; Robichaud *et al.*, 1980), endogenous ABA is often high in dormant seeds (Taylorson and Hendricks, 1977), treatments which remove endogenous ABA can release dormancy (Ackerson, 1984) and application of exogenous ABA at physiological concentrations can inhibit germination (Black, 1983). Therefore, one would predict that mutants defective in ABA synthesis or response could be selected at the level of germination. Initially, ABA deficient mutants were isolated using an elegant selection taking advantage of the non-germinating phenotype of some of the GA deficient mutants. Because ABA and GA have been shown to have antagonistic effects on many processes, including germination, Koornneef *et al.* (1982) reasoned that germinating revertants of the non-germinating GA-sensitive dwarves might be altered in ABA metabolism or response. When these secondary mutations were allowed to segregate from the mutation producing the GA-deficiency, they were found to map to a single locus, designated *aba.* Seeds of the *aba* mutants contain from 2—10 % of wild type ABA levels and leaves contain 15—30 % of wild type levels, depending on the allele (Koornneef *et al.,* 1982). In addition to being nondormant the *aba* mutants had poor regulation of water relations and tended to be wilty and stunted — effects consistent with the known role of ABA in regulating stomatal closure. Application of exogenous ABA reduced the observed wilting but did not restore dormancy.

The ABA-deficient mutants have been used to analyze the source and role of ABA in seed dormancy (Karssen *et al.,* 1983). Endogenous ABA levels peak twice during development of wild type seeds, but remain low throughout seed development in the *aba* mutant. Comparison of the developmental timing of these peaks and the capacity for precocious germination showed that the onset of dormancy correlated with the presence of ABA in wild type seeds and that dormancy was maintained even after the endogenous ABA decreased. These observations show that high levels of ABA need not be continuously present to induce and maintain the inhibition of germination associated with dormancy. It is not known how this temporal separation between the presence of ABA and its effect is mediated. Analysis of plants in which either the maternal tissue or the embryo, but not both, were homozygous for the *aba* mutation indicated that the two peaks of ABA measured in wild type seeds corresponded to a large maternal ABA fraction midway through seed development and a smaller embryonic fraction extending somewhat later in development. Tests of germination capacity showed that the embryonic ABA was required for induction of dormancy. Neither maternal ABA nor applied ABA could induce dormancy in embryos of the *aba* mutant, but the levels of maternal ABA correlated with the degree of dormancy, reflected in the period of after-ripening required to increase germination in light.

More recently, ABA-insensitive mutants were selected for their ability to germinate on perlite saturated with 10 μM ABA (Koornneef *et al.,* 1984). These could be broken into 3 complementation groups: *abi1*, *abi2* and *abi3*. ABA sensitivity was characterized in terms of effects on seed dormancy, seedling growth and water relations of the mature plant. Mutations at all 3 loci result in reduced dormancy and reduced inhibition of seedling growth by applied ABA. As in the *aba* mutants, although all 3 classes of *abi* mutants exhibit reduced dormancy, none are viviparous under normal conditions (i. e., all desiccate and enter developmental arrest). However, analysis of double mutants with the *aba* mutation have shown that in one of these combinations, *aba abi3,* the seeds fail to complete maturation in that they neither dry nor become desiccation tolerant (Karssen, pers.

comm.). Interestingly, it is maternal ABA which is required for induction of desiccation and developmental arrest, indicating that although dormancy and developmental arrest have the same phenotype (non-germination), the tissue which induces the response is different. The *abi1* and *abi2* mutants are prone to wilting, but the *abi3* mutants appear to have normal water relations. Measurements of endogenous ABA in seeds showed that levels in the *abi* mutants are at least as high as in wild type (Koornneef *et al.*, 1984), indicating that increased turnover is probably not the cause of insensitivity. Surprisingly, although application of ABA can reduce withering (wilting of leaves and stems, shriveling of pods) of wild type and ABA-deficient plants, it actually increased withering of the *abi1* and *abi2* mutants. (The *abi3* plants, having apparently normal water relations, were not tested in these experiments.) The reason for this anomalous response to ABA is not known. One possibility is that exogenous ABA could stimulate turnover of endogenous ABA, which is marginally effective in regulating stomatal aperture, and the applied ABA never reaches a compartment where it could induce stomatal closure. This hypothesis is consistent with the observations that applied ABA does not equilibrate with endogenous ABA in isolated mesophyll cells of *Xanthium* (Bray and Zeevaart, 1986) and is metabolized more rapidly than endogenous ABA in turgid *Xanthium* leaves (Creelman *et al.*, 1987).

Recently, Karssen and Lacka (1985) examined the interaction between gibberellin and ABA effects on seed dormancy by using genotypes with various combinations of GA and ABA deficient alleles. They found that endogenous ABA is required for induction of dormancy and that the ability to synthesize GA is irrelevant for this process. Futhermore, although GA is required to permit germination of dormant (wild type) seeds, it is GA sensitivity which controls the release from dormancy, not the ability to synthesize GA. Finally, measurements of endogenous ABA and GA showed that the endogenous levels of these two hormones peak at different stages in seed development. The fact that these hormones are neither present nor active simultaneously in seeds led the authors to conclude that the hormone balance theory of GA/ABA antagonism is incorrect and that the ABA effect on GA requirement for germination is indirect.

III. Environmental Regulation of Growth and Development

A. Introduction

The environment regulates many aspects of plant growth and development. Light and gravity affect the orientation of growth and, in many species, photoperiod regulates flowering, allowing these plants to use the least variable environmental cue as a signal for committment to reproductive development. Both the quantity and quality of light affect germination and hypocotyl elongation during early seedling growth. Extended low temperature treatments may be required to break seed dormancy (stratification) or permit/induce flowering (vernalization). Although various photoreceptors and phytohormones have been implicated, in no case have all components of the transduction pathway been indentified. As with phytohormones, mutational analysis could help clarify the functions of those components already identified and perhaps identify the intermediates in the transduction pathway.

B. Tropic Responses

Over 100 years ago Darwin observed and experimentally manipulated plants tropic responses to light and gravity (1881). Since then, much effort has gone into identifying the receptors for these environmental cues as well as the mechanisms of differential growth which produce the tropic curvature.

Although it is generally thought that gravity perception involves displacement of some subcellular component(s), these statoliths have not been unequivocally identified. A popular candidate for this function is the starch grain since amyloplast sedimentation has been correlated with geotropic responses of many plants (Volkmann and Sievers, 1979). However, the strong geotropic response of a starchless mutant of *Arabidopsis* indicates that starch grains are not required for the perception of gravity (Caspar and Somerville, 1986).

In the case of phototropism, plants must perceive the presence, quantity and location of a light source. Measurements of action spectra have indicated that the receptor involved absorbs blue light, but the pattern shows characteristics of absorption spectra for both carotenoids and flavins (Curry and Gruen, 1959). Studies with carotenoid deficient mutants of maize (Bandurski and Galston, 1951) and the phototropic fungus *Phycomyces* (Presti *et al.*, 1977) show that phototropism is not reduced at all or not to the same extent as the carotenoids, indicating that a carotenoid is probably not the photoreceptor. However, the carotenoid-like aspect of the action spectrum can be at least partly attributed to a secondary role of carotenoids in phototropism, giving information about the location of a light source by acting as an internal light screen to create a light gradient across a photostimulated organ (Vierstra and Poff, 1981). Initial biochemical and genetic studies with *Phycomyces* suggested that riboflavin was the receptor (Delbruck *et al.*, 1976). However, more recent experiments indicate that more than one photoreceptor is involved (Galland and Lipson, 1985).

Although the initial environmental signal is different for photo- and gravitropism, it is presumed that much of the transduction pathway involved in producing differential growth across a shoot or root is shared by these responses. A great deal of research has been devoted to examining the roles of various regulatory and messenger molecules in this process, especially auxin, ABA and more recently Ca^{++} (Pickard, 1985). In spite of much correlative evidence between asymmetric distribution of these substances and differential growth, the question remains whether redistribution of any of these occurs sufficiently rapidly to induce differential growth (Firn and Digby, 1980). Genetic evidence of a role for auxin in mediating geotropism is provided by ageotropic behavior of auxin resistant mutants of *Arabidopsis* and reduced auxin sensitivity of ageotropic mutants of tomato and barley. A role for ABA in root geotropism seems unlikely because of the normal geotropic response of maize in which ABA biosynthesis was blocked either genetically or chemically (Moore and Smith, 1984).

Another way to approach this question is to select mutants with reduced photo- or geotropism (photo$^-$ and geo$^-$, respectively) and attempt to determine the epistatic relationships between them. Working with *Arabidopsis,* Poff *et al.* (1987) have isolated mutants in 3 classes: photo$^-$ geo$^+$, photo$^+$ geo$^-$, and photo$^-$ geo$^-$. Presumably the first two classes affect early steps in transduction of the light or gravity stimulus, respectively, while the last class affects steps common to both processes. The fluence response curve for one of the photo$^-$ geo$^+$ mutants shows a shift to a higher threshold fluence, suggesting that the block in this isolate is at or near the photoreceptor. An interesting feature of one of the photo$^+$ geo$^-$ isolates is that only shoot geotropism is affected, indicating that control of shoot and root geotropism can be distinguished genetically (Bullen and Poff, 1987). Finally the existence of some mutants blocked in both photo- and geotropism supports the assumption that these responses share common elements.

C. Phytochrome

The best known photoreceptor, phytochrome, absorbs either red or far red light, depending on its conformation. Phytochrome mediated effects are characterized by their red, far-red reversibility which allow plants to sense both photoperiod and light quality. Many physiological studies have indicated that phytochrome is involved in regulating light-requiring seed germination, light-inhibited hypocotyl elongation in seedlings, "greening" of etiolated plants and photoperiodically controlled flowering. Some mutants of *Arabidopsis,* which were isolated on the basis of having hyper-elongated hypocotyls when growing in white light, have no spectrophotometrically detectable phytochrome in either seeds or hypocotyls (*hy1* and *hy2*), while a third class (*hy3*) shows reduced phytochrome levels only in seeds (Spruit *et al.,* 1980). Measurements of fluence response curves for seed germination in these mutants showed that they were less sensitive to red light induction and far red inhibition of germination (Cone and Kendrick, 1985). However the action spectra for induction and inhibition of germination were still characteristic of phytochrome response suggesting that the photoreceptor itself was not altered and that the leaky light responsiveness might result from a low residual level of phytochrome, sufficient to stimulate germination but not inhibit hypocotyl elongation. None of the *hy* mutants map near the phytochrome structural gene, which has been mapped, by linkage to restriction fragment length polymorphisms (RFLPs), to the proximal end of chromosome 1 (Chang *et al.*, 1987). Thus the *hy* mutants are more likely to affect regulation of phytochrome synthesis or stability or may alter a post-receptor step in the transduction pathway.

D. Flowering Induction

The transition from vegetative to reproductive growth is the most conspicuous change in plant development. Because reproductive success depends largely on flowering under the optimal internal and environmental conditions, the flowering responses of plant species reflect the degree of variability in their habitats. Thus, plants in temperate regions tend to be more dependent on environmental signals than those in tropical regions (Murfet, 1977). Both photoperiod and low-temperature (vernalization) can affect the onset of flowering and much work has been done to characterize the way in which plants sense and respond to the environmental factors inducing flowering (reviewed by Lang, 1965; Evans, 1971; Zeevart, 1976; Bernier *et al.*, 1981). However, the molecular processes involved are still unknown. Progress in this area has been hindered by the lack of a biological assay for the induced state, the long lag period between recognition of the environmental signal inducing flowering and the response at the meristem, and the fact that many different experimental conditions can induce flowering as a stress response.

The flowering process can be divided into three general steps: sensing of the inductive conditions (e. g. appropriate temperature or photoperiod), production and transport of a transmissible stimulus and response to the stimulus at the apical meristem. Low temperature seems to be sensed at the apical meristem itself in a process that requires cell division but nothing is known about the sensing mechanism (Lang, 1965). Photoperiod is sensed in the leaves (Lang, 1965) by a mechanism that involves phytochrome (Borthwick *et al.*, 1952). Experiments with interrupted light or dark cycles have shown that plants respond to the length of the dark period, presumably by sensing the change in levels of the P^{fr} form during this period (Lang, 1965). Production of a transmissible stimulus after photoperiodic induction has been demonstrated by grafting experiments (Lang, 1965) leading Chailakhyan (1936) to postulate the existence of a floral hormone called florigen, but a substance with the properties expected of florigen has yet to be identified. Production of a graft transmissible stimulus after cold induction, "vernalin", has been shown in only a few species (Melchers, 1939; Lang, 1965). In most of the cold-requiring species, no transmissible stimulus has been found and the induced condition seems to be a state of the vernalized meristem that can be transmitted only through cell division (Schwabe, 1954; Lang, 1965; Bernier *et al.*, 1981). In plants with both temperature and photoperiodic requirements, vernalization increases the sensitivity to photoperiod or eliminates its requirement (Lang, 1965; Bernier *et al.*, 1981).

In *Arabidopsis,* vegetative and reproductive phases are temporally separated. Flowering is induced by both long-day photoperiod and low temperature (reviewed by Napp-Zinn, 1969 and 1985) but neither requirement is absolute and *Arabidopsis* will eventually flower even under unfavorable conditions. All geographical races of *Arabidopsis* behave as long-day plants, flowering earlier under long day conditions (Napp-Zinn, 1985). The chilling requirement shows a wide genetic variation among races of *Arabidopsis:* some require vernalization to flower (e. g. Stockholm, St), while others show no response to cold treatment (Napp-Zinn, 1985). Most races used in the laboratory, such as Columbia and Landsberg have no chilling requirement and flower early. Several features of the physiology of flowering in *Arabidopsis,* such as the fact that it flowers under unfavorable conditions or even in complete darkness (Redei *et al.*, 1974), suggest that it may have a mechanism to induce flowering independent of the environmental conditions. Photoperiod and, in the vernalization requiring genotypes, cold temperatures could decrease the time preceding flowering through the same or additional mechanisms. Thus, the flowering process in *Arabidopsis* might be represented better as a complex network than as a single chain of events. Consistent with this idea, non-flowering mutants, which would be expected to arise frequently if flowering induction were a single pathway, have never been reported.

Because we have no clues about the chain of events leading to flowering, a mutant analysis alone will not reveal the molecular mechanisms underlying this process. However, by obtaining appropriate mutants we may be able to isolate and characterize some of the genes involved and analyze the function of their products. A number of mutations affecting flowering in *Arabidopsis* have been isolated and partially characterized. Many mutants in which the onset of flowering is delayed have been isolated following mutagenesis of early-flowering races of *Arabidopsis* (McKelvie, 1962; Hussein and Van der Veen, 1965, 1968; Hussein, 1968; Vetrilova, 1973; Koornneef *et al.*, 1983). The majority of these mutations behave as monogenic recessives *(fb, fca, fe, ft, fy, e1, e4, e6)* or semidominants *(f1* through *f5, fg, e3)* (McKelvie, 1962; Hussein, 1968; Koornneef *et al.*, 1983), but one is dominant *(f)* (McKelvie, 1962). In some of the late flowering mutants (e. g. *f, fca, e1, e6*) vernalization causes the mutants to flower at the same time as the wild-type parent of the original race. This indicates that these plants have a normal cold inducible mechanism and that the delayed flowering is not simply a non-specific effect of a mutation which reduces growth rate or vigor. The fact that many early-flowering races do not require vernalization may reflect the presence of an alternative way to induce flowering. This alternative pathway could be the same as that which makes *Arabidopsis* flower under non-inductive conditions. Variability for the different steps of this pathway could be responsible for the wide variation found in the cold requirements of natural populations (Napp-Zinn, 1985). A prediction of this hypothesis is that mutations that eliminate the cold requirement in winter-annual (late) races would represent a gain of this alternative pathway and thus should be relatively infrequent and probably dominant or semidominant over their normal alleles. Although no such mutant analysis has been done in late flowering races, Hussein (1968) subjected the late flowering mutant *fca* to a second cycle of mutagenesis and selected mutants which flowered earlier than *fca*. Consistent with the hypothesis mentioned above, he found two early revertant plants, one dominant and one semidominant with respect to the *fca* mutation.

Photoperiodic induction of flowering is believed to be mediated by phytochrome, in its active P^{fr} form. However, several observations indicate that, at least in *Arabidopsis,* the mechanism of photoperiodic induction could involve other receptor molecules. The analysis of the light induction spectrum shows a strong inductive effect of blue light (Meijer, 1959; Brown and Klein, 1971), similar to that found in other crucifer species (Funke, 1948). This led Brown and Klein (1971) to suggest the existence of a blue light photoreceptor that could transfer excitation energy to phytochrome. The role of phytochrome in this process is brought into question by the observation that mutants of *Arabidopsis* lacking almost any detectable phytochrome in etiolated seedlings, (*hy1* and *hy2*) (Koornneef *et al.*, 1980), do not show any alteration in the flowering induction. However, the biochemical lesions in the *hy* mutants have not been identified and it is not known whether they contain phytochrome in green tissue. These results indicate that either the levels of phytochrome required for flowering induction are very low or alternative mechanisms which can induce flowering do not depend on the phytochrome present in etiolated tissue.

The possibilities of interactions with other control circuits have complicated mutant analysis of photoperiodic requirements for flowering. Although photoperiodic mutants, whose flowering times differ from that of wild type under a variety of photoperiodic regimes have been isolated (Redei, 1962; Kranz, 1981), the different photoperiodic regimes provided different amounts of photosynthetically active light and may have been exerting a nutritional effect. In addition, it is not yet known whether any of these mutants show a cold requirement. The isolation of meaningful photoperiodic mutants in *Arabidopsis* will require testing of temperature requirements as well as careful control of both photosynthetically active and photoinductive light. Under these conditions it should be possible to select mutants that do not respond to light wavelengths known to induce flowering in normal plants. These mutants will help to characterize the genes involved in the process of photoperiodic induction.

Mutant analysis can also be used to study the role of different hormones in the flowering response. Gibberellic acid (GA) can substitute for long-days (Langridge, 1957) or cold treatment (Napp-Zinn, 1969) to induce flowering. Therefore it was thought to be involved in the flowering response of long-day plants. However, mutants of *Arabidopsis* with no detectable GA (e. g. *ga1*) (Koornneef and Van der Veen, 1980) flower at the same time as the wild-type under long-day photoperiod, and the double mutant with the genotype *(ga1, fca)* shows the same response to vernalization as the mutant *fca* (JMZ unpublished observations). These observations indicate that either GA is not required for these responses or very low levels are sufficient for the response. Other hormones and metabolites have been proposed to have a role in flowering induction (Lang, 1965) and the construction and analysis of the appropriate mutants could be very useful in testing those hypotheses.

The isolation and characterization of the *Arabidopsis* mutants described above reveals a complexity of the flowering process that the simple florigen concept is unable to explain. Similarly, the complexity revealed by physiological studies with a variety of species cannot be reduced to a single pathway of flowering induction. Although the variability seen in the flowering responses of many species indicates that the mechanisms involved are probably not universal among flowering plants, they are likely to share some common elements which can be identified by the combination of genetic and molecular genetic analyses. These approaches can contribute to a theory capable of explaining the entire phenomenology.

IV. Conclusions and Future Directions

Many years of physiological and genetic studies in a number of species have identified various components of a number of regulatory processes associated with specific developmental phenomena, but in no case is an entire transduction pathway understood. In *Arabidopsis,* mutations affecting sensitivity to phytohormones have been recovered at a relatively low frequency and most appear to be leaky in that sensitivity is reduced but not lost. Physiological analysis of these mutants has permitted us to examine the effects of each hormone separately and double mutants have been used to analyze interactions between hormones (e. g. GA and ABA effects on seed dormancy and germination). In addition to providing support for models suggesting roles for these hormones in regulating a variety of processes, these studies have also raised many questions about the mechanism of hormone action. For example, why do some of the insensitive mutants (*axr1* and *er1*) have extremely pleiotropic effects, while others (*aux1, gai, abi1, abi2* and *abi3*) have lost sensitivity for only a subset of the hormonally-inducible responses? Do these differences in sensitivity profiles reflect different threshold sensitivities for the different responses, different receptors for tissue- or stage-specific processes, or mutations further downstream in the various transduction pathways observed? We still do not know the nature of any of the lesions affecting sensitivity, but this should be accessible with molecular genetic techniques. Isolation of the affected genes and localization of their products should allow us to answer some of these questions. In addition, it may help to explain why so many of the insensitive mutations are dominant *(dwf, er1, gai, abi1),* which in turn may give us some clue about the mechanism of hormone action.

In principle it should be possible to identify all components of a pathway by mutant analysis, but the rarity of some classes of mutations may make mutant selection a rate-limiting step. The rapid life-cycle and small size of *Arabidopsis* make it possible to screen large numbers of plants and thus should permit isolation of some of the rarer classes of mutants.

Furthermore, some of these mutants might be viable only if conditional and the relative ease of handling large numbers of *Arabidopsis* may permit the isolation of temperature sensitive lines which can be subsequently screened for mutations of interest. Although it is often not possible to screen directly for mutations affecting developmentally regulated processes, recombinant DNA technology has been used to fuse selectable marker genes, such as alcohol dehydrogenase, to control regions of developmentally regulated genes with the goal of selecting mutations affecting trans-acting factors involved in regulating gene expression (Bonner *et al.*, 1984). This strategy is currently being employed to identify genes involved in light-induced (Chory and Ausubel, 1987) and seed-specific (Pang and Meyerowitz, 1987) gene expression.

Using these mutants to understand the mechanism of developmentally regulated gene expression or the determination process itself will ultimately require identification of the affected genes and their products. The small genome of *Arabidopsis* is ideal for molecular genetic analysis and intensive efforts by a number of laboratories should soon make chromosome walking, cloning by complementation, and transposon tagging feasible. Currently, both an RFLP map (Chang *et al.*, 1987) and a complete physical map composed of overlapping cosmid clones (Hauge *et al.*, 1987) are being constructed and correlated with the genetic map of visible markers. Additionally, a plant-transformable *Arabidopsis thaliana* genomic library is being constructed in *Agrobacterium* for eventual use in shotgun cloning experiments (Ludwig *et al.*, 1987; Olszewski and Ausubel, 1987). Finally, although no active transposable elements have yet been found in *Arabidopsis*, numerous groups are either looking for endogenous transposable elements (Mossie and Meyerowitz, 1987; Martinez-Zapater *et al.*, 1987) or introducing them from other organisms (Zhang *et al.*, 1986). The fact that the maize Ac element is mobile in tobacco (Baker *et al.*, 1986) bodes well for its utility in *Arabidopsis*. The combination of classical and molecular genetic approaches to developmentally interesting mutants should allow us to identify and isolate genes involved in development. At that point, analysis of the developmental process can be pursued in whatever organism is best-suited for it.

Acknowledgements

We thank C. Karssen and M. Koornneef for sharing unpublished results with us. In addition, we thank T. Caspar, S. Gilmour, T. Lynch and B. Moffatt for helpful discussions in the preparation of this manuscript.

V. References

Ackerson, R. C., 1984: Abscisic acid and precocious germination in soybeans. J. Exp. Bot. **35,** 414—421.

Addicott, F. T., Lyon, J. L., Ohkuma, K., Thiessen, W. E., Carns, H. R., Smith, O. E., Cornforth, J. W., Milborrow, B. V., Ryback, G., Wareing, P. F., 1968: Abscisic acid: a new name for abscisin II (dormin). Science **159,** 1493.

Anderson, L. W. J., 1978: Abscisic acid induces formation of floating leaves in the heterophyllous aquatic angiosperm *Potomageton nodosus.* Science **201,** 1135—1138.

Baker, B., Schell, J., Lorz, H., Federoff, N., 1986: Transposition of the maize controlling element "Activator" in tobacco. Proc. Natl. Acad. Sci. U.S.A. **83,** 4844—4848.

Bandurski, R. S., Galston, A. W., 1951: Phototropic sensitivity of coleoptile of albino corn. Maize Genet. Coop. Newsletter **25,** 5.

Bender, W., Spierer, P., Hogness, D. S., 1983: Chromosomal walking and jumping to isolate DNA from the *Ace* and *rosy* loci and the bithorax complex in *Drosophila melanogaster.* J. Mol. Biol. **168,** 17—34.

Bernier, G., Kinet, J.-M., Sachs, R. M., 1981: The physiology of flowering. Vol. 1, 149pp. Boca Raton, Fla.: CRC Press.

Bingham P. M., Levis, R., Rubin, G. M., 1981: Cloning of DNA sequence from the *white* locus of *Drosophila melanogaster* by a novel and general method. Cell **25,** 693—704.

Black, M., 1983: Abscisic acid in seed germination and dormancy. In: Abscisic Acid. Addicott, F. T. (ed.), New York: Praeger Publishers.

Blakely, L. M., Radaway, S. J., Hollen, L. B., Croker, S. G., 1972: Control and kinetics of branch root formation in cultured root segments of *Haplopappus ravenii.* Plant Physiol. **50,** 35—42.

Bleeker, A. B., Estelle, M., Somerville, C. R., Kende, H., 1987: Characterization of an ethylene-resistant mutant in *Arabidopsis thaliana.* Third International Meeting on *Arabidopsis,* Abstract **#77.**

Bonner, J. J., Parks, C., Parker-Thornburg, J., Mortin, M. A., Pelham, H. R. B., 1984: The use of promoter fusions in *Drosophila* genetics: Isolation of mutations affecting the heat shock response. Cell **37,** 979—992.

Borthwick, M. A., Hendricks, S. B., Parker, M. W., 1952: The reaction controlling floral initiation. Proc. Natl. Acad. Sci. U.S.A. **38,** 924—934.

Bray, E. A., Zeevaart, J. A. D., 1986: Compartmentation and equilibration of abscisic acid in isolated *Xanthium* cells. Plant Physiol. **80,** 105—109.

Brenner, M. L., Burr, B., Burr, F., 1977: Correlation of genetic vivipary in corn with abscisic acid concentration. Plant Physiol. Suppl. **59,** 76.

Brian, P. W., Elson, G. W., Hemming, H. G., Radley, M., 1954: The plant-growth promoting properties of gibberellic acid, a metabolic product of the fungus *Gibberella fujikoroi.* J. Sci. Food Agr. **5,** 602—612.

Brown, J. A. M., Klein, W. H., 1971: Photomorphogenesis in *Arabidopsis thaliana.* Plant Physiol. **47,** 393—399.

Bullen, B., Poff, K. L., 1987: Physiological characterization of mutants of *Arabidopsis* with altered geotropism. Third International Meeting on *Arabidopsis,* Abstract **#149.**

Caspar, T., Somerville, C., 1985: Geotropic roots and shoots of a starch-free mutant of *Arabidopsis.* Plant Physiol. **77S,** 105.

Chailakhyan, M. K., 1936: New facts in support of the hormonal theory of plant development. Dokl. Acad. Sci. U.S.S.R. **13,** 79—83.

Chang, C., DeJohn, A. W., Pruitt, R. E., Meyerowitz, E. M., 1987: A restriction fragment length polymorphism map of the *Arabidopsis* genome. Third International Meeting on *Arabidopsis,* Abstract **#31.**

Chen, T. H. H., Gusta, L. V., 1983: Abscisic acid-induced freezing resistance in cultured plant cells. Plant Physiol. **73,** 71—75.

Cheng, T.-Y., 1972: Induction of indoleacetic acid synthesis in tobacco pith explants. Plant Physiol. **50,** 723—727.

Chory, J., Ausubel, F., 1987: Genetic analysis of photoreceptor action pathways. Third International Meeting on *Arabidopsis.* Abstract **#56.**

Cone, J. W., Kendrick, R. E., 1985: Fluence-response curves and action spectra for promotion and inhibition of seed germination in wild type and long-hypocotyl mutants of *Arabidopsis thaliana* (L.). Planta **163,** 43—54.

Creelman, R. A., Gage, D. A., Stults, J. T., Zeevaart, J. A. D., 1987: Abscisic acid biosynthesis in leaves and roots of *Xanthium strumarium.* Plant Physiol., in press.

Curry, G. M., Gruen, H. E., 1959: Action spectra for the positive and negative phototropism of *Phycomyces* sporangiophores. Proc. Natl. Acad. Sci. U. S. A. **45,** 797—804.

Darwin, C., 1881: The Power of Movement in Plants. New York: D. Appleton and Co.

Davies, W. J., Mansfield, T. A., 1983: The role of abscisic acid in drought avoidance. In: Abscisic Acid, Addicott, F. T. (ed.), New York: Praeger Publishers.

Delbruck, M., Katzir, A., Presti, D., 1976: Responses of *Phycomyces* indicating optical excitation of the lowest triplet state of riboflavin. Proc. Natl. Acad. Sci. U.S.A. **73,** 1969—1973.

Estelle, M. A., Somerville, C. R., 1987: Auxin resistant mutants of *Arabidopsis thaliana* with an altered morphology. Mol. Gen. Genet. **206,** 200—206.

Evans, L. T., 1971: Flower induction and the florigen concept. Ann. Rev. Plant Physiol. **22,** 365—394.

Federoff, N. V., Furtek, D. B., Nelson, O. E., Jr., 1984: Cloning of the *bronze* locus in maize by a simple and generalizable procedure using the transposable controlling element *Activator (Ac).* Proc. Natl. Acad. Sci. U.S.A. **81,** 3825—3829.

Firn, R. D., Digby, J., 1980: The establishment of tropic curvatures in plants, Annu. Rev. Plant Physiol. **31,** 131—148.

Funke, G. L., 1948: The photoperiodicity of flowering under short day with supplemental light of different wavelengths. In: Vernalization and photoperiodism. Vol. **1,** pp. 79—82. Murneek, A. E., Whyte, R. O. (eds.). Waltham, Mass.: Lotsya.

Galland, P., Lipson, E. D., 1985: Action spectra for phototropic balance in *Phycomyces blakesleeanus:* Dependence on reference wavelength and intensity range. Photochem. Photobiol. **41,** 323—329.

Hauge, B., Fritze, C., Nam, H.-G., Paek, K.-H., Goodman, H. M., 1987: Progress in constructing a physical map of the *Arabidopsis thaliana* genome. Third International Meeting on *Arabidopsis,* Abstract **#32.**

Haughn, G. W., Somerville, C. R., 1987: Genetic control of morphogenesis in *Arabidopsis.* Dev. Genet., in press.

Henikoff, S., Tatchell, K., Hall, B. D., Nasmyth, K. A., 1981: Isolation of a gene from *Drosophila* by complementation in yeast. Nature **289,** 33—37.

Hussein, H. A. S., 1968: Genetic analysis of mutagen-induced flowering time variation in *Arabidopsis thaliana* (L.) Heynh. 88 pp. Thesis, Wageningen, The Netherlands.

Hussein, H. A. S., van der Veen, J. H., 1965: Induced mutations for flowering time. *Arabidopsis* Inf. Serv. **2,** 6—8.

Hussein, H. A. S., van der Veen, J. H., 1968: Genotypic analysis of induced mutations for flowering time and leaf number in *Arabidopsis thaliana. Arabidopsis* Inf. Serv. **5,** 30.

Jablonski, J. R., Skoog, F., 1954: Cell enlargement and cell division in excised tobacco pith tissue. Physiol. Plant **7,** 16—24.

Jacobs, M., Ray, P., 1976: Rapid auxin-induced decrease in free space pH and its relationship to auxin-induced growth in maize and pea. Plant Physiol. **58,** 203—209.

Jones, R. L., 1973: Gibberellins: Their physiological role. Ann. Rev. Plant Physiol. **24,** 571—598.

Karssen, C. M., Brinkhorst-van der Swan, D. L. C., Breekland, A. E., Koornneef, M., 1983: Induction of dormancy during seed development by endogenous abscisic acid: studies on abscisic acid deficient genotypes of *Arabidopsis thaliana* (L.) Heynh. Planta **157,** 158—165.

Karssen, C. M., Lacka, E., 1985: A revision of the hormone balance theory of seed dormancy: studies on gibberellin and/or abscisic acid-deficient mutants of *Arabidopsis thaliana.* In: Plant Growth Substances, pp. 315—323, Bopp, M. (ed.). Berlin: Springer-Verlag.

Kaufman, T. C., Abbott, M. K., 1984: Homoeotic genes and the specification of segmental identity in the embryo and adult thorax of *Drosophila melanogaster.* In: Molecular Aspects of Early Development, pp. 182—218. Malacinski, G. M., Klein, W. H. (eds.). New York: Plenum Press.

Kelly, M. O., Bradford, K. J., 1986: Insensitivity of the *Diageotropica* tomato mutant to auxin. Plant Physiol. **82,** 713—717.

Koornneef, M., Rolff, E., Spruit, C. J. P., 1980: Genetic control of light-inhibited hypocotyl elongation in *Arabidopsis thaliana* (L.) Heynh. Z. Pflanzenphysiol. **100,** 147—160.

Koornneef, M., van der Veen, J. H., 1980: Induction and analysis of gibberellin sensitive mutants in *Arabidopsis thaliana* (L.) Heynh. Theor. Appl. Genet. **58,** 257—263.

Koornneef, M., Dellaert, L. W. M., van der Veen, J. H., 1982: EMS- and radiation-induced mutation frequencies at individual loci in *Arabidopsis thaliana* (L.) Heynh. Mutation Res. **93,** 109—123.

Koornneef, M., Jorna, M. L., Brinkhorst van der Swan, D. L. C., Karssen, C. M., 1982: The isolation of abscisic acid (ABA) deficient mutants by selection of induced revertants in non-germinating gibberellin sensitive lines of *Arabidopsis thaliana* (L.) Heynh. Theor. Appl. Genet. **61,** 385—393.

Koornneef, M., van Eden, J., Hanhart, C. J., Stam, P., Braaksma, F. J., Feenstra, W. J., 1983: Linkage map of *Arabidopsis thaliana.* J. of Hered. **74,** 265—272.

Koornneef, M., Reuling, G., Karssen, C. M., 1984: The isolation and characterization of abscisic acid-insensitive mutants of *Arabidopsis thaliana.* Physiol. Plant. **61,** 377—383.

Koornneef, M., Elgersma, A., Hanhart, C. J., van Loenen-Martinet, E. P., van Rijn,

L., Zeevaart, J. A. D., 1985: A gibberellin insensitive mutant of *Arabidopsis thaliana.* Physiol. Plant **65,** 33—39.
Kranz, A. R., 1981: Genphysiologie lichtinduzierter Entwicklungsprozesse bei *Arabidopsis thaliana* (L.) Heynh. Ber. Dtsch. Bot. Ges. **94,** 181.
Kurosawa, E., 1926: Experimental studies on the secretion of *Fusarium heterosporum* on rice plants. Trans. Natl. Hist. Soc. Formosa **16,** 213—227.
Lang, A., 1965: Physiology of flower initiation. In: Encyclopedia of plant physiology. Vol. **15-1,** pp. 1380—1536. Ruhland, W. (ed.). Berlin: Springer-Verlag.
Langridge, J., 1957: Effect of day-length and gibberellic acid on the flowering of *Arabidopsis.* Nature **180,** 36—37.
Leutwiler, L. S., Hough-Evans, B. R., Meyerowitz, E. M., 1984: The DNA of *Arabidopsis thaliana.* Mol. Gen. Genet. **194,** 15—23.
Ludwig, B., Stein, P., Olszewski, N., Ausubel, F., 1987: Construction of a plant-transformable *Arabidopsis thaliana* genomic library in *Agrobacterium.* Third International Meeting on *Arabidopsis,* Abstract **#47.**
Maher, E. P., Martindale, S. J. B., 1980: Mutants of *Arabidopsis thaliana* with altered response to auxin and gravity. Biochem. Genet. **18,** 1041—1053.
Malmberg, R. L., McIndoo, J., 1983: Abnormal floral development of tobacco mutant with elevated polyamine levels. Nature **305,** 623—625.
Martinez-Zapater, J. M., Somerville, S., Somerville, C. R., 1987: Variegated mutants of *Arabidopsis.* Third International Meeting on Arabidopsis, Abstract **#107.**
McKelvie, A. D., 1962: A list of mutant genes in *Arabidopsis thaliana* (L.) Heynh. Radiat. Bot. **1,** 233—241.
Meijer, G., 1959: The spectral dependence of flowering and elongation. Acta Bot. Neerl. **8,** 189.
Melchers, G., 1939: Die Bluhhormone. Ber. Dtsch. Bot. Ges. **57,** 29—48.
Mertens, R., Weiler, E. W., 1983: Kinetic studies on the redistribution of endogenous growth regulators in gravireacting plant organs. Planta **158,** 339—345.
Moore, R., Smith, J. D., 1984: Growth, graviresponsiveness and abscisic-acid content of *Zea mays* seedlings treated with fluridone. Planta **162,** 342—344.
Mossie, K. G., Meyerowitz, E., 1987: In search of transposable genetic elements in *Arabidopsis thaliana.* Third International Meeting on Arabidopsis, Abstract **#41.**
Murfet, I. C., 1977: Environmental interaction and the genetics of flowering. Ann. Rev. Plant Physiol. **28,** 253—278.
Napp-Zinn, K., 1969: *Arabidopsis thaliana* (L.) Heynh. In: The induction of flowering: some case hostories. pp. 291—304. Evans, L. T. (ed.). Melbourne: Macmillan.
Napp-Zinn, K., 1985: *Arabidopsis thaliana.* In: Handbook of flowering. Vol. **1,** pp. 492—503. Halevy, A. H. (ed). Boca Raton, Fla.: CRC Press.
Olszewski, N., Ausubel, F., 1987: A potential vector for the cloning of plant genes by phenotypic complementation. Third International Meeting on *Arabidopsis,* Abstract **#45.**
Pang, P., Meyerowitz, E., 1987: Seed specific gene expression in *Arabidopsis thaliana.* Third International Meeting on *Arabidopsis,* Abstract **#53.**
Phillips, I. D., 1975: Apical dominance. Ann. Rev. Plant Physiol. **26,** 341—367.
Phinney, B. O., 1984: Gibberellin A_1, dwarfism and the control of shoot elongation in higher plants. In: The Biosynthesis and Metabolism of Plant Hormones, pp. 17—42. Crozier, A., Hillman, J. R. (eds.). Cambridge University Press.
Pickard, B., 1985: Roles of hormones, protons and Ca^{++} in geotropism. In:

Hormonal Regulation of Development III, Encyclopedia of Plant Physiol., New Series, vol. **11**, pp. 193—281. Pharis, R. P., Reid, D. M. (eds.). Berlin: Springer-Verlag.

Poethig, R. S., 1985: Homeotic mutations in maize. In: Plant Genetics, pp. 33—44. Freeling, M. (ed.). New York: Alan R. Liss, Inc.

Poff, K. L., Best, T., Gregg, M., Ren, Z., 1987: Mutants of *Arabidopsis thaliana* with altered phototropism and/or altered geotropism. Third International Meeting on *Arabidopsis*, Abstract **#79.**

Presti, D., Hsu, W. J., Delbruck, M., 1977: Phototropism in *Phycomyces* mutants lacking β-carotene. Photochem. Photobiol. **26,** 403—405.

Radley, M., 1956: Occurrence of substances similar to gibberellic acid in higher plants. Nature (London) **178,** 1070—1071.

Redei, G. P., 1962: Supervital mutants of Arabidopsis. Genetics **47,** 443—460.

Redei, G. P., Acedo, G., Gavazzi, G., 1974: Flower differentiation in *Arabidopsis.* Stadler Symp. **6,** 135—168.

Robichaud, C. S., Wong, J., Sussex, I. M., 1980: Control of *in vitro* growth of viviparous embryo mutants of maize by abscisic acid. Dev. Genet. **1,** 325—330.

Rogler, C. E., Hackett, W. P., 1974: Phase change in *Hedera helix:* stabilization of the mature form with abscisic acid and growth retardants. Physiol. Plant. **34,** 148—152.

Schnieke, A., Harbers, K., Jaenisch, R., 1983: Embryonic lethal mutation in mice induced by retrovirus insertion into the α1(I) collagen gene. Nature **304,** 315—320.

Schwabe, W. W., 1954: The site of vernalization and translocation of the stimulus. J. Exp. Bot. **5,** 389—400.

Shen-Miller, J., 1973: Rhythmic differences in the basipetal movement of indoleacetic acid between separated upper and lower halves of geotropically stimulated corn coleoptiles. Plant Physiol. **52,** 166—170.

Sponsel, V., 1987: Gibberellin biosynthesis and metabolism. In: Plant Hormones and Their Role in Plant Growth and Development. Davies, P. J. (ed.), pp. 43—75. Dordrecht: Martinus Nijhoff Publishers.

Spruit, C. J. P., van der Boom, A., Koornneef, M., 1980: Light induced germination and phytochrome content of seeds of some mutants of *Arabidopsis.* Arabidopsis Inf. Service **17,** 137—141.

Stodola, F. H., Raper, K. B., Fennell, D. I., Conway, H. F., Sohns, V. E., Langford, C. T., Jackson, R. W., 1955: The microbiological production of gibberellins A and X. Arch. Biochem. Biophys. **54,** 240—245.

Tagliana, L., Nissen, S., Blake, T. K., 1986: Comparison of growth, exogenous auxin sensitivity, and endogenous indole-3-acetic acid content in roots of *Horduem vulgare* L. and an agravitropic mutant. Biochem. Genet. **24,** 839—848.

Taylorson, R. B., Hendricks, S. B., 1977: Dormancy in seeds. Ann. Rev. Plant Physiol. **28,** 331—354.

Trewavas, A., 1981: How do plant growth substances work? Plant Cell Envir. **4,** 203—228.

Van Stevenick, R. F. M., Van Stevenick, M. E., 1983: Abscisic acid and membrane transport. In: Abscisic Acid. Addicott, F. T. (ed.), pp. 171—236. New York: Praeger Publishers.

Vetrilova, M., 1973: Genetic and physiological analysis of induced late mutants of *Arabidopsis thaliana* (L.) Heynh. Biol. Plant. **15,** 391—397.

Vierstra, R. D., Poff, K. L., 1981: Role of carotenoids in the phototropic response of corn seedlings. Plant Physiol. **68,** 798—801.

Volkmann, D., Sievers, A., 1979: Graviperception in multicellular organs. In: Physiology of Movements. Encyclopedia of Plant Physiol., new series, vol. **7**, Haupt, W., Feinleib, M. E. (eds.), pp. 573—600. Berlin: Springer-Verlag.
Walbot, V., 1978: Control mechanisms for plant embryogeny. In: Dormancy and Developmental Arrest: Experimental Analysis in Plants and Animals. Clutter, M. E. (ed.), pp. 114—166. New York: Academic Press.
Wareing, P. F., Phillips, I. D. J., 1983: Abscisic acid in bud dormancy and apical dominance. In: Abscisic Acid. Addicott, F. T. (ed.), pp. 301—330. New York: Praeger Publishers.
Went, F. W., Thimann, K. V., 1937: Phytohormones. New York: Macmillan.
West, C. A., Phinney, B. O., 1956: Properties of gibberellin-like factors from extracts of higher plants. Plant Physiol. **31** (Suppl.) XX (Abstr.).
Yabuta, S. F., Sumiki, Y., 1938: Communication to the editor. J. Agric. Chem. Soc. Japan **14,** 1526.
Yang, S. F., Hoffmann, N. E., 1984: Ethylene biosynthesis and its regulation in higher plants. Ann. Rev. Plant Physiol. **35,** 155—189.
Zeevaart, J. A. D., 1976: Physiology of flower formation. Ann. Rev. Plant Physiol. **27,** 321—348.
Zeevaart, J. A. D., 1983: Endogenous gibberellins in dwarf mutants of *Arabidopsis thaliana.* In: Plant Research '83, Annual Report of the MSU-DOE Plant Research Laboratory, pp. 157—158. East Lansing, Michigan.
Zeevaart, J. A. D., 1986: Characterization of three dwarf mutants of *Arabidopsis thaliana.* In: Plant Research '86, Annual Report of the MSU-DOE Plant Research Laboratory, pp. 130—131. East Lansing, Michigan.
Zhang, H., Somerville, C. R., 1987: Transfer of the maize transposable element *mu1* into *Arabidopsis thaliana.* Plant Sci. **48,** 165—173.

Chapter 2

Regulation of Gene Expression During Seed Germination and Postgerminative Development

John J. Harada, Robert A. Dietrich, Lucio Comai, and Catherine S. Baden

Department of Botany, University of California, Davis, CA 95616, U.S.A.

With 3 Figures

Contents

I. Introduction

Seed germination is a pivotal stage in the sporophytic life cycle of higher plants during which growth and differentiation of the primary plant body resumes following a period of quiescence imposed late in embryogeny. Many of the specific biochemical and physiological processes which characterize germinating seeds, particularly those occurring in storage organs, are unique to this stage (reviewed by Bewley and Black, 1983, and summarized below). From the viewpoint that differential gene expression underlies plant development, the relative specificity of these processes suggests that distinct gene sets are activated and repressed during this stage. Identifying these genes and defining mechanisms involved in regulating their expression will aid in understanding the control of germination-specific processes.

In this chapter, we will describe the regulation of gene expression related to germination, emphasizing recent studies from our laboratory. We have focused on the dicotyledonous oilseed plant *Brassica napus*, in part, because it can be transformed and regenerated, and is, therefore, amenable to studies of genetic mechanisms involved in gene regulation. Although germination is defined as the period beginning with dry seed imbibition and terminating with radicle emergence through the testa (Bewley and Black, 1983), we will also consider aspects of seed maturation and postgerminative development as well. We define postgermination as the period after germination during which events uniquely related to the reinitiation of sporophyte growth occur. Our intent is to provide an overview of the spatial and temporal regulation of gene expression in germinating seeds and seedlings and to indicate how study of these genes may aid in identifying signals which control processes unique to this major transition point in the plant life cycle.

II. Differential Gene Expression Underlies Seed Germination

Because the starting point of germination is usually the dry seed, characteristics of mature embryos are relevant to germination. Two morphologically distinct embryo parts, the root-shoot axis and cotyledons, are initiated during the first 15 days of embryogenesis (Tykarska, 1979; 1980). The axis consists of a shoot apical meristem with the first leaf primordium (epicotyl), a hypocotyl, and a root primordium with an apical meristem. Cotyledons are the major storage organ in *Brassica napus* seeds as the endosperm is compressed to a thin layer completely surrounding the embryo (Norton *et al.*, 1976). Lipids and proteins are abundant in *Brassica* cotyledons, comprising 30—50 % and 10—45 % of seed mass, respectively (Weiss, 1983). These storage macromolecules accumulate at rapid rates during the maturation stage of embryogenesis and are stored in specific subcellular organelles. Late during maturation stage, embryo dehydration occurs, roughly coinciding with a peak of abscisic acid accumulation and the onset of desiccation tolerance in *Brassica napus* (Crouch and Sussex, 1981; Finkelstein *et al.*, 1985). Desiccation is accompanied by decreased rates of protein and RNA synthesis as evidenced by lower concentrations of polyribosomes and polyadenylated RNAs, and metabolic rates are low (reviewed by Raghavan, 1986). Both abscisic acid and desiccation may play a role in promoting maturation processes and inhibiting premature germination of immature embryos (reviewed by Quatrano, 1986).

With non-dormant seeds such as *Brassica napus*, imbibition is a sufficient condition to initiate the dramatic changes in embryo morphology and physiology which occur during and following germination. By analogy to other plants, seed imbibition initiates rapid increases in the rates of respiration and protein and RNA synthesis in both axes and storage organs (reviewed by Bewley and Black, 1985). Polyribosome formation can commence in the presence of RNA synthesis inhibitors, implying that translational components as well as mRNA are stored in the dry seed (reviewed by Payne, 1976). The physiological changes are mirrored by corresponding changes in cellular morphology. Many organelles, e. g., mitochondria, plastids, endoplasmic reticulum, Golgi bodies, which are

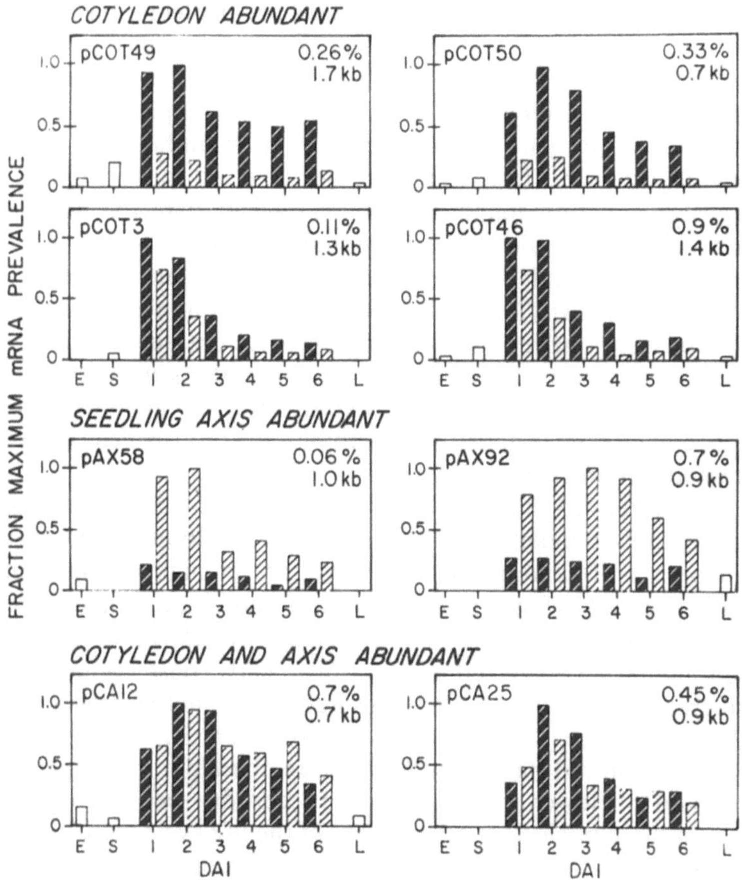

Fig. 1. Developmental regulation of genes encoding postgermination-abundant mRNAs. The graphs summarize the accumulation of mRNAs which reacted with the indicated postgermination-abundant cDNA clones (see text). Numbers in the upper right corner of the boxes show the maximal mRNA level given as percent of polyribosomal, polyadenylated RNA (top) and mRNA size as determined by denaturing agarose gel electrophoresis (bottom). Abbreviations and symbols: E, 23—27 days after flowering embryos; S, dry seed; L, leaf; 1-6 DAI, days after the start of imbibition; dark cross-hatching, mRNA level in cotyledons; light cross-hatching, mRNA level in seedling axes. Method: A sequence excess of the indicated postgermination-abundant cDNA clone was reacted with 0.2 µg of polyribosomal, polyadenylated RNA applied directly onto nitrocellulose filters. Following hybridization, the RNA dots were cut from the filter, dried, and counted by liquid scintillation counting

either cytologically difficult to discern or undetectable in dry seeds become rapidly activated and morphologically distinct following imbibition, with some organelles, e. g., glyoxysomes in oilseeds, proliferating extensively (Jacobsen, 1984). Storage organelle contents are depleted during the postgermination period, and protein bodies appear to coalesce to form the vacuole.

The distinct physiology of germinating seeds appears to reflect the expression of unique gene sets. *In vitro* translation studies have shown that relatively abundant mRNAs which are not detected in developing or dry seeds (postgermination-abundant mRNAs) accumulate during and after germination in both cotyledons and seedling axes (Dure *et al.*, 1981; Aspart *et al.*, 1984; Misra and Bewley, 1985 A; Sanchez-Martinez *et al.*, 1986). However, the sensitivity of these experiments cannot eliminate the possibility that postgermination-abundant mRNAs are present at lower levels in developing embryos or dry seeds. For example, the kinetic hybridization studies of Galau and Dure (1981) showed that the vast majority of mRNAs in cotton cotyledons 24 hours after imbibition are also detected in early and late maturation stage embryos.

We have extended these studies by showing that a gene class encoding postgermination-abundant mRNAs is highly developmentally regulated. cDNA clones representing mRNAs which are at high concentrations in germinating seeds (14 hours after the start of imbibition) but less prevalent in dry seeds were identified in differential hybridization experiments. Fourteen hours was selected as a critical early postgermination stage because greater than 80 % of seeds had newly emerged radicles protruding through the seed coat at that time. Figure 1 summarizes representative studies which showed that the mRNAs were abundant in postgerminative seedlings and at a lower level in immature seeds, dry seeds, and leaves. The detection of these mRNAs at other stages of the life cycle indicated that the gene families are not exclusively expressed during the postgermination stage. Nevertheless, the differential mRNA accumulation pattern as well as the other studies cited above suggest that an mRNA set(s) accumulates and presumably encodes proteins which function primarily during postgerminative development.

III. Spatial Regulation of Postgermination-Abundant Genes

The two major seedling components, cotyledons and seedling axes, have diverse primary roles. Following germination, cotyledons serve as a primary site of storage reserve hydrolysis and mobilization, whereas, during seed maturation, the synthesis and accumulation of storage macromolecules occurs. This dramatic transition in cotyledon genetic activity and physiology takes place in the absence of cell division. The mobilized metabolites, primarily sucrose and amino acids, serve as a major source of nutrients for the growing seedling until autotrophic growth commences. Mobilization of food reserves is mediated by enzymes involved in the hydrolysis of storage macromolecules, e. g., proteinases, lipases, as well as those responsible for converting metabolites into translocated forms, e. g., exopeptidases, beta-oxidation enzymes, glyoxylate cycle enzymes (reviewed by Ching, 1972). Studies have shown that many of these enzymes are synthesized *de novo* following germination while others are present in dry seeds (reviewed by Marcus and Rodway, 1982). Concomitant with the physiological changes is development of the vascular system for the translocation of sucrose and amino acids from cotyledons to the axis (Esau, 1977).

In contrast, activity in the root-shoot axis is devoted largely to the growth and development of the mature plant body. Emergence of the axis through the testa provides the most obvious morphological change in germinating seeds. Several studies have suggested that radicle and/or hypocotyl elongation first results by cell expansion, although in certain seeds, cell division may also contribute to initial axis growth (reviewed by Bewley and Black, 1983). Following germination, organogenesis and accompanying histodifferentiation occur primarily at the apices. Although the processes have been well characterized morphologically, the specific physiological mechanisms involved have not been defined.

The differences in cotyledon and seedling axis physiology suggest that alternate gene sets may be expressed in the two seedling parts. To confirm this possibility, we showed that many postgermination-abundant mRNAs accumulated preferentially in a particular seedling part. Figure 1 summarizes representative studies in which the cDNA clones were reacted with mRNA isolated from dark-grown seedling axes and cotyledons at one through six days after the start of imbibition. Based on the spatial accumulation pattern, the postgermination-abundant mRNAs could be grouped into three sets: cotyledon-abundant mRNAs (pCOT clones), seedling axis-abundant mRNAs (pAX clones), and mRNAs abundant in both seedling parts (pCA clones). Although mRNAs in the two former groups were clearly more prevalent in one seedling component, they were also detected in the other part. Recent *in situ* hybridization studies in seedling tissue verified that the results were not due to cross-contamination of dissected tissues of RNA preparations (R. A. D. and J. J. H., unpublished results).

The majority of characterized postgermination-abundant clones, eight of fourteen, represented mRNAs which were prevalent in seedling cotyledons. All of the cloned cotyledon-abundant mRNAs displayed similar temporal and spatial accumulation patterns, suggesting coordinate expression of this gene set. Because the primary physiological processes in postgerminative cotyledons are involved with storage macromolecule mobilization, it is possible that the cloned mRNAs may encode enzymes which mediate the hydrolysis of reserves or conversion of metabolites into sucrose or translocated amino acids.

We have obtained evidence to support this hypothesis by characterizing the expression of two genes encoding enzymes involved in lipid mobilization, isocitrate lyase and malate synthase. Both are key enzymes of the glyoxylate cycle which effect the net conversion of two molecules of acetyl

CoA generated from fatty acid beta-oxidation into succinate. The metabolite can then be converted into sucrose for translocation to the axis (Beevers, 1979). *Brassica napus* isocitrate lyase and malate synthase cDNA clones were obtained from the germinating seed cDNA library using heterologous probes from *Aspergillus nidulans* (kindly provided by G. Turner; Ballance and Turner, 1986) and cucumber (kindly provided by C. Leaver; Smith and Leaver, 1986). We found that isocitrate lyase and malate synthase genes were expressed similarly with cotyledon-abundant genes; the mRNAs accumulated preferentially in cotyledons, increased then decreased during postgerminative development, and were at very low levels in leaves and immature (23—27 days after flowering) seeds (J. J. H., L. C., and C. S. B., manuscript in preparation). Thus, genes encoding glyoxylate cycle enzymes and cotyledon-abundant mRNAs appear to be coordinately expressed. Regardless of whether the other cotyledon-abundant mRNAs encode mobilization enzymes, the results suggest that a relatively large gene set expressed during postgerminative development is similarly regulated.

Further support for the coordinate expression of cotyledon-abundant genes came from studies which showed that the mRNAs were similarly distributed in seedling tissue. Using *in situ* hybridization protocols, we found that isocitrate lyase, malate synthase, and pCOT1 mRNA were prevalent in the storage parenchyma cells of dark-grown seedling cotyledons two days after the start of imbibition (R. A. D. and J. J. H., unpublished results). More strikingly, all three mRNAs were relatively abundant in upper hypocotyl cortex cells and less prevalent in the basal hypocotyl and upper root. The finding that reserve mobilization enzyme mRNAs are present in the axis is not surprising since lipid and protein bodies are detected in these tissues in the dry seed, although at lower amounts than in cotyledon cells. In addition to emphasizing that cotyledon-abundant genes are similarly expressed during the postgermination period, the results also showed that they are differentially regulated in axes. It is presently unclear if this pattern of expression reflects the distribution of storage reserves, cellular activation patterns, or distinct requirements of the cells.

In contrast to cotyledon-abundant mRNAs, the two cloned axis-abundant mRNAs accumulated in temporally unique patterns (Fig. 1). Lalonde and Bewley (1986) also showed that distinct protein sets are synthesized in pea axes at different times after imbibition. At least one axis-abundant gene is spatially regulated in the axis, but its expression pattern is opposite to that observed for cotyledon-abundant genes. *In situ* hybridization studies showed that pAX92 mRNA is more prevalent in the basal hypocotyl than in the upper root, the upper hypocotyl, and cotyledons. Datta *et al.* (1987) reported that prevalent mRNAs which accumulate in soybean axes during and after germination were spatially regulated. By reacting cDNA clones with RNA isolated from different parts of the axis at various postgermination stages, they showed that mRNAs which accumulated within 9 hours after imbibition were distributed throughout the seedling axis while a second group, most abundant 24 hours after imbibition, were primarily localized in the hypocotyl region. Thus, in contrast to cotyledons, regulation of postgerminative gene expression in axes appears to be complex, with different mRNA sets showing divergent temporal and spatial accumulation programs.

These varied expression patterns most likely reflect the fact that the axis is a collection of organs and cell types, each different morphologically and physiologically. In part, this diversity makes it difficult to speculate on the identity of seedling axis-abundant mRNAs. For similar reasons, it is also difficult to assess cotyledon and axis-abundant gene function. Although these genes are expressed in both seedling parts, their expression differs from that of constitutively expressed genes in that the mRNAs are most prevalent after germination.

IV. Activation of Postgermination-Abundant Genes

To define mechanisms involved in the control of germination-specific processes, it is necessary to determine the precise developmental stage at which postgermination-abundant genes are activated. Figure 1 shows that the cloned mRNAs accumulated to near maximal levels within one day after imbibition. We further showed that the concentration of many post-germination-abundant mRNAs increased significantly within eight to sixteen hours after imbibition by examining the kinetics of mRNA accumulation (Fig. 2). Assuming that mRNA synthesis and degradation occur at minimal rates in dry seeds, the results suggest that postgermination-abundant mRNA accumulation probably results from increased transcript synthesis. Thus, gene activation appears to be an early response after imbibition.

However, higher sensitivity experiments than those summarized in Figures 1 and 2 showed that all postgermination-abundant mRNAs except that represented by pAX58 were stored in dry seeds, indicating that they accumulated during seed maturation. Stored mRNAs have been shown to consist of residual seed maturation specific mRNAs which are rapidly degraded in germinating seeds, e. g., storage proteins, late-embryogenesis abundant proteins (Mori *et al.*, 1978; Galau *et al.*, 1986), and germination-specific mRNAs which become prevalent in postgerminative storage organs, e. g., carboxypeptidase, isocitrate lyase, and malate synthase (Ihle and Dure, 1969; Weir *et al.*, 1980).

To determine whether stored postgermination-abundant mRNA accumulation results from constitutive expression throughout embryogeny or gene activation at a specific stage, we measured mRNA prevalence during seed maturation. As summarized in Figure 3, the results showed that the spatially regulated postgermination-abundant mRNA sets displayed opposite accumulation patterns. Cotyledon-abundant mRNA concentration increased during late maturation beginning at 35—40 days after flowering but did not attain levels found in postgerminative seedlings. We also showed that isocitrate lyase and malate synthase mRNA accumulation patterns during embryogeny were identical to that of other cotyledon-abundant mRNAs (J. J. H., L. C., and C. S. B., manuscript in preparation). In contrast, seedling axis-abundant mRNAs were relatively prevalent in young embryos, but declined at later stages reaching minimal levels by 35—40 days after flowering. The results accentuate additional differences in the expression of the two gene classes and underscore the apparent coordinate expression of cotyledon-abundant genes.

The differential accumulation of cotyledon-abundant and seedling axis-abundant mRNAs implicates 35—40 days after flowering as a critical stage in seed development. This embryonic stage coincides with the initiation of maturation drying and with a peak of abscisic acid accumulation. The correlation with the beginning of dehydration is interesting in relation to the hypothesis that desiccation is a signal which shifts plant development from an embryogenic to a germinative program. Bewley and his colleagues (Dasgupta and Bewley, 1982; Misra and Bewley, 1985B) demonstrated that rehydration of prematurely desiccated embryos resulted in the accumulation of mRNAs which are characteristic of germinating rather than developing seeds. In addition, α-amylase, a postgermination abundant enzyme which is not normally induced by gibberellic acid in

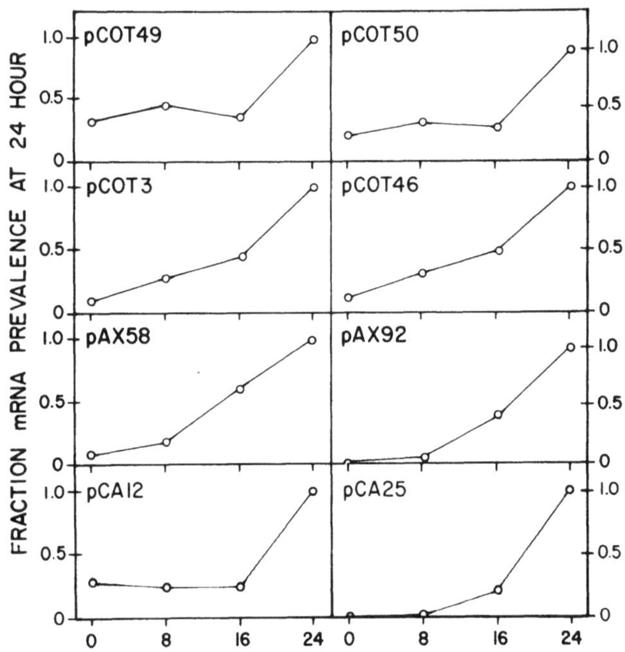

Fig. 2. Early accumulation of postgermination-abundant mRNAs. mRNAs corresponding to each of the cDNA clones were quantified in total, polyadenylated RNA obtained from dry seeds imbibed for the indicated times. Results are given as fraction of mRNA present in seedlings 24 hours after imbibition. Zero time shows mRNA level in dry seeds

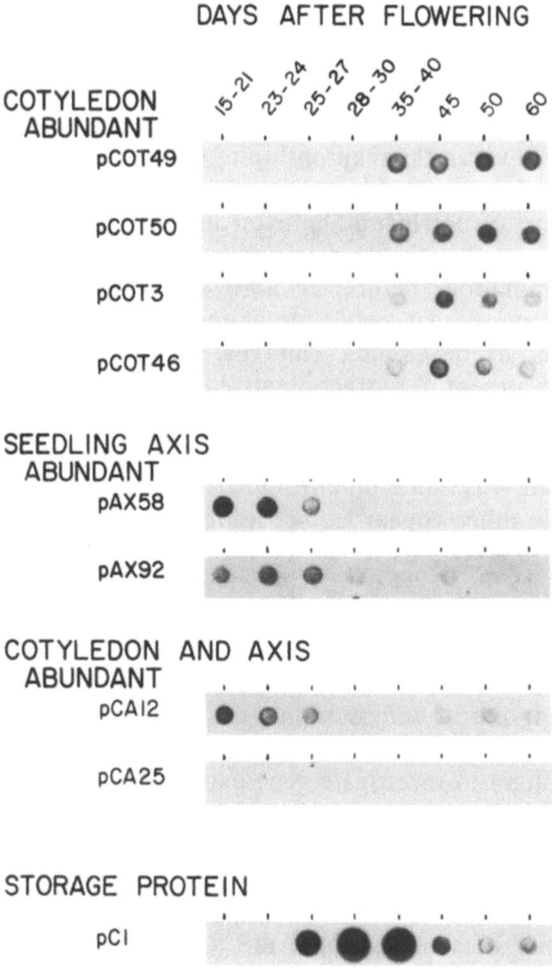

Fig. 3. Activation of genes encoding postgermination-abundant mRNAs during seed maturation. Autoradiograms show the accumulation of mRNAs corresponding to the indicated cDNA clones during embryogeny in RNA dot blot hybridization studies. Methods employed were the same as those described in Figure 1, except that total, polyadenylated RNA was used. The storage protein, cruciferin, cDNA clone pC1 was generously provided by Dr. M. L. Crouch and provides a developmental control for embryonic stages

immature developing wheat aleurone, was produced in prematurely desiccated grains treated with the growth regulator (Armstrong *et al.*, 1982). However, the elegant studies of Crouch and her colleagues (Finkelstein and Crouch, 1986) showed that immature embryos cultured on high osmoticum continue to accumulate embryo-specific proteins and mRNAs. Thus, desiccation does not appear to be a sufficient stimulus to reprogram development completely to a germinative mode. The significance of the correlation between the accumulation of cotyledon-abundant mRNAs and abscisic acid is unclear. Dure *et al.* (1981) showed that an mRNA set which becomes detectable during late embryogeny in cotton cotyledons and persists after germination accumulated in immature embryos treated with the hormone. However, the relationship of these mRNAs to the cotyledon-abundant mRNAs described here is not known.

The presence of cotyledon-abundant mRNAs and/or proteins in dry seeds may suggest that they aid in activating embryos during the initial stages of germination, before *de novo* synthesis becomes significant. However, we found that cotyledon-abundant mRNAs accumulated to different levels in developing embryos, ranging between 2—45% of maximal levels present in postgerminative cotyledons. Differential mRNA and, presumably, protein accumulation could reflect different requirements of seeds for the mRNA/proteins early during germination, and may possibly explain why some mobilization enzyme activities are present in dry seeds while other appear to be synthesized *de novo* following germination.

V. Future Directions

From the standpoint of understanding the control of germination-specific processes, genes encoding cotyledon-abundant mRNAs appear to be an excellent paradigm to investigate gene expression during germination for several reasons. First, isocitrate lyase and malate synthase genes which have major roles following germination, have already been identified, and the other cotyledon-abundant mRNAs may also encode enzymes which participate in the mobilization of storage reserves. We may gain insight into the identity of some of these mRNAs in DNA sequence analysis studies and by determining the intracellular location of the corresponding proteins using antibodies generated with synthetic peptides deduced from these sequences. Second, the apparent coordinate expression of these diverse genes may suggest that they are regulated by a common mechanism, thus facilitating studies designed to identify *cis*-acting regulatory sequences and *trans*-acting factors. Identifying *trans*-acting molecules involved in controlling the expression of these genes and determining how the activity of the factor(s) is regulated should aid in defining the signals which activate germination-specific physiological processes. Finally, the accumulation of cotyledon-abundant mRNAs during both embryogeny and postgerminative development raises an important question; are these genes regulated by distinct or common mechanisms during these two developmental stages? The former alternative may suggest that in preparation for germination, these genes are activated and "stored" in a potentially active transcriptional state. The failure to accumulate large amounts of these mRNAs during seed maturation may simply reflect lower RNA and protein synthetic activities in desiccating embryos. On the other hand, the genes may be controlled differently during these two stages. We have shown that four cotyledon-abundant mRNAs are encoded by gene families, an expected result because *Brassica napus* is an allotetraploid (U, 1935). While it is possible that different gene family members may be expressed at the two stages, preliminary results suggest that the same malate synthase gene class is most highly expressed during both embryogenesis and postgermination (L. C. and J. J. H., unpublished observation). Definitive answers to these questions await characterization of the relevant regulatory sequences and factors involved in controlling postgermination-abundant genes.

Acknowledgements

We would like to thank Drs. Bob Fischer and Tom Rost for their comments on this manuscript. This research was supported, in part, by the National Science Foundation, grant number DCB8518182 awarded to J. J. H. R. A. D. was supported by a graduate fellowship from the McKnight Foundation.

VI. References

Armstrong, C., Black, M., Chapman, J. M., Norman, H. A., Angold, R., 1982: The induction of sensitivity to gibberellin in aleurone tissue of developing wheat grain. I. The effect of dehydration. Planta **154,** 573—577.

Aspart, L., Meyer, Y., Laroche, M., Penon, P., 1984: Developmental regulation of the synthesis of proteins encoded by stored mRNA in radish embryos. Plant Physiol. **76,** 664—673.

Ballance, D. J., Turner, G., 1986: Gene cloning in *Aspergillus nidulans:* isolation of the isocitrate lyase gene *(acuD).* Mol. Gen. Genet. **202,** 271—275.

Beevers, H., 1979: Microbodies in higher plants. Ann. Rev. Plant Physiol. **30,** 159—193.

Bewley, J. D., Black, M., 1983: Physiology and Biochemistry of Seeds in Relation to Germination. Vol. 1. Development, Germination, and Growth. Springer-Verlag, Berlin, 306 pp.

Bewley, J. D., Black, M., 1985: Seeds: Physiology of Development and Germination. Plenum Press, New York. 367 pp.

Ching, T. M., 1972: Metabolism of germinating seeds. In: Seed Biology. Vol. 2, Germination control, metabolism, and pathology, pp. 103—218. Kozlowski, T. T. (ed). Academic Press, New York.

Crouch, M. L., Sussex, I. A., 1981: Development and storagesprotein synthesis in *Brassica napus* L. embryos *in vivo* and *in vitro.* Planta **153,** 64—74.

Dasgupta, J., Bewley, J. D., 1982: Dessication of axes of *Phaseolus vulgaris* during development of a switch from a development pattern of protein synthesis to a germination pattern. Plant Physiol. **70,** 1224—1227.

Datta, K., Parker, H., Averyhart-Fullard, B., Schmidt, A., Marcus, A., 1987: Gene expression in the soybean seed axis during germination and early seedling growth. Planta **170,** 209—216.

Dure, L. III., Greenway, S. C., Galau, G. A., 1981: Developmental biochemistry of cottonseed embryogenesis and germination: changing messenger ribonucleic acid populations as shown by *in vitro* and *in vivo* protein synthesis. Biochem. **20,** 4162—4168.

Esau, K., 1977. Anatomy of Seed Plants, Second Edition. John Wiley and Sons, New York. 550 pp.

Finkelstein, R. R., Tenbarge, K. M., Shumway, J. E., Crouch, M. L., 1985: Role of ABA in maturation of rapeseed embryos. Plant Physiol. **78,** 630—636.

Finkelstein, R. R., Crouch, M. L., 1986: Rapeseed embryo development in culture on high osmoticum is similar to that in seeds. Plant Physiol. **81,** 907—912.

Galau, G. A., Dure, L. III., 1981: Developmental biochemistry of cottonseed embryogenesis and germination: changing messenger ribonucleic acid populations as shown by reciprocal heterologous complementary deoxyribonucleic acid-messenger ribonucleic acid hybridization. Biochem. **20,** 4169—4178.

Galau, G. A., Hughes, D. W., Dure, L. III., 1986: Abscisic acid induction of cloned cotton late embryogenesis-abundant *(Lea)* mRNAs. Plant Molec. Biol. **7,** 155—170.

Ihle, J. N., Dure, L. S. III., 1969: Synthesis of a protease in germinating cotton cotyledons catalysed by mRNA synthesised during embryogenesis. Biochem. Biophys. Res. Commun. **36,** 705—710.

Jacobsen, J. V., 1984: The seed: germination. In: Embryology of Angiosperms, pp. 611—646. Johri, B. M. (ed.). Springer-Verlag, Berlin.

Lalonde, L., Bewley, J. D., 1986: Patterns of protein synthesis during the germination of pea axes, and the effects of an interrupting desiccation period. Planta **167,** 504—510.

Marcus, A., Rodway, S., 1982. Nucleic acid and protein synthesis during germination. In: The Molecular Biology of Plant Development, pp. 337—361. Smith, H. (ed.). University of California Press, Berkeley.

Misra, S., Bewley, J. D., 1985 A: The messenger RNA population in the embryonic axes of *Phaseolus vulgaris* during development and following germination. J. Exp. Bot. **36,** 1644—1652.

Misra, S., Bewley, J. D., 1985 B: Reprogramming of protein synthesis from a developmental to a germinative mode induced by desiccation of the axes of *Phaseolus vulgaris*. Plant Physiol. **78,** 876—882.

Mori, T., Wakabayashi, Y., Takagi, S., 1978: Occurrence of mRNA for storage protein in dry soybean seeds. J. Biochem. (Tokyo) **84:** 1103—1111.

Norton, G., Harris, J. F., Tomlinson, A., 1976: Development and deposition of proteins in oilseeds. In: Plant Proteins, pp. 59—79. North, G. (ed.). Butterworth, London.

Payne, P. I., 1976: The long-lived messenger ribonucleic acid of flowering-plant seeds. Biol. Rev. **51,** 329—363.

Quatrano, R. S., 1986: Regulation of gene expression by abscisic acid during angiosperm embryo development. In: Oxford Surveys of Plant Molecular and Cell Biology. Vol. 3, pp. 467—477. Miflin, B. J. (ed.). Oxford University Press, Oxford.

Raghavan, V., 1986: Embryogenesis in Angiosperms: A Developmental and Experimental Study. Cambridge University Press, Cambridge, 303 pp.

Sanchez-Martinez, D., Puigdomenech, P., Pages, M., 1986: Regulation of gene expression in developing *Zea mays* embryos. Plant Physiol. **82,** 543—549.

Smith, S. M., Leaver, C. J., 1986: Glyoxysomal malate synthase of cucumber:

molecular cloning of a cDNA and regulation of enzyme synthesis during germination. Plant Physiol. **81,** 762—767.

Tykarska, T., 1979: Rape embryogenesis. II. Development of the embryo proper. Acta Soc. Bot. Pol. **48,** 391—421.

Tykarska, T., 1980: Rape embryogenesis. III. Embryo development in time. Acta Soc. Bot. Pol. **49,** 369—385.

U., N., 1935: Genome analysis in *Brassica* with special reference to the experimental formation of *B. napus* and peculiar mode of fertilization. Jap. J. Bot. **7,** 389—452.

Weir, E. M., Riezman, H., Grieneberger, J.-M., Becker, W. M., Leaver, C. J., 1980: Regulation of glyoxysomal enzymes during germination of cucumber. Eur. J. Biochem. **112,** 469—477.

Weiss, E. A., 1983: Rapeseed. In: Oilseed Crops, pp. 161—215. Longman, London.

Chapter 3

Genes Involved in the Patterns of Maize Leaf Cell Division

Michael Freeling[1], Deverie K. Bongard-Pierce[1], Nicholas Harberd[1], Barbara Lane[1], and Sarah Hake[2]

[1] Department of Genetics, University of California, Berkeley, CA 94720, [2] Plant Gene Expression Center, USDA, 800 Buchanan St., Albany, CA 94710, U.S.A.

With 6 Figures

Contents

I. Introduction

Beginning with a mature leaf, it is relatively easy to discover the size, location, and at least some of the specialized molecular and physiological properties of each component cell and tissue. The origin and growth of the leaf clearly preceeds function. Many of the physiological functions of the tissues and cells of a corn leaf are known, but such knowledge has little to do with how these cells came to be located where they are and will not be reviewed here. Classical techniques of fixation, sectioning, and histochemical staining, coupled with more recent methods of in situ cellular RNA and antigen localizations, can describe in detail the end result of development. Detailed anatomical and histological descriptions of the maize leaf are available (Sharman, 1942; Esau, 1943). Figure 1 depicts various parts of a mature maize leaf. The histological section passing transversely through the blade denotes only the anatomical components and cell types necessary to follow this discussion. A typical corn plant generates about 20 near-identical leaves, the leaf being the most visible of the four components that comprise a vegetative, shoot segment (called the "phytomer"; see Galinat, 1959).

How do the cells of the mature leaf get there? Unlike animal cells, plant cells are instantly cemented into place following division as new extracellular cell-wall material is deposited. Cell membranes do not touch except via cytoplasmic continuities, plasmodesmata, caused by bundles of spindle microtubules that may persist as new membranes and walls form along the plane of mitosis. Plant cells do not migrate or rotate upon one another to make new associations. The cellular pattern of the maize leaf is the

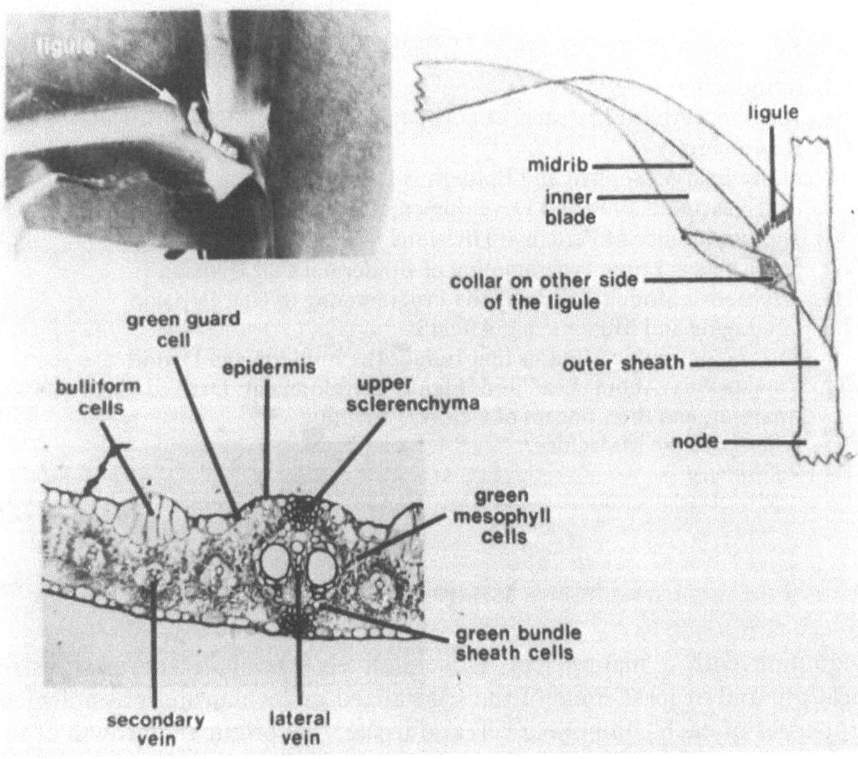

Fig. 1. Identification of the parts of a maize leaf pulled away from the stem. In the upper photograph the arrow indicates the mature ligule. The lower photograph shows a toluidine-blue-stained cross-section through a mature blade in the region of a lateral vein (c. f. Freeling and Hake, 1985). The upper or abaxial surface is at the top

summation of the number and planes (directions) of cell divisions. Differential cell expansion affects the overall size of the organ without altering cellular pattern directly. Judging from observations of the Ragged *(Rg)* maize mutant and of leaves with holes punched in them, young leaf cells do not change their planes of cell division to "fill in" voids caused by cell death. Because of this fixed nature of plant organ development, genetic mosaic analyses of leaf cell lineages have been particularly informative.

II. The Shoot Apical Meristem as a Self-Regulating Unit

As stated previously, when a cell lineage in a primordial leaf of maize stops dividing, or is ablated, nearby cells do not (completely) fill in the void. In contrast, meristems are self-regulating. The shoot apical meristem of many plants not only fill in ablated regions, but isolated meristem pieces regenerate whole meristems. Pinkington (1929) found that vertically bisected apices of *Vicia* and *Lupinus* generated two complete apices. Sussex (1953) found that one-sixth of a vertically cut potato shoot meristem was sufficient to regulate; that is, this piece grew into a complete meristem. Chapter VII of C. W. Wardlaw's (1968) text entitled "Morphogenesis in Plants" details the evidence from angiosperms and ferns that the shoot apical meristem is a self-regulated population of cells. It is not known how the information of the whole is stored in the parts; cell-cell biophysical stresses and positional fields have been suggested, as will be discussed briefly.

Working with maize, a historically difficult species for tissue culture, E. Irish and T. Nelson (Yale University, persl. comm.) have succeeded in culturing shoot apices cut just below the next to the newest leaf (plastochron 2). These meristems initiate new, apparently normal leaves. Even when the apical dome is scrambled with a needle, normal function is eventually regained. Although this is the result expected from studies on other plants, Irish and Nelson's success permits the surgical investigations so important in order for maize to become a developmental system-one in which histology, allometry, cell biology, experimental embryology, molecular biology and, of course, genetics can all be focused on a single phenomenon.

III. Heterochrony

The term "heterochrony" is used by comparative morphologists studying both animals (Raff and Kaufman, 1983) and plants (see Kaplan, 1971; Lord and Hill, 1987) to denote the general rule that relatively small changes in the *timing* of developmental events often lead to large changes in form and consequent behavior. As argued and documented clearly by Lord and Hill (1987), plant shoots have proved particularly useful for the study of heterochronic events because their indeterminant meristems generate a series of determinant segments (including organs like leaves) that often differ in form. Dramatic examples include "phase changes" from juvenile to adult leaf form or in flower morphology. If a group of organ initials begins to differentiate sooner (smaller) or later (larger), dramatically different patterns of cell division can ensue. A "heterochronic gene" would be involved in either the timing of the differentiation signal or the ability for cells to accurately know their own ages. Hundreds of morphological mutants have occurred in plants (Hilu, 1983; Gottlieb, 1984). It is well known that many more exist in genetically well studied plant species maize or *Arabidopsis,* but have never been described in print. The comparative approach has been valuable. However, to move from a general rule such as heterochrony to a causal understanding of plant morphogenesis will certainly require focusing experiments on a few model systems. As knowledge of one system accumulates, the importance of available mutants can be determined. One will be able to ask: How do cells keep track of age and time? How are the "boundary conditions" of primordial size transduced to meaningful instructions about the planes and rates of subsequent cell division? In short, how do cells in groups talk to one another? We see in the maize leaf, and especially in the ligule of the leaf, a potentially useful model system for answering such questions. The leaf morphology mutants *Kn1* and *Hsf*-family are alterations in the age-identity of certain leaf primordial cells; these mutants are "heterochronic".

IV. Maize Leaf Mesophyll and Epidermis Lineage Maps

Because leaf mesophyll, bundle sheath, and the guard cells of the epidermis contain chloroplasts (Fig. 1), the cellular lineages of these tissues can be determined using somatic mosaics marked by the uncovering of an albino allele. Steffenson (1968) X-irradiated +/albino heterozygotic kernels and recorded the plants that had white meosphyll stripes (sectors) occurring on more than one leaf. Recurring sectors resulted from mutational events that occurred during proliferation of the meristem itself. Steffenson estimated that there were 16—32 leaf initials (sometimes called leaf "founder cells") forming a slightly overlapping ring around the apical dome because the narrowest of these recurring stripes was 1/16 to 1/32 of the corrected leaf width. Steffenson found good agreement between his "fate map" and the histological results of previous workers.

Sharman's (1942) histological observations suggested that the leaf primordium arose from initial cells in both the outer layer (epidermis, LI) and the adjacent internal layer (LII) of the shoot apical meristem. Poethig (1984) devised a leaf fate-mapping strategy that has generated a detailed and experimentally useful reconstruction of the meristem's leaf initials. Poethig irradiated germinated albino heterozygotes that had initiated a known number of leaf primordia. Poethig first recognized his sectors as narrow white (mesophyll) stripes in the blade, but then went on to evaluate whether surrounding epidermis was also white. Figure 2 redraws the four most general types of sectors found; Poethig's LI or LII interpretations are indicated on the figure and in its legend. The careful design of this experiment permitted several conclusions. Among the most important for our discussion are: (1) Individual initial cells have no strict tissue fate. (2) The existence of sector types B, C and D implies that at least two layers of meristem (LI and LII) contribute to the leaf. (3) Reasoning similarly to Steffenson, Poethig calculated that the LI initials were approximately 42 cells in circumference. However, Poethig emphasizes that the existence of sectors like Type C (Fig. 2) suggests that the initials are not single-cell rings but, rather, ridges comprising at least three cells. In summary, Poethig's studies implicate a strip of cells at least 3 cells "high" (longitudinal axis), at least 2 cells deep and approximately 42 cells around; thus, at least 252 meristem initial cells originate a leaf at an approximately 90° angle from the surface of the apex.

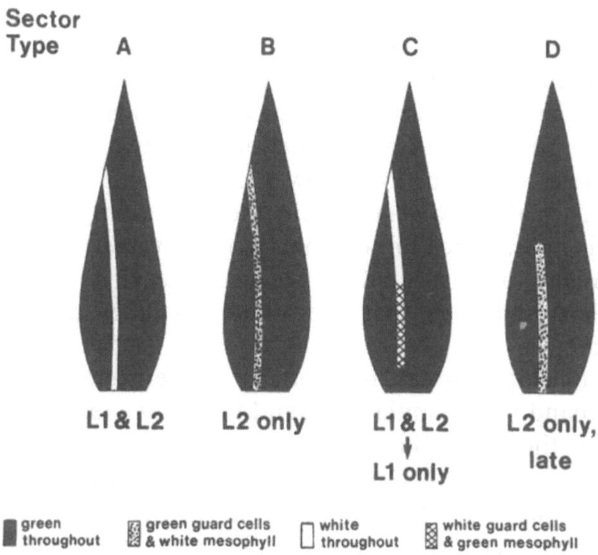

Fig. 2. Poethig's (1984) classes of leaf sectors arising from single, white meristem initial cells. "L1" refers to cells in the outermost layer (epidermis) of the meristem. "L2" refers to the layer of cells just beneath L1

V. Leaf Vascularization and Developmental Compartments

The tip of the maize leaf matures first. Leaves at all stages of maturity, from plastochron 1 (first leaf from the apex) to senescence display a tip to base developmental gradient. Both Sharman (1942) and Esau (1943) describe the time course of vascularization. In one inbred line, the plastochron 1 leaf has no discernable median procambial strand (pre-midrib), but the procambial strand is present in the shoot apex. The front of procambial differentiation will move from the apex up into the leaf. The laterals, which run parallel to the midrib, differentiate at the base of Esau's plastochron 2 leaf and differentiation proceeds in both directions. Smaller procambial strands are first seen in the blade and differentiate down toward the stem. Clearly, the bilateral symmetry and repeated longitudinal subdivisions of the leaf, as marked by the pattern of veins, are established early in the leaf primordium or in the meristem. In the absence of genetic lineage studies where cells that will form cambial tissue are followed with a biochemical marker, it remains possible that the midrib, and perhaps some laterals, have their own initial cells in the meristem. It is not clear if the branching patterns of the smaller veins of the leaf are the consequence of cell lineages.

In a variety of dicots, if a vascular strand in the stem internode is broken, new xylem cells (an easily recognized vascular cell type) develop in a downward direction by the differentiation of extant parenchyma cells (Simon, 1908; see Jacobs, 1952, and the text by Wardlaw, 1968). If procambial differentiation in leaves behaves similarly, there is no reason to assume that the vascular system in leaves arises from initial cells with strictly determined fates. Sachs (1969) has shown that auxin alone is sufficient to cause the differentiation of new xylem strands in pea root. It seems likely that the position and differentiation of the vascular system follow pre-existing lines of auxin flow. How such lines might space themselves out so symmetrically and regularly in the very young leaf is unknown. While there is no evidence for a clonal basis for veination pattern, there is no direct evidence against it either; because parenchyma can differentiate into vascular strands does not necessitate that they do originate from uncommitted cells.

The relationship between the pattens of meristem cell division that generate the band of leaf initial cells and the subsequent division patterns that engender the leaf's shape is not known. By analogy with similar problems encountered in *Drosophila,* the pertinent question is apparent: Is the midrib or any other longitudinal line on the leaf a developmental compartment boundary? (See Brower, 1985, for review and citations.) A compartment boundary is a line beyond which a cell lineage may not grow. In practice, compartment boundaries are usually discovered by using x-rays to promote mitotic recombination leading to twin sectors, one slow growing and the other faster growing and marked by an obvious phenotype. If no compartment boundary exists, the faster-growing cells "fill-in" the deficit made by the slower-growing cells. However, if a compartment boundary does exist, then the faster cells grow up to this arbitrary line and no further. Compartments (or bipartite parasegments) are the functional domains of several developmentally important genes in *Drosophila.*

In our own work, we have never seen a white sector from a single initial, like those diagrammed in Figure 2, that crossed a midrib or lateral vein (we have seen large sectors that include most of a leaf, but these certainly reflect a mosaic apex). However, since the prevailing polarity of cell division in the developing leaf is the same as that of the differentiating veins, this is to be expected.

Twin-sector compartment analyses should be possible for the developing leaf if Green is correct in his prediction that walled plant cells alter their planes of cell division in response to physical stress (Green, 1980; Green and Lang, 1981; this idea will be discussed further). A twin sector of slow-dividing and fast-dividing cells should create physical stress that would be alleviated if the faster-growing cells "fill in the difference". Such experiments have not been done in plants, but should be possible to do in the meristem since it is self-regulating and *may* be possible to do in very young leaves as well.

Hake and Freeling (1986a) found several narrow sectors that were represented in upper leaves and in a few tassel branches as well. These data make it unlikely that the tassel is a developmental compartment.

VI. The Importance of Periclinal Divisions

Cells, when dividing within the plant, tend to divide in a direction. The result of a single epidermal cell dividing in this way (anticlinally) would be a sheet of cells one cell thick. The result of an internal cell dividing in this way would be a column of cells. Because the maize leaf is long and narrow, the typical lineage is a file of cells. For cells on the surface, divisions that generate cells out of the surface plane are called "periclinal" (one can imagine "periclinal" divisions for internal cells as well if the prevailing division pattern is oriented with respect to the surface, such as procambial divisions). For example, the single LI initial cell that generated the completely white Type A sector of Figure 2 must have first divided anticlinally so that its daughters could buckle out from the surface of the apex, and these daughters must have then undergone extensive periclinal divisions to contribute all the cells of the mesophyll. Had the daughters of these LI initials not divided periclinally, they would have contributed epidermal cells only.

Organogenesis in plants generally occurs by forming a new axis of growth at right-angles to the preceeding axis. Lateral roots grow out at a 90° angle from primary roots. Branches – tillers, ear shoots or tassel branches in maize – grow from the right-angle axillary buds associated with each shoot segment (phytomer). Leaves and ligules (as will be discussed) originate as primordial ridges rising perpendicular to the plane of extant cells. The involvement of periclinal divisions in organogenesis is a rule of plant development (see references in Green, 1980).

Why do new organs arise at particular places on old organs or in the apical dome? Next to nothing is known about this pattern-formation problem in plants. Once at least one periclinal division is initiated, one might study the mechanics of division, recruitment of adjacent cells and other aspects of the massive reorientation of division and expansion planes. Green and coworkers (Green and Brooks, 1978; Green and Lang, 1981; Green, 1984) have studied the growth of buds from the leaves of a succulent plant, *Graptopelum paraguayense*. The polarity of individual bud initial cells, as measured by cellulose hoop reinforcement and cortical microtubule arrays, changes gradually, both in the interior of the leaf where periclinal divisions first occur, and in the epidermis as well. Green (1984) explains bud growth largely in terms of repolarization of cells in response to physical stretching. Green's work has certainly shown that once a few cells begin dividing and expanding in a new direction, the repolarization of adjacent cells is gradual, sporatic and probabilistic, not strictly programmed on an individual cell basis.

VII. Strict Versus Loose Programming of Epidermal Cell Divisions

In theory, it is possible to know the exact sequence of cell divisions from an LI cell of the meristem to an adult cell. For example, a two cell leaf hair might be derived from a total of 14 divisions, 13 generating daughters in the plane of the leaf surface — 4 longitudinal, 2 across, 6 longitudinal, 1 across — followed by a periclinal division involved in the second cell of the hair. Given such knowlegde, just how fixed are these sequences of epidermal cell division?

The leaf epidermis originates as a uniform sheet of cells in easily discernable, longitudinal files. At a preset developmental time, from tip to base, some of these cells participate in a strictly programmed "dance" of cell division that culminates in a pair of guard cells (Stebbins and Shah, 1960). Figure 3 shows the steps to this dance, identifies the individual cells in the legend and shows the sort of data that can be obtained by focusing a light microscope on the epidermis of wholemounted leaves that have been fixed and stained with hematoxylin. These results were obtained in monocots with the use of paraffin sections by Stebbins and Shaw (1960). Galatis (1980) has accomplished a definitive electron microscopic description of maize guard-cell morphogenesis with special reference to microtubules and their apparent organizing centers. Stebbins and Shaw (1960) put forth an interesting, testable, but circumstantial hypothesis that the guard cell mother cell (GMC of Fig. 3) induces competent, adjacent epidermal cells to divide and produce subsidiary cells; this is among the first suggestions of cell-cell induction in plants. Although the steps of cell division to stomata formation are strictly programmed, the placement of the stomata in the leaf is more flexible. Sachs (1978) uses the phrase "differentiation-dependent pattern formation" in reference to guard cell placement. This phrase connotes the possibility that information affecting guard cell placement may originate from the expression of previously differentiated cells; in this case, the major procambial strands of the leaf are already in place and may serve to organize subsequent developmental events.

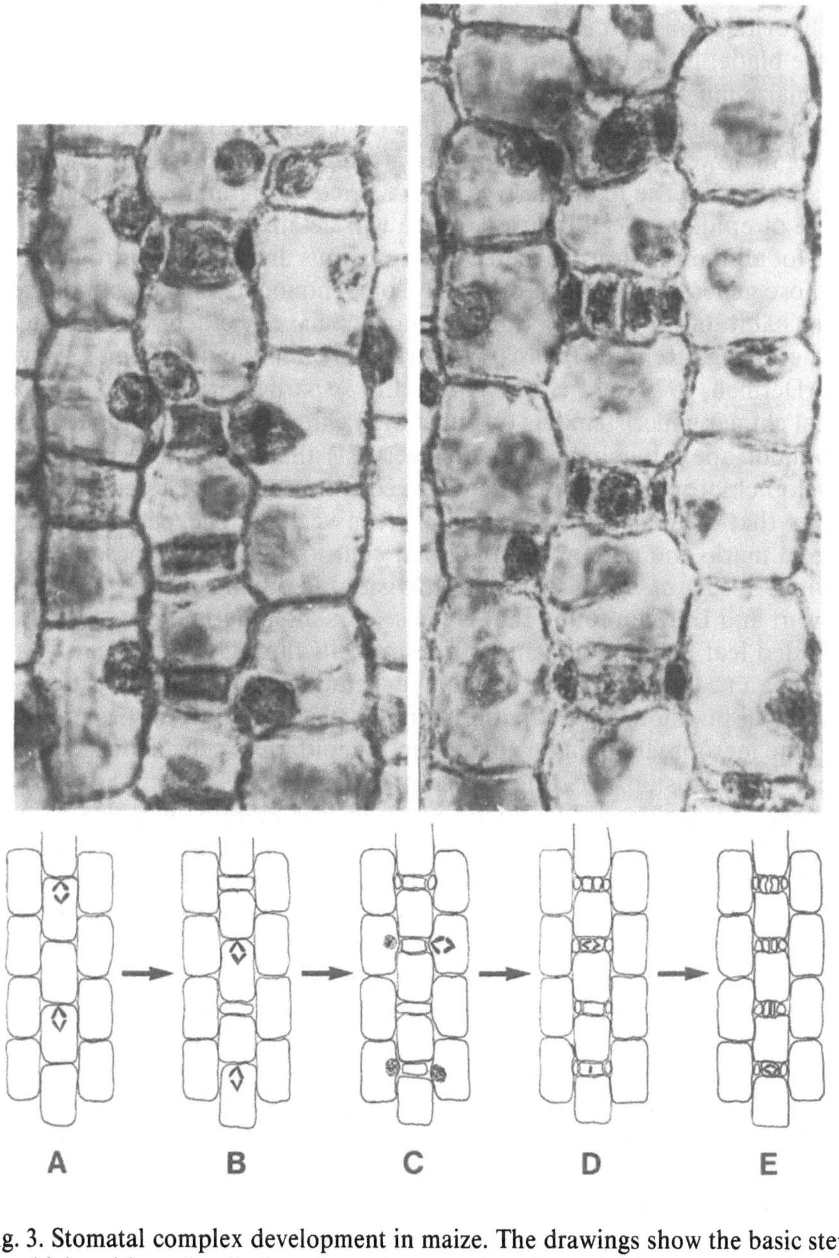

Fig. 3. Stomatal complex development in maize. The drawings show the basic steps by which epidermal cells form stomata. A and B show the first asymmetric divisions which give rise to files of small guard cell mother cells (GMC), separated by larger epidermal cells. C and D depict the bulging of the cells lateral to the GMC and their subsequent asymmetric divisions to produce the subsidiary cells (SC). Once the SC are formed, the GMC then divide transversely to produce pairs of guard cells as shown in D and E. The photos show whole mounts of hematoxylin-stained epidermal cells in various stages of stomatal development. The photo on the left corresponds to stages A, B and C; the photo on the right to stages D and E. Magnification = 1000 ×

The epidermis also generates various types of hairs, which are especially evident on the outer surface of the sheath and inner (upper) surface of the blade. Little is known of their pattern or differentiation, although a few maize mutants are "hairy". The most unique and experimentally useful leaf epidermal derivatives are the ligule (inner or upper surface) and collar (outer or lower surface), which separate the mature leaf into blade and sheath (Fig. 1). The ligule derives from a strictly programmed change in the plane of epidermal cell division, as we will detail.

Not all the divisions of maize leaf epidermis are as strictly programmed as those generating stomata or ligule. Some monocots differ from maize in their habit of branching. Such monocots have yielded to cell-lineage studies using rare buds that carry "periclinal chimeric meristems" (Stewart and Dermen, 1979, and citations therein). A particularly informative periclinal chimeric meristem has dysfunctional chloroplasts in all inner cells of the shoot apex, but has an outer layer (LI) that retains the potential to develop chloroplasts. The branch derived from such a meristem puts forth leaves that are basically white with green sectors. The pattern of green sectors marks the pattern of epidermal contribution to the mesophyll, the obvious result of single periclinal divisions. Looking at maize relatives, Stewart and Dermen found that green sectors of epidermal origin generally included leaf margin and some stripes of cells throughout the leaf. Leaves on the same branch have similar but certainly not identical patterns of epidermal invasion. Such a pattern might be called "loosely" programmed, as if the mesophyll might be able to recruit epidermal cells in areas of more rapid cell division. Poethig's sector Type D (Fig. 2) demonstrates the tendency for epidermis to divide into the leaf margin in maize as well. Harberd, Hake and Freeling (1987) reported on a periclinal chimeric sector in an unfertilized ear of maize that had been exposed to light to induce greening. The sector was essentially white with a green epidermis, the latter identified by autofluorescence of the chlorophyll in the guard cells. Each ovule was surrounded by leaf-like homologues called glumes, and dozens were included in the sector. Each glume had a few green streaks, and they all looked basically the same, but were not identical; the pattern reminded Scott Poethig (persl. comm.) of the loose pattern of epidermal invasion expected at the very tip of monocot blades.

We conclude that some sequences of cell division in the developing epidermis are strictly programmed (stomatal complex and ligule). Periclinal divisions that contribute a daughter to the mesophyll are under looser control. It seems likely that different rules and mechanisms will apply to strict and loose programs.

VIII. Alternative Models Involving the Programming of Cell Division

Taking leaf epidermal cell division patterns as exemplary, there certainly must be some causal mechanism that tells cells at particular positions to change the direction of cell division before dividing again. There are at least two very different models that have been advanced to explain such proximal causes. The first, argued eloquently by Paul Green (Green, 1985; Green and Lang, 1981) emphasizes how the extant structure of an apex or organ or tissue imposes physical stress on nearby cells. Presumably, cells tend to respond to stretching and twisting by dividing in new, more comfortable directions. Green's biphysical explanation is supported by his laboratory's results on the haphazard recruitment of neighboring initial cells into new axilary buds (discussed previously), by the mathematically precise rules of leaf initiation in many plant species (phyllotaxy), and especially by the recent demonstration of physical stretch-induced opening of ion channels in animal cells (Guharay and Sachs, 1984, 1985). There is at least one result from voltage patch-clamp studies on plant protoplasts and protoplast membranes that suggests that pressure-sensitive anion-channels exist in plants as well (Falke *et al.,* 1986). How stress at the cell wall might be transduced to the membrane is not clear. Gene action in Green's view is relegated to the nebulous quantitative realm of establishing the fundamental architecture of the extant apex or organ. Green recognizes that the differences in morphology of different species are certainly genetic, but sees no need to ivoke genes to explain *observable* cases of organogenesis since the new organ always occurs in a pre-existing physical context.

Alternatively, pattern formation can be seen as a consequence of a cellular-level phenomenon called "positional information" by Lewis Wolpert (1968, 1971). In its simplest sense, a cell's aquisition of position information is a two-step process. First, cells in a group have their positions specified in relation to boundary regions. Any number of mechanisms might, in theory, suffice, but ideas involving chemical gradients (e. g. Meinhardt and Gierer, 1974) or waves (e. g. Goodwin and Cohen, 1969) are the most popular. Second, the cells consult their genetic program to transduce this positional data into behaviors, such as shifts in polarity, change of cell type, and the like. One might expect organ-, tissue- or cell-specific gene products to be involved in the second step. Plant developmental systems are not yet experimentally disciplined enough to test for the applicability of the positional information concept. For this discussion, it is important to contrast the positional information concept with Green's more specialized biophysical theory. While both require cells to interpret information from their surroundings, the biophysical model sees this information as mechanical stress only.

The positional information concept is a response to self-regulation data: a part can regenerate the whole. On the other hand, the biophysical theory may connote a "house of cards" prediction; remove one card and the entire structure may remain, but remove many and the entire pattern disappears in strict accordance to the laws of physics. It is not clear how Green's biophysical explanations of phylotaxy accomodate the fact that even small parts of the meristem can regenerate the entire meristem.

We will discuss these concepts further in the context of the ligule, a leaf organ that is particularly useful from an experimental viewpoint.

IX. The Ligule and Mutants that Affect It

We have chosen the developing maize ligule as an experimental system because it is a dispensible organ, it is morphologically simple, it is of epidermal origin and can therefore be observed, and several maize mutants are known to affect it. The dispensibility and the consequent opportunities for mutational analysis are especially important because it is now possible to clone genes for which transposon-induced mutants exist. Using this system, we hope to study the rules and mechanisms of the genesis, programmed cell divisions and differentiation of a plant organ.

The ligule of a mature leaf is a thin fringe of cells that runs all the way across the leaf on its inner-upper surface at the point where the leaf extends outward from the stalk (Fig. 1 and inset). The sheath wraps around the stalk, the blade tilts out and the ligule, which delineates sheath from blade, presses against the stalk, perhaps functioning to keep water and insects away from the rapidly growing parts of the shoot. On the outer-lower surface of the leaf in the region of the ligule is the collar. This is a thickening that mechanically reinforces the sheath-blade tilt. I maize, the two recessive liguleless mutants have lost both ligule and collar; thus, the entire region is often called the ligule. However, in other grass species, ligule fringe and collar have sometimes evolved separately.

Sharman (1942) showed that the ligule derives its cells entirely from the epidermis. The periclinal divisions initiating the ligule occur in leaves 4-5 plastochrons old in a thin strip near the base of the leaf. Sharman has seen these first periclinal divisions clearly in paraffin sections. Using a method of hematoxylin staining of fixed, whole leaves unrolled from the meristem, one of us (D. B.-P.) also observed these periclinal divisions and their growth into the ligule itself. Figure 4A shows a plastochron 6 leaf surface; notice the band of epidermal cells positioned at a right angle to the veins. The inset (Fig. 4) shows a lower magnification of the younger part of this leaf. Focusing beneath the epidermis does not uncover any extra divisions. Figure 4B shows a comparable epidermal surface of a liguleless mutant (*lg1/lg1*); notice the band of larger cells in the position where the ligule should be. These epidermal cells might be arrested at a particular stage in the process of changing planes; it will be interesting to visualize the microtubule arrays within these cells and, perhaps, also the preprophase band in order to assess polarity (Wick *et al.*, 1981). Although the products of the epidermal divisions are absent in the liguless mutant, everything else about the ligular region seems normal: a band of translucence, an anastomosis of smaller veins, and a flaring at the margin of the leaves (inset, Fig. 4). The pattern of the ligule remains in the recessive liguleless mutants; the simplest interpretation is that the epidermis of the liguleless mutant does not respond to an underlying pattern.

A plastochron 5 leaf is about 1 cm long and the ligule can be seen dividing about 5 mm from the line of attachment. Division begins in the ligule at the midrib and proceeds to the margin (Hake *et al.*, 1985, and D. B.-P., unpublished). The ligule seems to be positioned at a 90° angle to each vein transversed. We know that the information, be it in the form of a molecule or a wave (as in mechanical stress), that tells the ligule where to divide comes from the apical meristem and not the blade. We know this from two observations. First, in some leaves of maize there is no blade at all. The husks are occasionally composed entirely of sheath and the ligule is the oldest part of the leaf. Second, dominant mutants that cause *de novo* ligule formation (*Kn1* and *Hsf* mutants, as will be discussed) involve unexpected mitotic activity and often generate *de novo* sheath as well.

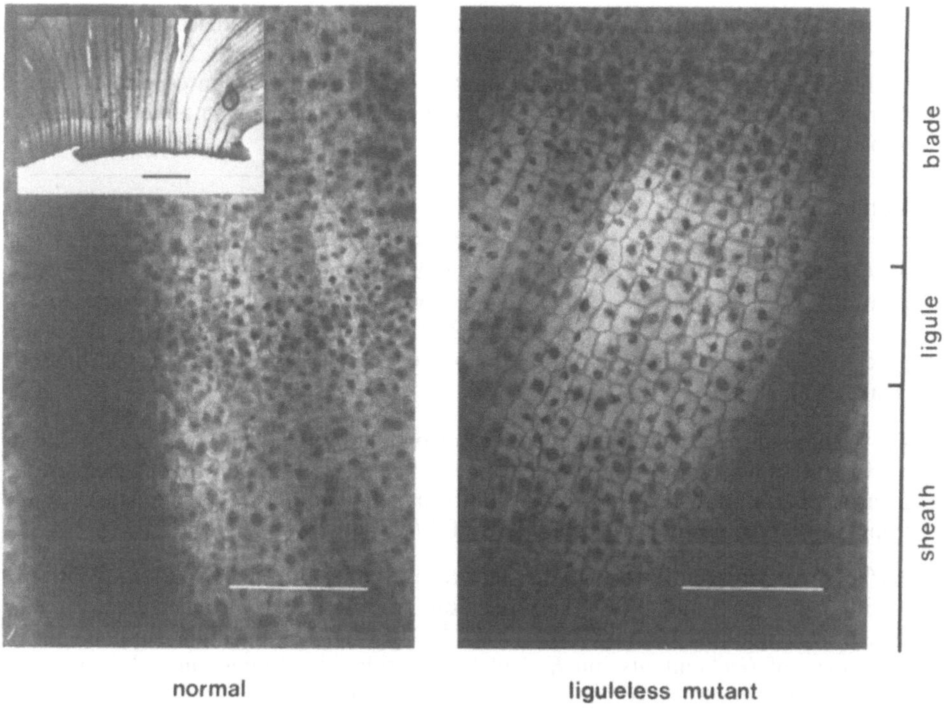

Fig. 4. Comparison of ligule development of *lg1* mutant homozygotes and wild-type siblings. The inset shows the overall view of the base of a plastochron 6 leaf. Bar is 1 mm
The left panel shows the surface of the developing ligule in plastochron 6. Magnification = 250×. Bar is 0.1 mm
The right panel shows the corresponding area presented on the left but in the liguleless mutant. Note that the cells at the blade-sheath boundary are not dividing and are larger. Data from D. B.-P. Magnification = 250X. Bar is 0.1 mm

Because the ligule is dispensible, any mutant that specifically alters or obliterates it can be recovered. Table 1 describes some of these mutants in detail. Some are obviously involved in the disruption of processes essential to ligule formation or pattern. Note that those mutants that remove the epidermal components of the ligule are recessive (*lg1* and *lg2*) while those mutants that alter its shape (*Rs, VG, Lxm, Kn2, Lg3*), position on the leaf (*Rld*), or *de novo* occurence (*Kn1, Hsf*) are all dominant. Formal genetic analyses of the mutants in the clases above are underway. Since such formal analyses are all about the same, we will review our results for the best known among them: Knotted-1 (*Kn1*).

Table 1. Mutants Affecting the Ligule

Recessive Mutants that Remove the Ligule.

lg1 and *lg2* (see Emerson, Beadle and Fraser, 1935). These liguleless mutants are fully penetrant and map to 11 on 2S and 83 on 3L, respectively. They originated spontaneously. See text for histology; mutants probably have nonresponding epidermises. Liguleless mutants remove ligule even when it arises in abnormal locations due to mutant action (unpublished).

Dominant Mutants that Add Extra Ligule.

Kn1. Fingerlike projections from the blade (Bryan and Sass, 1942; Gelinas, Postlethwait and Nelson, 1969). Some of the more severe, developmentally late *Kn1* mutants (127 on 1L) — like *Kn1-O* and *-Z2* (Freeling and Hake, 1985) — specify patches of *de novo* ligule well out on the blade without affecting the normal ligule (Hake et al., 1985).

*Rld**. Three unplaced EMS mutants were recovered by Bird and Neuffer (1985) as having rolled leaves and some ligule fringe on the abaxial (under) side of leaf. One of us (B. L., unpublished) has found bulliform cells and hairs characteristic of the upper epidermis on the lower epidermis of *Rld** blades.

*Hsf**. Three unplaced EMS mutants were recovered by Bird and Neuffer (1985) as having many hairs and sheathlike projections from the blade delineated by a new ligule. The ligule is normal (Hake *et al.*, 1985). Many inbred lines suppress the phenotype of *Hsf** mutants, but *Kn1* alleles override this suppression and display a strong interaction with *Hsf** mutants (M. F., unpublished).

Dominant Mutants that Change the Pattern of Ligule Formation.

Kn1 (Bryan and Sass, 1942). There are seven morphological mutants that map to the *Kn1* locus (127 on 1L; Freeling and Hake, 1985). Among other phenotypes, mutants *Kn1-O, -N1, -N2* and *-Z2* often show "ligule displacement", where the ligule is spread out on the blade in patches. Often, this "displaced" ligule is in pairs over lateral veins (see Fig. 3). *Kn1* induces the epidermis to grow into a ligule. An "age-aquisition" model is discussed in text. *Kn1* displaces the adaxial ligule induced by *Rld** (S. H., unpublished) and the new ligules present with *Hsf** (M. F. and B. L., unpublished).

Lg3. First described by H. S. Perry as a dominant, liguleless mutant near the centromere on chromosome 3 (c. f. Coe and Neuffer, 1977). Actually, the ligule exists, is "displaced" up the blade, but not in pairs as it is with *Kn1* (M. F., unpublished).

Table 1. Continued

Kn2. Recovered and described by M. Zuber (c. f. Freeling and Hake, 1985) as a mutant that generates finger-like projections of leaf just at the ligule. Unlinked to *Kn1* (Freeling and Hake, 1985). Distorts ligule.

*Lxm**. Two unplaced, EMS-induced mutants have been recovered and characterized by Bird and Neuffer (1985) that flatten the midrib of the leaf. This structural alteration somehow results in a ligule that arcs from the margin to run parallel with the midrib.

*Rs**. Several unplaced, dominant rough sheath mutants from a variety of sources (c. f. Emerson, Beadle and Fraser, 1935; Bird and Neuffer, 1985; M. F., unpublished) are characterized by too much sheath with prominant veins. Ligule fringe is distored and collar is sometimes removed or altered (M. F., unpublished).

Dominant Mutant that Slows Ligule Growth.

Vg1 (Sprague, 1939). Not only are the glumes surrounding both male and female flowers shortened, but the ligule is shortened to a ridge (W. Galinat, persl. comm.). It is difficult to imagine what the glume and ligule might have in common.

* Denotes mutants that appear to be similar but have not been tested for allelism or placed on a chromosome.

X. *Kn1*: Neomorphic Mutants that Induce the Epidermis to Divide

The original mutant at the Knotted-1 locus (*Kn1-O;* "O" for "original" mutant allele) was recovered by geneticist A. A. Bryan in a field of corn in 1922. Bryan and Sass (1941) described its most dramatic, dominant phenotype: grotesque fingerlike projections of anatomically normal leaves growing in the region of the lateral veins of the blades. Gelinas, Postlethwait and Nelson (1969) showed that the ligule of *Kn1-O* plants was often displaced up the blade, and prepared histological sections to show that the first cellular manifestation of a knot was an unexpected periclinal division in the epidermis – a division reminiscent of ligule initiation (Sharman, 1942).

Thanks to our maize geneticist colleagues and some good luck of our own, we now have seven independent mutants that map to the *Kn1* locus at position 127 near *Adh1* on the long arm of chromosome 1 (Freeling and Hake, 1985). Not all of these mutants consistently specify knots, nor displaced ligule, nor paired, *de novo* ligule fringes. However, they do share the phenotype of a tendency for the mesophyll, sclerenchyma and bundle sheath cells normally associated with lateral veins (see transverse section of Fig. 1) to remain nongreen and parenchyma-like. In a dramatic mutant like *Kn1-O* in a permissive genetic background, pairs of *de novo* ligules form around these nongreen stripes, lining up parallel to them. Figure 5 shows

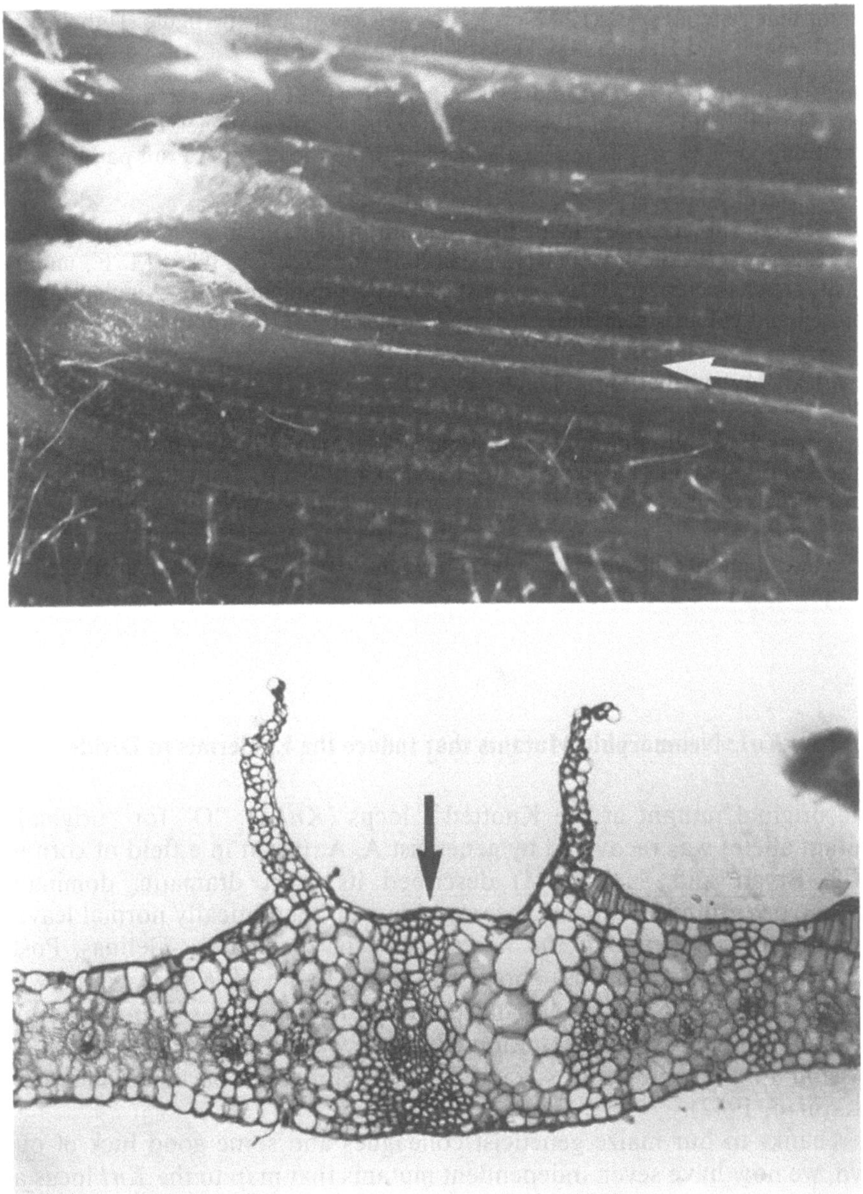

Fig. 5. *Kn1-O* mutants often generate *de novo* ligule as paired fringes paralleling lateral veins. The upper photograph is the surface of a knotted blade; the lower photograph is a stained cross-section through a pair of ligules. Our working hypothesis predicts that the upper epidermis between the ligules, denoted by the arrow, is actually sheath. Note the parenchyma-like cells around the lateral vein. Data from Freeling and Hake (1985); reproduced with permission

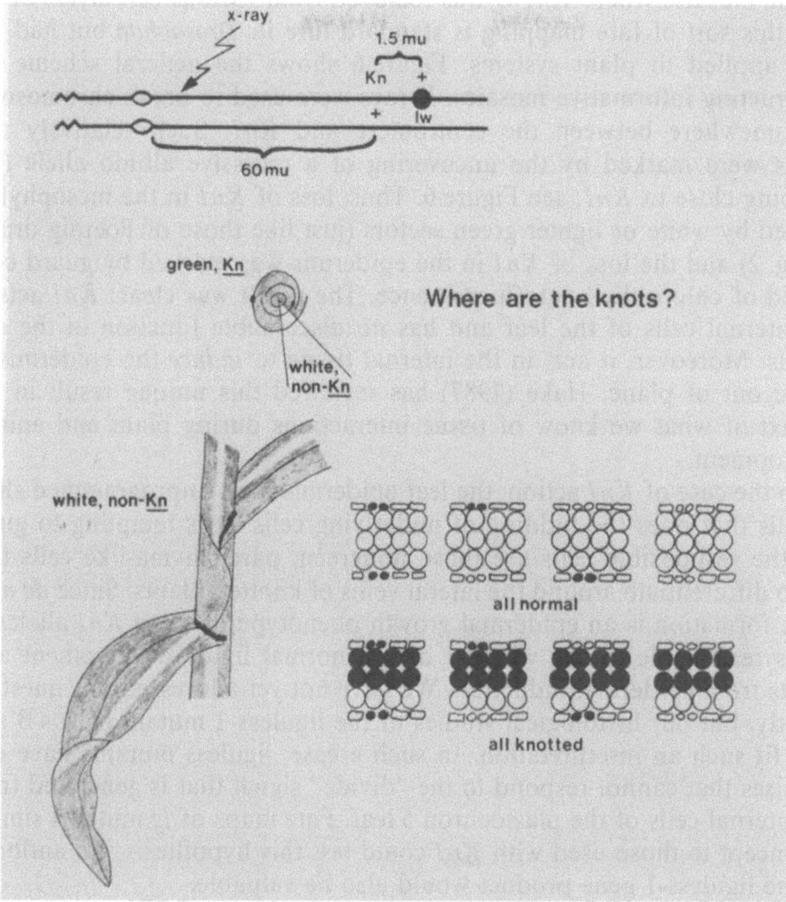

Fig. 6. Hake and Freeling's (1986) scheme for evaluating the tissue-layer of blade in which the *Kn1-O* mutant functions. Shaded or dark areas denote green tissue. White cells or tissue derives from the loss of chromosome 1L carrying *Lw*, uncovering the *lw* albino mutant. See text for how these data, diagrammed in the 8 leaf sections, lead to our conclusion that *Kn1* acts in the internal cells of the blade to induce the epidermis to divide periclinally

such a case; note the parenchyma-like cells near the lateral vein. Further out on the blade, knots predominate over *de novo* ligule.

The use of various translocations with chromosome 1L breakpoints near *Kn1* permitted a routine aneuploid study in which it was shown that a wild-type allele of *Kn1* has no effect on the severity of the *Kn1* mutant phenotype (Freeling and Hake, 1985). This result suggested that these mutants are neomorphic. That is, they encode a new function in the leaf or so vastly overproduce a normal product that the normal allele's contribution to the phenotype is unmeasurable.

In which cells does *Kn1* act? Hake and Freeling (1986 b) analysed genetic mosaics where *Kn1-O* was removed from various cell layers of the leaf; this sort of fate mapping is standard fare in *Drosophila* but had not been applied to plant systems. Figure 6 shows the general scheme for constructing informative mosaics. X-rays were used to break chromosome 1L somewhere between the centromere and *Kn1*. Such relatively rare events were marked by the uncovering of a recessive albino allele (*lw*) mapping close to *Kn1*; see Figure 6. Thus, loss of *Kn1* in the mesophyll is marked by white or lighter green sectors (just like those of Poethig drawn in Fig. 2) and the loss of *Kn1* in the epidermis was marked by guard cells devoid of chlorophyll autofluorescence. The result was clear: *Kn1* acts in the internal cells of the leaf and has no discernable function in the epidermis. Moreover, it acts in the internal tissue to *induce* the epidermis to divide out of plane. Hake (1987) has reviewed this unique result in the context of what we know of tissue interactions during plant and animal development.

In the case of *Kn1* action, the leaf epidermis is an unprogrammed sheet of cells that does the bidding of underlying cells. It is tempting to guess that the responsible cells are those nongreen, parenchyma-like cells that fail to differentiate around the lateral veins of knotted plants. Since *de novo* ligule formation is an epidermal growth phenotype of some *Kn1* alleles, it seems reasonable to ask whether or not normal ligule development also results from epidermal induction. We have not yet addressed this question directly, but our histological studies of the liguless-1 mutant (Fig. 4B and text) fit such an interpretation. In such a case, liguless mutants have epidermises that cannot respond to the "divide" signal that is generated from the internal cells of the plastochron 5 leaf. Fate maps of *lg* mutants similar in concept to those used with *Kn1* could test this hypothesis. An antibody for the liguless-1 gene product would also be valuable.

XI. Conclusions About Leaf and Ligule Development Derived from Mutant Analyses, and the Concept of Cell Age Identity

A major point of the biophysical theory of organogenesis is to obviate the need for specific genes controlling the pattern of periclinal divisions. For the specific case of the maize ligule, four of the mutants of Table 1 — *lg1, lg2, Lg3* and *Kn2* — are specific for ligule alterations. *Vg* arrests cell divisions both in the ligule and, for some reason, the leaf-like glumes in the flowers. The mere existence of such organ-*specific* mutants fits poorly with a purely biophysical model for ligule initiation.

Enough is known about *Kn1* action to support a rather detailed working hypothesis. *Kn1* probably encodes a new protein which is expressed in the developing parenchyma cells near lateral veins. The function of this protein is to delay normal differentiation by a set amount of time, measured in cell generations, without altering the rate of primordial cell division itself. The result is that developmentally younger cells find themselves out-of-place in older tissue. In regions far out in the blade, these younger cells go on dividing even when surrounding cells have stopped; thus, knots (tube-like projections) of perfectly normal tissue develop. In regions nearer the ligule, these younger cells are so young they retain the identity of the sheath (the younger half of the leaf). They may induce strips of sheath above lateral veins, and *de novo* ligule develops as a programmed consequence of a sheath-blade boundary; thus the peculiar paired ligules shown in Figure 5. It seems likely from the *Kn1* fate mapping that the sheath or blade identity of a leaf tissue is centered within the internal cells of the leaf and not in the epidermis. However, it is not clear whether ligule is induced directly from an underlying blade-sheath boundary or indirectly by a secondary blade-sheath epidermal interaction. To answer this question will require markers — like specific message probes or monoclonal antibodies — that are specific to the various leaf cell types.

Even though *Kn1* mutants act to alter the pattern of ligule in the leaf, our studies demonstrate that *Kn1* does not change any leaf pattern at all. *Kn1* somehow retards the internal parenchyma cells' ability to recognize their chronological age. Age aquisition could involve a cell's position in a morphogenetic field. Age converts directly to cell-type in the mature leaf. Our studies have disconnected a cell's age identity from the program of its cell divisions and from its neighboring cells. It follows that a cell's age identity is aquired early in leaf development and is somatically inherited and autonomously expressed thereafter. Epidermal phenotypes result from normal tissue-tissue induction, but the internal leaf cells doing the inducing are developmentally retarded.

A previous section of this chapter emphasized heterochrony as a basic rule of morphogenesis. Because *Kn1* mutants alter developmental age-identity without altering the number or rate of cell divisions, it may now be possible to approach specific questions as to how development time is reckoned.

One of us (S. H., unpublished) has recently recovered several phenotypic revertants of *Kn1-O* from lines carrying transpositionally active *Mu* elements. Some map to the *Kn1* locus and are likely to be *Mu* transposon insertions, which should permit the cloning of *Kn1* by transposon tagging (Lillis and Freeling, 1986). Others map away from *Kn1* and could be dominant suppressors of *Kn1* expression; these could provide leads to other genes that are involved in the early aquisition of cellular age and identity.

Kn1 is the best understood mutant (Table 1) but may not be the most important one. We see the *Kn1* story as a first, albeit small, dividend from our continuing investment in a system powerful enough to eventually supply rules and mechanisms for pattern formation (programmed cell divisions), organogenesis and causal steps of cellular differentiation in a higher plant.

XIII. Where Are the Molecules?

We are well aware that the mainstream of "developmental genetics" in plants is concerned with cell- and organ-specific enhancer or silencer sites that tie coding sequences into the plant's development. Indeed, we are actively pursuing this sort of research in other projects. Perhaps by working backwards from enhancers to the regulatory proteins that bind them, then to their enhancers, and so forth, a developmental master gene will be discovered. However, in the words of Arthur Kessler (1967), living systems operate by "fixed rules and flexible strategies". Following a tradition that has been amply validated by breakthroughs in our understanding of *Drosophila* segment identity and embryo polarities, we have more confidence in the genetic approach practiced on well-understood cellular systems. Eventually, the mutants listed in Table 1, and others yet to be discovered, will be cloned, and their products will be followed *in situ*. We can only trust that this biological system will prove powerful enough to disclose the meaning of these molecules.

Acknowledgements

We thank Scott Poethig for helpful criticism. We thank Elizabeth Montgomery for editing and help with graphics. N. H. is an U. K. SERC postdoctoral fellow. This research was supported by an NSF (U. S.) grant to M. F. and S. H.

XIII. References

Bird, R. McK., Neuffer, M. G., 1985: Odd new dominant mutations affecting the development of the maize leaf. In: Plant Genetics. Freeling, M. (ed.), pp. 818—821. New York: Alan R. Liss.

Brower, D., 1985: The sequential comparmentalization of Drosophila segments revisited. Cell **41,** 361—364.

Bryan, A. A., Sass, J. E., 1941: Heritable characters in maize. J. Hered. **32,** 343—346.

Coe, E. H., Jr., Neuffer, M. G., 1977: The genetics of corn. In: Corn and Corn Improvement. Sprague, G. F. (ed.), pp. 111—223. Madison: American Society of Agronomy.

Emerson, R. A., Beadle, G. W., Fraser, A. C., 1935: A summary of linkage studies in maize. Cornell Univ. Agric. Exp. Stn. Mem. **180.**

Esau, K., 1943: Ontogeny of the vascular bundle in zea mays. Hilgardia **15,** 327—368.

Falke, L., Edwards, K. L., Misler, S., Pickard, B., 1986: A mechanotransductive ion channel in patches from cultured tobacco cell plasmalemma. Plant Physiol. Suppl. **80,** 9 (Abstr.).

Freeling, M., Hake, S., 1985: Developmental genetics of mutants that specify knotted leaves in maize. Genetics **111,** 617—634.

Galatis, B., 1980: Microtubules and guard-cell morphogenesis in *Zea mays* L. J. Cell Sci. **45,** 211—244.

Galinat, W. C., 1959: The phytomer in relation to the floral homologies in the american *Maydeae.* Harvard University Museum Leaflets **19,** 1—32.

Gelinas, D., Postlethwait, S. N., Nelson, O. E., 1969: Characterization of development in maize through the use of mutants. II. The abnormal growth conditioned by the knotted mutant. Am. J. Bot. **56,** 671—678.

Goodwin, B. C., Cohen, M. H., 1969: A phase-shift model for the spatial and temporal organization of developing systems. J. Theoret. Biol. **25,** 49—107.

Gottlieb, L. D., 1984: Genetics and morphological evolution in plants. Am. Nat. **123,** 681—709.

Green, P. B., 1980: Organogenesis — a biophysical view. Ann. Rev. Plant Physiol. **31,** 51—82.

Green, P. B., 1984: Shifts in plant cell axiality: histogenetic influences on cellulose orientation in the succulent, *Graptopetalum.* Develop. Biol. **103,** 18—27.

Green, P. B., 1985: Surface of the shoot apex: a reinforcement-field theory for phyllotaxis. J. Cell Sci. Suppl. **2,** 181—201.

Green, P. B., Brooks, K. E., 1978: Stem formation from a succulent leaf: its bearing on theories of axiation. Am. J. Bot. **65,** 13—26.

Green, P. B., Lang, J. M., 1981: Toward a biophysical theory of organogenesis: birefringence observations on regenerating leaves in the succulent *Graptopetalum paraguayense* E. Walther. Planta **151,** 413—426.

Guharay, F., Sachs, F., 1984: Stretch-activated single ion channel currents in tissue-cultured embryonic chick skeletal muscle. J. Physiol. **352,** 685—701.

Guharay, F., Sachs, F., 1985: Mechanotransducer ion channels in chick skeletal muscle: the effects of extracellular pH. J. Physiol. **363,** 119—134.

Hake, S., 1987: Tissue interactions in plant development. Bioessays **6,** 58—60.

Hake, S., Bird, R. McK., Neuffer, M. G., Freeling, M., 1985: The maize ligule and mutants that affect it. In: Plant Genetics. Freeling, M., pp. 61—72. New York: Alan R. Liss.

Hake, S., Freeling, M., 1986a: Is the tassel a developmental compartment in the young meristem? Maize Genet. Coop. News Lett. **60,** 23—24.

Hake, S., Freeling, M., 1986b: Analysis of genetic mosaics shows that the extra epidermal cell divisions in *Knotted* mutant maize plants are induced by adjacent mesophyll cells. Nature **320,** 621—623.

Harberd, N., Hake, S., Freeling, M., 1987: Programmed periclinal divisions of epidermal cells during glume development. Maize Genet. Coop. News Lett. **61,** 23—24.

Hilu, K. W., 1983: The role of single-gene mutations in the evolution of flowering plants. Evol. Biol. **16,** 97—128.

Jacobs, W. P., 1952: The role of auxin in differentiation of xylem around a wound. Am. J. of Bot. **39,** 301—309.

Kaplan, D. R., 1971: On the value of comparative development in phytogenetic studies — a rejoiner. Phytomorphology **21,** 134—140.

Kessler, A., 1967: The Ghost in the Machine. New York: Macmillan.

Lillis, M., Freeling, M., 1986: *Mu* transposons of maize. Trends Genet. **2,** 183—188.

Lord, E. M., Hill, J. P., 1987: Evidence for heterochrony in the evolution of plant form. In: Development as an Evolutionary Process. Raff, R. A., Raff, E. C. (eds.), pp. 47—70. New York: Alan R. Liss.

Meinhardt, H., Gierer, A., 1974: Applications of a theory of biological pattern formation based on lateral inhibition. J. Cell Sci. **15,** 321—346.

Pinkington, M., 1929: The regeneration of the stem apex. New Phytol. **28,** 37—53.
Poethig, R. S., 1984: Cellular parameters of leaf morphogenesis in maize and tobacco. In: Contemporary Problems in Plant Anatomy. White, R. A., Dickison, W. C. (eds.), pp. 235—259. New York: Academic Press.
Raff, R. A., Kaufman, T. C., 1983: Embryos, Genes and Evolution. New York: Macmillan.
Sachs, T., 1969: Polarity and the induction of organized vascular tissues. Ann. Bot. **33,** 263—275.
Sachs, T., 1978: Patterned differentiation in plants. Differentiation **11,** 65—73.
Sharman, B. C., 1942: Developmental anatomy of the shoot of *Zea mays* L. Ann. Bot. **6,** 245—282.
Simon, S., 1908: Experimentelle Untersuchungen über die Entstehung von Gefäßverbindungen. Ber. Deutsch. Bot. Ges. **26,** 364—396.
Sprague, G. F., 1939: Heritable characters in maize. 50 — Vestigal glume. J. Hered. **30,** 143—145.
Stebbins, G. L., Shah, S. S., 1960: Developmental studies of cell differentiation in the epidermis of monocotyledons. II. Cytological features of stomatal development in the Gramineae. Develop. Biol. **2,** 477—500.
Steffensen, D. M., 1968: A reconstruction of cell development in the shoot apex of maize. Am. J. Bot. **55,** 354—369.
Stewart, R. N., Dermen, H., 1979: Ontogeny in monocotyledons as revealed by studies of the developmental anatomy of periclincal chloroplast chimeras. Am. J. Bot. **66,** 47—58.
Sussex, I. M., 1953: Regeneration of the potato shoot apex. Nature **171,** 224—225.
Wardlaw, C. W., 1968: Morphogenesis in Plants. London: Methuen.
Wick, S. M., Seagull, R. W., Osborn, M., Weber, K., Gunning, B. E. S., 1981: Immunofluorescence microscopy of organized microtubule arrays in structurally stabilized meristematic plant cells. J. Cell Biol. **89,** 685—690.
Wolpert, L., 1968: The French flag problem: a contribution to the discussion on pattern development and regulation. In: Towards a theoretical biology, Vol. 1. Waddington, C. H. (ed.), pp. 125—133. Edinburgh: Edinburgh Univ. Press.
Wolpert, L., 1971: Positional information and pattern formation. Curr. Top. Develop. Biol. **6,** 183—224.

Chapter 4

Molecular Analysis of Genes Determining Spatial Patterns in *Antirrhinum majus*

Enrico S. Coen, Jorge Almeida, Tim P. Robbins, Andrew Hudson, and Rosemary Carpenter

John Innes Institute, Colney Lane, Norwich NR4 7UH, U.K.

With 4 Figures

Contents

I. Introduction

Coloration in plants and animals provides favourable material for the analysis of pattern formation. The ease of observing differences in colour has lead to the documentation of an enormous amount of variation in the pattern and intensity of pigmentation. Genetic analysis of this variation has revealed large allelic series at loci such as *white* and *yellow* in *Drosophila melanogaster* (Judd, 1976; Nash and Yarking, 1974) *succinea* in the beetle *Harmonia axyridis* (Tan, 1945), *agouti* in mammals (Searle, 1968), *nivea (niv) and pallida (pal)* in *Antirrhinum majus* (Baur, 1924; Fincham and Harrison, 1967; Carpenter *et al.*, 1987), *c1, a1, a2,* and *R* in *Zea mays* (Coe and Neuffer, 1977) and *an1, an2* in *Petunia hybrida* (Bianchi *et al.*, 1978; Gerats *et al.*, 1983). A major reason for the variability in colour is that it usually does not affect the viability of an organism, although in natural populations coloration may have important consequences on ecological aspects such as the dispersal of pollen in flowering plants (Stanton *et al.*, 1986).

The variation in pigmentation patterns may be classified into two types: clonal and non-clonal. Clonal patterns arise from genetic changes in cells that occur during the development of the organism and which are subsequently passed on to daughter cells through mitosis. The resulting patterns consist of clonal sectors of altered pigmentation which reflect cell lineage during ontogeny. For example, variegated flowers of *A. majus* which carry the *pallida*^{recurrens}-2 (*pal*^{rec}-2) allele have spots and sectors of red pigment on an ivory background (Fig. 1b). Close inspection of these sectors reveals their clonal origin since the number of cells in a sector is normally 1, 2, 4, 8 . . ., as expected if they originated 0, 1, 2, 3 . . . divisions before complete flower development. The molecular basis of this clonal pattern is now understood (Martin *et al.*, 1985). The *pal*^{rec}-2 allele has a transposable element (Tam3) inserted in the *pal* gene, blocking its expression. Since the *pal* gene encodes an enzyme required for pigment biosynthesis, the presence of the element prevents the production of colour. However, during the development of the flower, Tam3 may excise from the *pal* locus to restore a functional *Pal*$^+$ gene which is then faithfully replicated in subsequent cell divisions to give red spots and sectors in the flower. Other mechanisms can also give rise to clonal pigmentation patterns, such as position effect variegation at the *white* locus of *D. melanogaster* (Demerec, 1940) or random X-chromosome inactivation in mammals (Lyon, 1961). The analysis of clonal patterns provides important information on the cell lineage of a structure and hence is an essential background to the study of non-clonal patterns.

Non-clonal patterns do not have a simple explanation in terms of cell lineage and are usually the more common type of pattern. They include cases in which distinct tissues or organs are differentially pigmented and also cases in which an apparently uniform structure contains differentially pigmented regions. The study of these types of pattern can therefore provide a general model for understanding differential gene expression in development. Although inter-cellular signalling must be involved at some stage in setting up non-clonal patterns, most mutations which alter these patterns of pigmentation are in genes whose action is cell-autonomous. These mutations therefore affect the realisation or execution of pre-existing patterns or pre-patterns. The term pre-pattern was first introduced by Stern (see Stern, 1968) and will be used here to indicate specific spatial distributions of molecules which regulate gene expression. The nature of pre-patterns and the diverse ways in which they can be revealed is now open to molecular analysis in a few genetically well-characterized systems and in this review we consider some of the recent findings in the study of spatial patterns in *A. majus*.

Three types of mutation may alter the spatial pattern or intensity of pigmentation. The first type consists of *cis*-acting mutations which affect genes encoding enzymes involved in pigment biosynthesis. In *A. majus* two such loci have been most extensively studied (Fig. 2); *nivea (niv)* which encodes the enzyme chalcone synthase (Spiribille and Forkmann, 1982) and *pallida (pal)* which encodes dihydroflavonol-4-reductase (E. Coen, J.

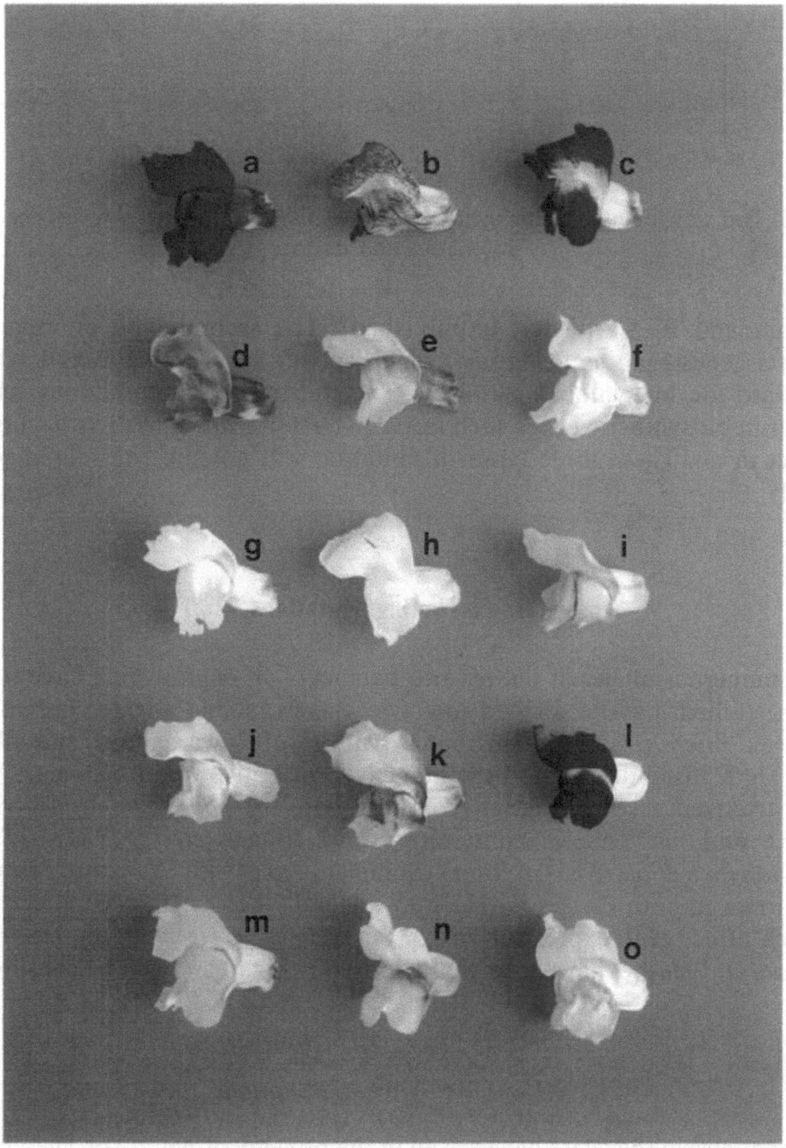

Fig. 1. Flowers illustrating various mutations in *Antirrhinum majus*. Flowers show: (a) full red wild-type (*pal*-501); (b) *pal*^{rec}-2; (c) *niv*^{rec}-98; (d) *pal*-518; (e) *pal*-33; (f) *pal*-32; (g) *pal*-15; (h) *niv*-540; (i) *pal*-510; (j) *pal*-42; (k) *pal*-41; (l) *delila;* (m) *sulfurea* which also carries the *pal*-15 mutation to reveal the yellow pigment; (n) *niv*-525; (o) *niv*-543

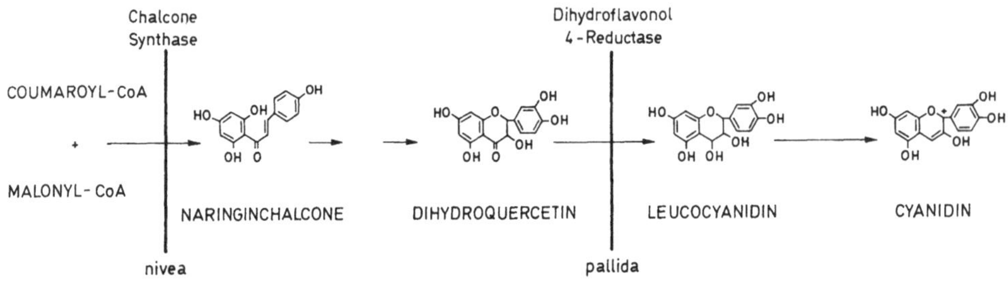

Fig. 2. Pathway to anthocyanin synthesis in *A. majus* showing the enzymes encoded by the *pal* and *niv* loci

Firmin and R. Carpenter, unpublished). The second type of mutation affects genes such as *delila (del)* which encode *trans*-acting factors which regulate the biosynthetic pathway. The third type are mutations which occur in biosynthetic genes such as *niv* but act in *trans* as well as having an effect in *cis*. These three types of mutation will now be considered separately.

II. *Cis*-Acting Mutations

The numerous alleles at the *niv* and *pal* loci of *A. majus* which have so far been studied may be divided into two classes: stable alleles, that confer reduced intensities or altered spatial patterns of flower colour, and alleles that show instability caused by the activity of transposable elements. Well characterized cases in the latter category are the pal^{rec}-2 allele described above and niv^{rec} alleles containing Tam1 (Bonas *et al.*, 1984a), Tam2 (Upadhyaya *et al.*, 1985) or Tam3 (Sommer *et al.*, 1985). A feature common to alleles in both classes is that they generally carry sequence alterations which affect gene expression in *cis*. Several stable *cis*-acting mutations at the *pal* and *niv* loci have been characterised in detail and will be considered first.

A. Stable Cis-Acting Mutations

Most alleles in this class have arisen in the progeny of pal^{rec}-2 or niv^{rec} lines. The majority of stable alleles produced from unstable lines confer full red colour on the flowers (revertants) and are due to excision of transposable elements in the germinal tissue (Bonas *et al.*, 1984b; Sommer *et al.*, 1985; Martin *et al.*, 1985; Upadhyaya *et al.*, 1985). Less frequently, a number of stable mutants conferring reduced intensities of coloration ranging from almost full red to very pale, or determining pigmentation confined to certain areas of the flower have been recovered in the progeny of pal^{rec}-2 and niv^{rec} (Fincham and Harrison, 1967; Carpenter *et al.*, 1984; Harrison and Carpenter, 1973; Carpenter *et al.*, 1987).

One of the best characterized series of alleles which alter the intensity or distribution of pigment are those derived from *pal*^{rec}-2 (see Fig. 3). This allele contains Tam3 located 70 bp upstream of the start of *pal* transcription (Coen *et al.*, 1986). The transposon is flanked by a 5 bp direct duplication of host sequence, the target duplication. Precise excision of a transposon will restore the wild type sequence of the locus at which it was inserted, a single copy of the target being retained. However, analysis of the sequences around the excision sites in full red revertants at *pal* and *niv* has shown that excision is not precise in general since, in the alleles examined the target duplication is retained in a modified form (Bonas *et al.*, 1984b; Sommer *et al.*, 1985; Coen *et al.*, 1986; Hudson *et al.*, 1987; Hehl *et al.*, 1987). For example, alleles *pal*-501 and *pal*-520, which confer a full red phenotype (Fig. 1a), contain insertions, relative to wild type, of 1 bp and 7 bp respectively at the site of Tam3 excision (Fig. 3). The 7 bp insertion in *pal*-520 is the result of deletion of 1 bp of each of the copies of the target duplication and an inverted duplication of one of the copies. This type of rearrangement has been postulated to arise either as a result of DNA polymerase strand switching (Nevers and Saedler, 1985) or following the resolution of intermediate hairpin structures generated on excision of Tam3 (Coen *et al.*, 1986).

The structures of several other stable alleles derived from *pal*^{rec}-2 are shown in Fig. 3. In *pal*-518, which confers reduced pigmentation (Fig. 1d), correlating with decreased levels of *pal* transcript (14 %) relative to full red flowers, 3 bp of the target duplication and 10 bp to its left have been deleted. It is very unlikely that the differences in transcript levels between *pal*-518 and *pal*-501 or *pal*-520 are due to differences in genetic background since the lines that carry these alleles have descended from a common *pal*^{rec}-2 progenitor which was inbred for many generations. Therefore, the sequence missing from *pal*-518 is necessary in *cis* for wild type levels of *pal* transcription. A reduction in transcription has also been shown in *niv* alleles carrying deletions generated by excision of Tam1 (Sommer, Bonas and Saedler, 1988).

The remaining alleles shown in Fig. 3 determine altered patterns of flower colour which may be divided into two classes. The first class includes alleles *pal*-33, *pal*-32 and *pal*-15, which cause pigmentation to be restricted mainly to the flower tubes (Fig. 1e, f, g). The second class contains *pal*-41 which gives a more complex pattern (Fig. 1k). In the alleles of the first class, excision of Tam3 has generated an inverted duplication of 4 or 5 bp of the target sequence. However, in contrast to the origin of *pal*-520, this rearrangement has been accompanied by deletions with end points approximately 10, 20 and 100 bp upstream of the target in *pal*-33, *pal*-32 and *pal*-15, respectively. The generation of these deletions can be explained by assuming that in the model of Coen *et al.* (1986) the endonuclease nicks that lead to resolution of the hairpins may occur not only with the target duplication, as in *pal*-520, but also at several sites beyond it. The second class (*pal*-41) shows the most radical alteration of the *pal* promoter: the entire region 5′ to the Tam3 excision site has been replaced by a foreign sequence. This is thought to have arisen from an aberrant transposition of Tam3 which, from recombination analysis, appears to have inverted a chromosome segment extending at least 6 map units from the *pal* locus (Robbins, Carpenter and Coen, unpublished).

One explanation for the generation of specific patterns of pigmentation by these sequence alterations is that the wild type *pal* promoter contains a set of sequences that respond to diverse regulatory signals spatially arranged as a pre-pattern in the flower. Novel spatial patterns are produced by mutations which change the interpretation of the pre-pattern by modifying the affinity of the *pal* promoter for different regulatory molecules.

Such a model, involving the interaction between transcription factors

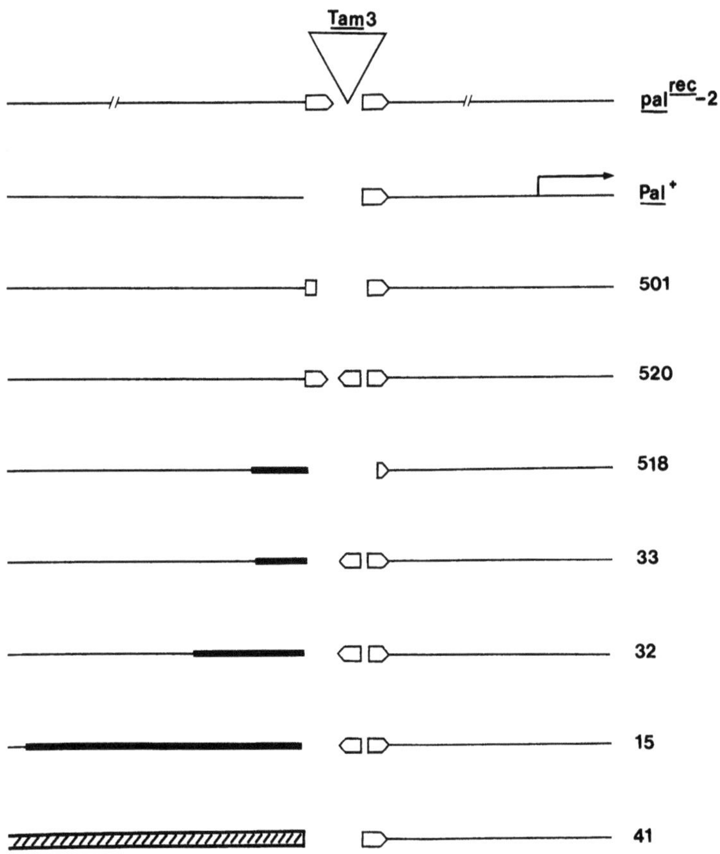

Fig. 3. Structures of *pal*^{rec}-2 and stable *pal* alleles in the region flanking the Tam3 excision site. The phenotypes confered by the alleles, the numbers of which are indicated on the right, are illustrated in Fig. 1. The boxed arrows represent the target duplication, the bold lines indicate deletions and the hatched box represents a sequence not present normally at the *pal* locus. The arrow over *Pal* shows the site of initiation of *pal* transcription (see text for details)

and the *pal* promoter can be invoked to explain the generation of patterns that are specific to the corolla tube. Although there is no direct evidence for such interactions, it is worth emphasizing some sequence features that might provide clues as to how modifications of DNA-protein and/or protein-protein interactions could arise as a result of the mutations described. The lobes of full red flowers are more intensely pigmented and contain significantly higher amounts of *pal* transcript than the tubes. Therefore, an overall reduction of *pal* transcription would be expected to give flowers which still have darker lobes than tubes. However, removal of 10 bp to the left of the target sequence in *pal*-518 results in a reduction of pigmentation that seems more accentuated in the lobes than in the tubes (compare Figs. 1a and d). This is more extreme in *pal*-33 which carries a deletion similar to that in *pal*-518. Other stable alleles derived from pal^{rec}-2 whose structures have not been investigated in detail, show a gradation of colour pattern between those of *pal*-518 and *pal*-33. In *pal*-32, a further deletion of about 10 bp to the left of the target sequence causes an almost complete inhibition of expression of this gene in the lobes (Fig. 1f). The region deleted in these alleles might therefore be a candidate for the binding of a hypothetical transcription factor, that differentially enhances *pal* gene expression in the lobes. Evidence for a *trans*-acting factor that plays a role in determining differential expression of *pal* in tubes and lobes is presented in section III. In *pal*-15, a deletion of approximately 100 bp to the left of Tam3 results in lack or pigmentation in the whole corolla except for a ring at the base of the tube (Fig. 1g). This ring of pigment is formed at a very early stage of flower development in both *pal* mutants and wild type, suggesting that the base of the tube represents an area where *pal* expression is subjected to different control mechanisms, presumably independently of sequences within 100 bp to the left of Tam3. It appears, therefore, from the analysis of this set of alleles, that separable *cis*-acting regulatory elements might confer spatial-specific expression on *pal*. The involvement of complex sets of *cis*-acting sequences in conferring cell-type-specific expression has also been postulated for the *yellow* gene of *Drosophila* (Chia *et al.*, 1986).

The pattern produced by *pal*-41 is quite distinct from that of the tube-specific alleles (Fig. 1k) and requires a different explanation. It is conceivable that this pattern partly reflects a pre-pattern different from the one that normally interacts with the *pal* promoter, since in this allele the entire region to the left of the Tam3 excision site has been replaced by foreign sequences. For example, in *Drosophila* an inversion between the *Antp* gene and a nearby gene results in the exchange of *cis*-acting sequences with the result that the *Antp* gene is brought under the control of the promoter of the other gene (Schneuwly *et al.*, 1987). However, the foreign sequence brought next to the *pal* gene in *pal*-41 does not promote transcription in flowers when in its original position (Robbins, Carpenter and Coen, unpublished). This indicates that the pattern observed in *pal*-41 may reflect a position effect of the adjacent foreign chromatin rather than the introduction of a new promoter. Such effects have been noted for particular transformants of the *white* gene in Drosophila resulting in novel distributions of the red eye pigment (Levis *et al.*, 1985).

B. Unstable Cis-Acting Mutations

Stable alleles showing altered gene expression can be produced by the excision of a transposable element from an unstable allele. An unstable allele can also give rise to new alleles with altered intensity or patterns of expression whilst retaining the transposable element at the locus. In these cases host gene expression can occur with the transposable element present, to produce a non-clonal background pigmentation with a superimposed clonal pattern of revertant cells. Such a mutation may owe its altered phenotype to changes within the transposable element, or to its reinsertion at a new position in the gene. These mutations are of interest for several reasons. Firstly, the transposable element alters the expression of its host gene and, like the mutations caused by excision, can reveal something of the wild-type function of the sequence into which it is inserted. Secondly, the frequency of reversion provides information about the behaviour of the transposable element after it has been altered or has transposed. Thirdly, the way in which the transposable element changes the expression of the host gene may be clarified by examining the effect of changing the element or moving it to a new position. Several unstable anthocyanin mutations containing the transposable elements Tam1, Tam2 and Tam3 have been studied in *A. majus* (see Coen and Carpenter, 1986, for review). The most important factor determining the effect of each transposable element on expression appears to be its location within the host gene. Alleles containing insertions in the promoter, in introns or exons, will each be considered in turn.

The *pal*^{rec}-2 allele of *A. majus* contains the 3.5 kb transposable element. Tam3 inserted into the promoter of the *pal* gene (Coen *et al.*, 1986). The flowers show a high frequency of dark red sites, caused by somatic excision of Tam3, against a colourless background, indicating that no expression of the gene occurs with Tam3 in position (Fig. 1b). This null phenotype might be expected to be due to the interruption of the promoter sequence, part of it being displaced over 3.5 kb away from the gene. However, several pieces of evidence suggest that Tam3 may not simply act as a passive spacer sequence, but may actively inhibit expression of the host promoter. Firstly, Tam3 completely prevents expression of *pal,* whereas even the large deletions of promoter sequences upstream of the insertion site in the alleles *pal*-15 and *pal*-41 do not. Further evidence is provided by a consideration of alleles derived from *pal*^{rec}-2, and of other alleles containing transposable elements in the promoter.

The allele *pal*-510 was produced from *pal*^{rec}-2. It shows a much reduced frequency of somatic excision, against a light red background pigmentation (Fig. 1i). This suggests that Tam3 no longer prevents expression of the gene completely, when in position. DNA studies show that Tam3 is in the same position as in its progenitor, but has been altered slightly at the end nearer

Table 1. Structural organisation of unstable alleles and their derivatives in *A. majus*. Large open triangles indicate the location of the elements and deletions in the elements are indicated by gaps in the triangles. Exons are shown by thicker lines and the start and direction of transcription of the host gene is indicated by a small solid triangle. The dotted line in *pal*-42 indicates a sequence which is not normally present at the *pal* locus

Allele	Element	Position	Molecular change from parental allele	Background expression	Reference
pal-2	Tam3			None	Coen *et al.*, 1986
pal-510	Tam3		3 bp loss from downstream end of element	Low	Unpublished
pal-42	Tam3		25 bp deletion of element and promoter, inversion upstream of element	Low, patterned	Unpublished
niv-53	Tam1	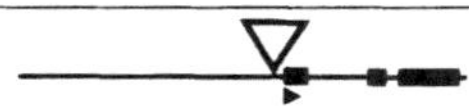		None	Bonas *et al.*, 1984
niv-46	Tam1		5 bp deletion from upstream end of element	Low	Hehl *et al.*, 1987
niv-44	Tam2			None	Upadhyaya *et al.*, 1985
niv-566	Tam2		Internal deletion of element	Low	Unpublished
niv-98	Tam3	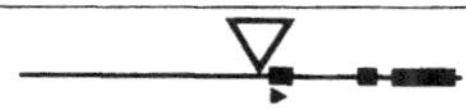		Low	Sommer *et al.*, 1985
niv-540	Tam3		Replication and transposition	None	Unpublished

the *pal* gene (Table 1; Hudson, Carpenter and Coen, unpublished results). This alteration is consistent with an excision of this end of Tam3, followed by immediate re-integration into its original position. That such a slight change can permit background *pal* expression suggests that the disrupting effect of the intact transposable element was not due entirely to interruption of the promoter, but was partly due to active inhibition by Tam3. Alternatively, the altered transposable element might still disrupt the promoter but now partially replace the missing function with its own internal sequences. Such expression of genes by promotion or enhancement from insertion sequences has been reported for the bacterial IS2, IS3 and IS5 (Glandsdorff *et al.*, 1981; Jund and Loison, 1982), *Ty1* of yeast (Roeder *et al.*, 1985), *copia* in *Drosophila* (Scott *et al.*, 1983) and many vertebrate retroviruses (e.g. Luciw *et al.*, 1983). In the case of *pal*-510, no transcript starting within Tam3 and running into the gene can be detected, and it is difficult to imagine how the altered transposable element could supply an enhancer or promoter function, when the intact Tam3 did not. This strongly suggests that the effect of the intact Tam3 is an active inhibition of *pal* gene expression.

The allele *pal*-42 was also derived from pal^{rec}-2, and it shows certain similarities to *pal*-510 in exhibiting a reduced frequency of somatic excision against a background pigmentation. However, whereas the background pigmentation produced by *pal*-510 has the same distribution as in the wild type flower, so that it is most intense at the base of the tube and in the face and lobes of the corolla, that caused by *pal*-42 shows a novel pattern (Fig. 1j). Molecular analysis of *pal*-42 reveals that Tam3 has undergone a small deletion at the end nearer the gene, similar to that in *pal*-510 but extending further into the promoter (Table 1). It differs dramatically at the other end of the element, where the same sequence as in *pal*-41 is found, suggesting that *pal*-41 and *pal*-42 are derived from a common progenitor (Robbins, Carpenter and Coen, unpublished results). It seems probable that *pal*-42 possesses background pigmentation for the same reasons as *pal*-510. The pattern of pigmentation in this allele may be due to the presence of the new sequence at the upstream end of Tam3, as has been discussed in section II A.

The behaviour of Tam1 in the promoter of the *niv*-53 allele is similar to that of Tam3 in pal^{rec}-2 (Bonas *et al.*, 1984). In this case Tam1 blocks expression when in position, but restores it on excision. The allele *niv*-46 is an apparently stable derivative of *niv*-53 showing a pale pigmentation and contains a 5 bp deletion in the end of Tam1 further from the gene (Table 1; Hehl *et al.*, 1987). This deletion therefore restores some gene expression whilst abolishing or greatly reducing the ability of Tam1 to excise.

There are a number of ways in which a transposable element in a promoter region might actively inhibit expression. One possible mechanism is that it might prevent expression by disrupting the secondary structure of the chromatin in its environment. A *Drosophila* transposable element, *HMS Beagle*, is known to increase the DNA'ase I sensitivity of a neighbouring DNA sequence, suggesting that the element alters the chromatin structure (Eissenberg *et al.,* 1985). However, such changes are usually associated with an increase in expression. Another possibility, proposed for inhibition of the *yellow* gene of *Drosophila* by *gypsy* is that the transposable element competes with the host promoter for transcription factors (Chia *et al.,* 1985).

The similarity of the alleles *pal*-510, *pal*-42 and *niv*-46, which show a reduced frequency of excision and background pigmentation, suggests another way in which Tam1 and Tam3 may inhibit expression. The three alleles all show alterations at one end of the transposable element. The ends are the proposed recognition sites for transposase molecules (Saedler and Nevers, 1985), and so the reduced frequency of excision presumably reflects a reduced affinity of transposase for the altered ends. Thus the inhibitory effect of the intact transposable elements may be due to transposase binding, possibly because both termini are brought together in a complex with transposase to produce a structure which physically restricts access of transcription factors to the host promoter. Alterations in the ends of the transposable elements may then reduce the affinity for transposase and the probability of forming the transposition complex, thus allowing some expression of the gene. A similar model has been put forward to explain the action of the maize transposable element *I* in the *a*1 gene (Schwarz-Sommer *et al.,* 1985). However, even an intact Tam3 does not always inhibit expression when located in the promoter. In the allele *niv*^{rec}-98, Tam3 both excises with a high frequency and allows background expression (Fig. 1c) (Sommer *et al.,*1985). It may be that the orientation of the element is also important in determining its phenotypic effects, since in *niv*^{rec}-98, Tam3 is in the opposite orientation to that found at *pal.* Alternatively the different structure of the promoters or sites of insertion of the element at *niv* and *pal* may affect the interaction between the element and the host gene.

In addition to insertions in the promoter, other transposon locations have been described in *A. majus.* The *niv*-44 allele contains Tam2 at the first exon-intron boundary (Upadhyaya *et al.,* 1985). In general, insertions in introns cause either a reduction or a complete abolition of expression of the host gene. No expression occurs when the transposable element contains poly-adenylation signals which terminate transcription. This is the case for the *copia* transposable element in the *white*^{apricot} allele of *Drosophila* (Zachar *et al.,* 1985) and the maize *Spm/En* transposable element in the *wx* gene (Gierl *et al.,* 1985). When the transposable element contains no polyadenylation signals a reduced level of expression usually results; probably because its inclusion in the pre-mRNA interferes with RNA processing. The maize *Adh1-S5340* allele containing *Mu* in the first intron appears to fall into this category (Bennetzen *et al.,* 1984). In the case of the *niv*-44 allele the insertion disrupts the splice-donor sequences. However, the absence of the original donor sequence may not be resposible for the phenotype since the new donor site is still consistent with the consensus sequence proposed by Mount (1982). The presence of Tam2 allows no expression, so it may either cause termination of transcription or provide a splice-acceptor site resulting in a mature transcript which contains Tam2 sequences. A reduced level of expression is observed in *niv*-567 which has been derived from *niv*-44, possibly by an internal deletion of Tam2. It shows a background pigmentation with both red and colourless sites superimposed. The background pigmentation suggests that whatever was preventing expression in *niv*-44 has now been deleted. A similar deletion in the *I* transposon of maize allows some gene expression by removing poly-adenylation signals (Schwartz-Sommer *et al.*, 1985). The sites seen on the corolla of *niv*-567 are probably caused by excision of the remainder of Tam2 to produce either mutant or wild-type splice-donor sequences.

The consequence of transposon insertions into coding regions are the most straightforward to understand. The allele *niv*-540 was derived from niv^{rec}-98, which contains Tam3 in the promoter. In contrast to its progenitor, *niv*-540 produces no background pigmentation and has a reduced frequency of red sites (Fig. 1h). The new allele contains two copies of Tam3 (Table 1). One copy, about 3 kb upstream of the promoter does not affect the phenotype of flowers, while the second is within the last exon of the *niv* gene and blocks *niv* expression completely. The reduced frequency of wild-type sites seen on *niv*-540 flowers is presumably a reflection of the location of Tam3 within a coding region, since many imprecise excisions will cause null mutations caused by amino acid substitutions and frame shifts.

Amongst all these alleles only *pal*-42 shows a novel spatial distribution of pigmentation. This may reflect the change in relative position on the chromosome of the *pal* gene in this allele as described earlier for *pal*-41 (see section II A).

III. *Trans*-Acting Mutations

A *trans*-regulatory gene may be defined as one which encodes a factor that interacts with one or more genes at other loci so as to produce altered transcriptional activity. The cumulative action of several such factors acting directly or indirectly on the structural genes of the anthocyanin pathway is thought to be responsible for the normal distribution of anthocyanin in the flower. *Cis*-alterations of structural genes may alter their interaction with particular *trans*-acting regulators to produce altered distributions of pigment that reveal underlying pre-patterns not visible in the wild type flower. It follows that mutations in the genes which encode *trans*-acting regulators might also be expected to reveal such pre-patterns.

In *A. majus,* the mutation of a single gene, *delila* (*del*), generates an altered spatial pattern of pigmentation by a *trans*-regulation of the anthocyanin pathway. The most dramatic change in plants homozygous for the recessive *del* allele is the loss of anthocyanin pigment from the corolla tube whilst apparently normal levels are maintained in the corolla lobe (Fig. 1l).

Several lines of evidence indicate that the *del* mutation alters the distribution of pigment synthesis, at least in the corolla, by acting in *trans* to change the expression of several structural genes required for anthocyanin biosynthesis. This has been most clearly demonstrated by the analysis of transcript levels for two structural genes of the pathway, *niv* and *pal*. By extracting RNA from dissected wild type (*Del*$^+$) and mutant (*del*) flowers it was shown that, whilst apparently normal levels of expression were maintained in the lobes of *del* flowers, the tubes revealed a 5—10 fold reduction in *niv* gene expression and an undetectable level of *pal* expression (Robbins, Carpenter and Coen, unpublished results). Additional evidence from the feeding of a leucocyanidin intermediate (Fig. 2) to flowers (Carpenter and Coen, unpublished data) and measuring levels of the late acting enzyme UDP-Glucose: flavonoid 3-0-glucosyltransferase (Martin *et al.*, 1987) suggests that subsequent steps in the pathway may also be altered at the transcriptional level in *del* mutants.

An unstable *del* mutation *del*^{rec}, which can restore pigment synthesis to single cells in the corolla tube, provides evidence that the *Del*$^+$ gene product can act in the final stages of flower development, in an intracellular or cell-autonomous fashion. This argues against any role for the Del$^+$ gene product in the establishment of pre-pattern but instead suggests a late acting intracellular role in the *trans*-activation of anthocyanin biosynthetic genes in the corolla tube. Because of the high levels of pigment synthesis in the lobes of *del* flowers it is not clear whether the *Del*$^+$ gene product also activates expression in the lobes to a small degree. Whether the *Del*$^+$ gene product itself might be differentially expressed between lobes and tubes is therefore uncertain. In addition to the corolla tube, a number of other normally pigmented structures lack colour in *del* plants, notably the lower stem and leaves. In specifically reducing the synthesis of pigment in a subset of normally pigmented structures the action of *del* is closely analogous to that of certain regulatory loci in maize. These loci appear to regulate the distribution of anthocyanin in the various tissues of the maize plant in a cell-autonomous manner (Coe and Neuffer, 1977). Two of these regulatory loci, *R* and *C*, alter the levels of at least two enzymes encoded by structural genes of the pathway (Dooner, 1983). However, whilst the action of these loci distinguishes between recognisable tissues or organs of the plant, the most dramatic result of *del* is to produce a spatial pattern resulting from differential expression within a single organ (the corolla). A pattern of pigmentation defined within thc corolla of *Petunia hybrida* has also been described (Mol *et al.*, 1983). Therefore it is possible for *trans*-regulatory factors to produce patterns of gene expression which need not correlate with the accepted divisions within a plant, of organ or tissue type. Gene expression which is specific to organs, tissues or regions within organs can therefore be considered to be particular cases of more general pre-patterns that are set up during development.

We have argued that the *Del*$^+$ gene product confers a pre-pattern which confers corolla tube pigmentation on the flower. It is important to distinguish this pre-pattern from the underlying pre-pattern of lobe-specific expression revealed in the absence of the *Del*$^+$ gene product. Presumably the latter pre-pattern is also present in the wild type flower but is not revealed until the *Del*$^+$ gene product is removed. Clearly this pre-pattern has a strict morphological correlation with the lobe and tube structures. However the analysis of sectors of cells in *pal*^{rec} flowers indicates that these two structures are not derived from separate cell lineages since clonal sectors of pigment can span both tubes and lobes. This is consistent with the view that the pre-pattern revealed by the *del* mutation reflects a division within the corolla ultimately determined by some form of intercellular signalling. Evidence for a pre-pattern conferring expression in the lobes also comes from a particular distribution of the yellow pigments termed aurones. These pigments are derived from chalcones which form the first step of the anthocyanin biosynthetic pathway. Aurones can be most readily observed in flowers blocked in the later stages of anthocyanin biosynthesis (e.g. *pal*-15). If such plants are also homozygous for the recessive *sulfurea* (*sulf*) mutation the spatial distribution of aurones in the corolla is similar to that of anthocyanins in a *del* plant. The yellow pigments are present throughout the lobes but are absent from the tube (cf. Figs. 1m and 1l). It is possible that this similarity arises from the response of both pigment pathways to the same lobe-specific pre-pattern. In contrast, the normal distribution of aurones in a dominant *Sulf*$^+$ flower is a restriction to the "lip" of the lower lobe (eg. Fig. 1g). This bears a limited resemblance to the distribution of anthocyanin that occurs in plants carrying a dominant *Eluta* (*El*$^+$) allele (not shown) or the *pal*-41 allele already described. This raises the possibility that both structural genes and *trans*-regulatory genes of pathways giving rise to two distinct floral pigments may be responding to the same pre-patterns within the corolla which can confine expression to the lobe or "lip" areas.

IV. Mutations Which Act Both in *Cis* and *Trans*

The majority of mutations in biosynthetic genes such as *niv* and *pal* described above reduce the overall quantity of gene expression. These mutations are recessive to wild-type since a single dose of the wild-type allele encodes sufficient gene product to confer the full red phenotype. However, several *niv* alleles have been described which are semi-dominant to wild-type (Carpenter *et al.*, 1987). Plants homozygous for the *niv*-525 allele have very pale flowers with pigment concentrated in the tubes and around the lower lip (Fig. 1n). When crossed to wild-type, the F_1 plants (*Niv*$^+$/*niv*-525) have a similar spatial distribution of pigment as *niv*-525/*niv*-525 but of a slightly darker intensity. Self-pollination of the F_1 plants gives F_2 progeny containing *niv*-525 homozygotes, *niv*-525/*Niv*$^+$ heterozygotes and *Niv*$^+$ homozygotes in a ratio of 1:2:1. The *niv*-543 allele is also semi-dominant but, unlike *niv*-525, it is unstable and has spots and sectors of pigment on the flower (Fig. 1o).

The chalcone synthase protein encoded by the *niv* locus is thought to be a multimeric enzyme, so the semi-dominance of *niv*-525 might be explained if this allele produces an altered chalcone synthase monomer which interacts with the wild-type protein to give heteromers with reduced enzyme activity. However, analysis of chalcone synthase protein and mRNA levels in different genotypes containing the *niv*-525 allele shows that *niv*-525 reduces the quantity of *niv* gene expression rather than altering the type of chalcone synthase produced (Coen and Carpenter, 1988). The *niv*-525 mutation therefore acts both in *cis* and in *trans* to alter the spatial distribution and quantity of *niv* gene expression. We do not fully undersand mechanism underlying this phenomenon.

The semi-dominant alleles *niv*-525 and *niv*-543 were both derived from *niv*^{rec}-98: Tam3, suggesting that they resulted from sequence changes induced by Tam3. This has been confirmed since an inverted duplication of about 200 bp occurs in the *niv*-525 allele at the site of Tam3 excision (Coen and Carpenter, 1988, Fig. 4). The structure of *niv*-525 is readily explained by the hairpin model of transposon excision (Coen *et al.*, 1986) in a similar manner to the explanation of the short inverted duplications produced by imprecise excision of Tam3 from the *pal* locus described above. The presence of an extra 200 bp in the *niv* promoter would be expected to have an effect in *cis* on the transcription of the *niv* gene. However, the effect of *niv*-525 on a *Niv*$^+$ allele in *trans* requires a different explanation.

A possible model for the semi-dominance of *niv*-525 is that it is due to the production of anti-sense RNA. Deletion analysis of the *Niv*$^+$ promoter showed that a region of 1.2 kb, which lies upstream of the TATA box, is required for expression of the promoter (Kaulen *et al.*, 1986). Most of this region is retained intact, upstream (to the left in Fig. 4) of the inverted duplication of *niv*-525, and would be expected to drive transcription to the right if provided with a suitable TATA box. The sequence TTATAAA,

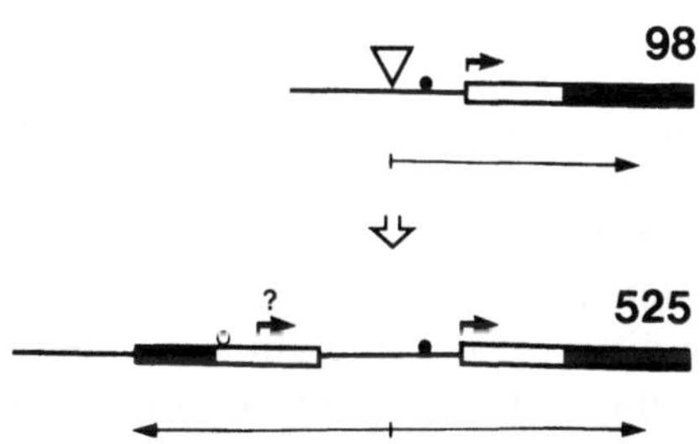

Fig. 4. Structure of the *niv*-525 allele and its progenitor, *niv*^{rec}-98. The short arrow pointing to the right in *niv*^{rec}-98 indicates the normal start site of *niv* transcription. The transcribed region comprises the non-translated leader (open bar) and the protein coding region (solid bar). The triangle indicates the location of Tam3 in *niv*^{rec}-98. The long arrows indicate a region of about 200 bp which are duplicated and inverted in *niv*-525. The usual TATA box is shown as a solid circle and the new potential TATA box in *niv*-525 is shown as an open circle

identical to that of the normal TATA box, is located within the inverted duplication near the junction of the inverted leader and coding sequence (open circle, Fig. 4). Transcription might therefore be driven from this new TATA box to produce an RNA with 40 bp at its 5′ end of the anti-sense strand of leader sequence. We propose that in a *niv*-525/*Niv*$^+$ heterozygote the *niv*-525 allele produces an anti-sense RNA molecule which hybridizes to the first 40 bp of the leader RNA from the *Niv*$^+$ allele and that the resulting double-stranded RNA is degraded in the nucleus. The spatial pattern of pigmentation observed in heterozygotes could be explained if greater levels of expression of anti-sense RNA occurred in regions of the flower with less colour. The distribution of anti-sense RNA would therefore reflect a novel pre-pattern, distinct from those described in previous sections.

One problem with the anti-sense model is that previous studies using DNA transfection indicate that inhibition by anti-sense RNA requires a 10 to 100-fold excess of DNA expressing anti-sense over DNA expressing sense RNA (Izant and Weintraub, 1985). However, in these experiments the DNA encoding anti-sense RNA is introduced at random locations in the genome whereas the *niv*-525 allele is located at a chromosome position which is homologous to the position of the *Niv*$^+$ allele which it inhibits. A strong effect of chromosome position on allelic interactions has been described in *D. melanogaster* and is termed transvection. For example, a dominant allele at the *white* locus, w^{DZL}, acts in *trans* to reduce expression of a w^+ allele only when the two alleles are at homologous chromosome positions (Bingham and Zachar, 1985). Such effects of chromosome position might be explained if the *trans*-effects are mediated by a labile RNA species (e. g. anti-sense RNA) which requires close proximity between interacting genes in order to maintain a high local concentration (Jack and Judd, 1979).

V. Conclusion

The analysis of mutations which alter pigmentation has revealed many non-clonal pre-patterns in the flower. *Cis*-acting mutations in pigment biosynthetic genes can reveal several different patterns by changing the affinity of the promoters for spatially distributed *trans*-acting transcription factors, hence giving different interpretations to the pre-patterns. Analysis of one such *trans*-acting factor, encoded by the *del* locus, shows that its own spatial distribution is in turn determined by a pre-pattern since it behaves in a cell-autonomous fashion. This provides a molecular model for Stern's notion that "development may be conceived as a consecutive series of patterned events in which a specific pre-pattern becomes expressed in a subsequent pattern, the subsequent pattern serving as a pre-pattern for the next patterned process" (Stern, 1968). Most mutations which produce non-clonal patterns in other organisms, such as *Z. mays* or *D. melanogaster,* act in a cell-autonomous way and hence affect genes which interpret a pre-pattern. This raises the question as to what determines pre-patterns.

Pre-patterns must ultimately derive from genes whose products act in an inter-cellular or non-autonomous manner. Mutations in this type of gene might be expected to have consequences on general aspects of morphogenesis as well as on pigment patterns. For example, the *pal*-41 allele produces a complex pattern of pigmentation: the upper lobes of the flower have little colour, most pigment being in a diffuse area around the lip of the lower lobes (Fig. 1k). The morphological mutation *cycloidea*^{radialis} results in radially symmetrical flowers in which all lobes resemble the lower middle lobe. If such a flower also contained the *pal*-41 allele, a relatively simple pattern of pigment would be expected in which a symmetrical diffuse band of pigment would be seen around the lip of the entire flower. The complex pattern observed with *pal*-41 in normal flowers therefore partly reflects a transformation of a simpler pattern through the action of the *Cycloidea*^{radialis} gene product which changes the flower morphology. This leads to the conclusion that our understanding of pre-patterns which determine the spatial distribution of pigment may ultimately depend on how they interact with genes involved in morphogenesis.

Acknowledgements

We would like to thank Joachim Bollmann, David McClelland and David Hopwood for helpful comments on the manuscript.

VI. References

Baur, E., 1924: Untersuchungen über das Wesen, die Entstehung und Vererbung von Rassenunterschieden bei *Antirrhinum majus*. Bibliotheca Genetica **4,** 1—70.

Bennetzen, J. L., Swanson, J., Taylor, W. C., Freeling, M., 1984: DNA insertion in the first intron of maize *Adh1* affects message levels: Cloning of progenitor and mutant *Adh1* alleles. Proc. Natl. Acad. Sci. U.S.A. **81,** 4125—4128.

Bianchi, F., Cornelissen, P. T. J., Gerats, A. G. M., Hogervorst, J. M. N., 1978: Regulation of gene action in *Petunia hybrida:* unstable alleles of a gene for flower colour. Theor. Appl. Genet. **53,** 157—167.

Bingham, P. M., Zachar, Z., 1985: Evidence that two mutations, w^{DZL} and z^1, affecting synapsis-dependent genetic behaviour of *white* are transcriptional regulatory mutations. Cell **40,** 819—825.

Bonas, U., Sommer, H., Harrison, B. J., Saedler, H., 1984a: The transposable element Tam1 of *Antirrhinum majus* is 17 kb long. Mol. Gen. Genet. **194,** 138—143.

Bonas, U., Sommer, H., Saedler, H., 1984b: The 17 kb Tam1 element of *Antirrhinum majus* induces a 3 bp duplication upon integration into the chalcone synthase gene. EMBO J. **3,** 1015—1019.

Carpenter, R., Coen, E. S., Hudson, A. D., Martin, C. R., 1984: Transposable genetic elements and genetic instability in *Antirrhinum*. Seventy-third Annual

Report of the Johan Innes Institute. (Norwich: Crowe and Sons Ltd.), pp. 52—64.
Carpenter, R., Martin, C., Coen, E. S., 1987: Comparison of genetic behaviour of the transposable element Tam3 at two unlinked pigment loci in *Antirrhinum majus*. Mol. Gen. Genet. **207,** 82—89.
Chia, W., Howes, G., Martin, M., Meng, Y., Moses, K., Tsubota, S., 1986: Molecular analysis of the *yellow* locus of *Drosophila*. EMBO J. **5**, 3597—3605.
Coe, E. H., Neuffer, M. G., 1977: The Genetics of Corn. In: Corn and corn improvement. Sprague, G. F. (ed), American Society of Agronomy, Madison, Wisconsin.
Coen, E. S., Carpenter, R., 1986: Transposable elements in *Antirrhinum majus:* generators of genetic diversity. Trends in Genet. **2,** 292—296.
Coen, E. S., Carpenter, R., Martin, C., 1986: Transposable elements generate novel spatial patterns of gene expression in *Antirrhinum majus*. Cell **47,** 285—296.
Coen, E. S., Carpenter, R., 1988: A semi-dominant allele, *niv*-525, acts in *trans* to inhibit expression of its wild-type homologue in *Antirrhinum majus*. EMBO J. **7** in press.
Demerec, M., 1940: Genetic behaviour of euchromatic segments inserted into heterochromatin. Genetics **25,** 618—627.
Dooner, H. K., 1983: Coordinate genetic regulation of flavonoid biosynthetic enzymes in maize. Mol. Gen. Genet. **189,** 136—141.
Fincham, J. R. S., Harrison, B., 1967: Instability at the *Pal* locus in *Antirrhinum majus*. II. Multiple alleles produced by mutation of one original unstable allele. Heredity **22,** 211—227.
Eissenberg, J. C., Kimbrell, D. A., Fristrom, J. W., Elgin. S. C. R., 1984: Chromatin structure at the 44D larval cuticle gene in *Drosophila:* the effect of a transposable element insertion. Nucl. Acids Res. **12,** 9025—9037.
Gerats, A. G. M., Farcy, E., Wallroth, M., Groot, S. P. C., Schram, A., 1983: Control of anthocyanin synthesis in *Petunia hybrida* by multiple allelic series of the genes *An1* and *An2*. Genetics **106,** 501—508.
Gierl, A., Schwarz-Sommer, Z., Saedler, H., 1985: Molecular interactions between the components of the En-I transposable element system of *Zea mays*. EMBO Journal **4,** 579—583.
Glandsdorff, N., Charlier, D., Zafarullah, M., 1980: Activation of gene expression by IS2 and IS3. Cold Spring Harbor Symp. Quant. Biol. **45,** 153—156.
Harrison, B. J., Carpenter, R., 1973: A comparison of the instabilities at the *nivea* and *pallida* loci in *Antirrhinum majus*. Heredity **31,** 309—323.
Hehl, R., Sommer, H., Saedler, H., 1987: Interaction between the Tam1 and Tam2 transposable elements of *Antirrhinum majus*. Mol. Gen. Genet. **207,** 47—53.
Hudson, A., Carpenter, R., Coen, E. S., 1987: De novo activation of the transposable element Tam2 of *Antirrhinum majus*. Mol. Gen. Genet. **207,** 54—59.
Izant, J. G., Weintraub, H., 1985: Constitutive and conditional suppression of exogenous and endogenous genes by anti-sense RNA. Science **229,** 345—352.
Jack, J. W., Judd, B. H., 1979: Allelic pairing and gene regulation: A model for the *zeste-white* interaction in *Drosophila melanogaster*. Proc. Natl. Acad. Sci. U.S.A. **76,** 1368—1372.
Judd, B. H., 1976: Genetic units of *Drosophila* — complex loci. In: The genetics and biology of *Drosophila* vol. *1b*. Ashburner, M., Novitski, E (eds.) pp. 767—799. London: Academic Press.
Jund, R., Loison, G., 1982: Activation of transcription of a yeast gene in *E. coli* by an IS*5* element. Nature **296,** 680—681.

Kaulen, H., Schell, J., Kreuzaler, F., 1986: Light induced expression of the chimeric chalcone synthase-NPTII gene in tobacco cells. EMBO J. **5,** 1—8.

Levis, R., Hazelrigg, T., Rubin, G. M. 1985: Effects of genomic position on the expression of transduced copies of the *white* gene of *Drosophila.* Science **229,** 558—561.

Luciw, P. A., Bishop, J. M., Varmus, H. E., Cappechi, M. R., 1983: Location and function of retroviral and SV40 sequences that enhance biochemical transformation after microinjection of DNA. Cell **33,** 705—716.

Lyon, M. F., 1961: Gene action in the X-chromosome of the mouse *(Mus musculus* L.). Nature **190,** 372—373.

Martin, C. R., Carpenter, R., Coen, E. S., Gerats, A. G. M., 1986: The control of floral pigmentation in *A. majus.* In: Soc. Expt. Biol. Symp. **32,** 19—52. Thomas and Grierson (eds.), Cambridge University Press.

Martin, C. R., Carpenter, R., Sommer, H., Saedler, H., Coen, E. S., 1985: Molecular analysis of instability in flower pigmentation of *Antirrhinum majus,* following isolation of the *pallida* locus by transposon tagging EMBO J. **4,** 1625—1630.

Mol, J. N. M., Schram, A. W., de Vlaming, P., Gerats, A. G. M., Kreuzaler, F., Hahlbrock, K., Reif, H. J., Veltkamp, E., 1983: Regulation of flavonoid gene expression in *Petunia hybrida:* Description and partial characterization of a conditional mutant in chalcone synthase gene expression. Mol. Gen. Genet. **192,** 424—429.

Mount, S. M., 1982: A catalogue of splice-junction sequences. Nucl. Acids Res. **10,** 459—472.

Nash, W. G., Yarkin, R. G., 1974: Genetic regulation and pattern formation: a study of the *yellow* locus in *Drosophila melanogaster.* Genet. Res. Camb. **24,** 19—26.

Roeder, G. S., Rose, A. B., Perlman, R. E., 1985: Transposable element sequences involved in the enhancement of yeast gene expression. Proc. Natl. Acad. Sci. U.S.A. **82,** 5428—5432.

Saedler, H., Nevers, P., 1985: Transposition in plants: a molecular model. EMBO J. **4,** 585—590.

Schwartz-Sommer, Zs., Gierl, A., Berntgen, R., Saedler, H., 1985: Sequence comparison of "states" of *a*1—*m*1 suggests a model for Spm (En) action. EMBO J. **4,** 2439—2443.

Schneuwly, S., Kuroiwa, A., Gehring, W. J., 1987: Molecular analysis of the dominant homoeotic *Antennapedia* phenotype. EMBO J. **6,** 201—206.

Scott, M. P., Weiner, A. J., Hazelrigg, T. I., Polisky, B. A., Pirotta, V., Scalenghe, F., Kaufman, T. C., 1983: The molecular organisation of the *Antennapaedia* locus of Drosophila. Cell **35,** 763—776.

Searle, A. G., 1968: Comparative genetics of coat colour in mammals. London: Academic Press.

Sommer, H., Carpenter, R., Harrison, B. J., Saedler, H., 1985: The transposable element Tam3 of *Antirrhinum majus* generates a novel type of sequence alteration upon excision. Mol. Gen. Genet. **199,** 225—231.

Sommer, H., Bonas, U., Saedler, H., 1988: Transposon-induced alterations in the promoter region affect transcription of the chalcone synthase gene of *Antirrhinum majus.* Mol. Gen. Genet. **211,** 49—55.

Spiribille, R., Forkmann, G., 1982: Genetic control of chalcone synthase activity in flowers of *Antirrhinum majus.* Phytochemistry **21,** 2231—2234.

Stanton, M. L., Snow, A. A., Handel, S. N., 1986: Floral evolution: attractiveness to pollinators increases male fitness. Science **232,** 1625—1627.

Stern, C., 1968: Developmental genetics of pattern. In: Genetic mosaics and other essays, pp. 130—173. Harvard Press.

Tan, C. C., 1945: Mosaic dominance in the inheritance of color patterns in the lady-bird beetle, *Harmonia axyridis*. Genetics **31,** 195—210.

Upadhyaya, K. C., Sommer, H., Krebers, E., Saedler, H., 1985: The paramutagenic line *niv*-44 has a 5kb insert, Tam2, in the chalcone synthase gene of *Antirrhinum majus*. Mol. Gen. Genet. **199,** 201—207.

Zachar, Z., Davidson, D., Garza, D., Bingham, P. M., 1985: A detailed developmental and structural study of the transcriptional effects of insertion of the *copia* transposon into the *white* locus of *Drosophila melanogaster*. Genetics **111,** 495—515.

Chapter 5

Isolation of Differentially Expressed Genes from Tomato Flowers

Charles S. Gasser, Alan G. Smith, Kim A. Budelier, Maud A. Hinchee, Sheila McCormick, Robert B. Horsch, Dilip M. Shah, and Robert T. Fraley

Plant Molecular Biology, Monsanto Company, 700 Chesterfield Village Parkway, St. Louis, MO 63198, U.S.A.

With 5 Figures

Contents

I. Introduction

The development of a higher plant is a complex process involving interactions between environmental cues and genetically determined programs for growth and differentiation that result in shifting patterns in cell division, and differential gene expression. The processes of flower induction and development provide a system in which all of these events can be studied.

We are interested in developmentally regulated genes, and in the role of such genes in flowering. The genes involved in development can be divided into two classes according to the function of their products. One class, the regulatory genes, consists of those genes whose primary function is to modulate the activity of other genes. Several genes of this class have been identified in animal systems. The genes for the glucocorticoid receptors of mammalian cells, which have been shown to mediate the regulation of other genes by glucocorticoids, are one example (Miesfeld *et al.*, 1986). A genetic regulatory function has also been hypothesized for several of the mammalian, and avian proto-oncogenes (Coppola and Cole, 1986; Verma *et al.,* 1985; Bishop, 1986), and for the homeotic genes of *Drosophila* (Gehring and Hiromi, 1986). In plants, at least one regulatory gene, encoding phytochrome, has been implicated in the control of flowering, as well as other light regulated processes (Wareing and Phillips, 1978). In those plants where the commitment to flower is regulated by photoperiod, phytochrome is the primary receptor responsible for the detection of day length (Wareing and Phillips, 1978). It is not known whether phytochrome acts by direct interaction with the genes, or through a series of second

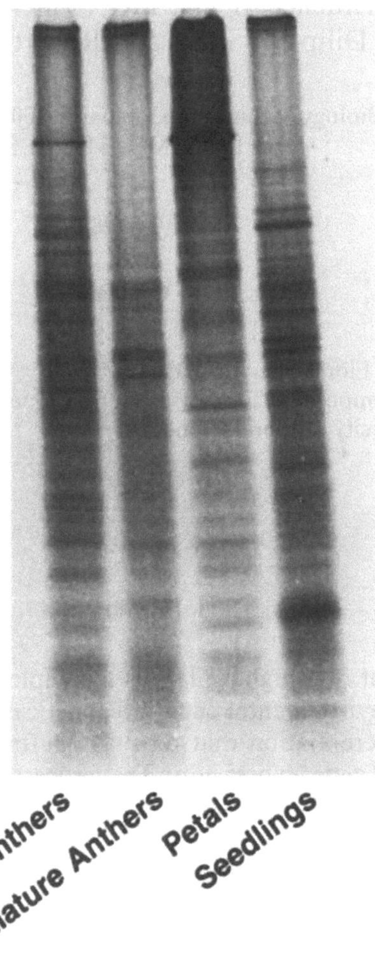

Fig. 1. Differential gene expression in tomato organs. Poly-A+ RNA was isolated from the indicated plant organs and translated *in vitro* in the presence of ^{35}S-methionine. The products were electrophoresed on a denaturing polyacrylamide gel. The gel was dried and exposed to X-ray film

messengers. It is possible that regulatory genes and their products are the primary agents governing the process of development. Currently there are intensive efforts in a number of labs to isolate specific factors responsible for gene regulation. Identification of these factors is difficult because they appear to be present in very small quantities, often for only short periods of time.

The second class of genes involved in development are those whose products directly form a functional part of the organism. This class includes genes for structural proteins, such as extensin, as well as genes for enzymes involved in organ specific pathways. In the formation of the flower this class would include genes for enzymes necessary for formation of the walls of pollen grains and floral pigments, genes for self incompatibility factors (Anderson *et al.,* 1986; Nasrallah *et al.,* 1985), and genes that are necessary for meiosis. Previous work suggests that the number of genes that are specific to floral organs is quite large. Kamalay and Goldberg (1980) have shown that the ovaries and anthers of tobacco plants each contain approximately 10,000 mRNA species that are unique to those organs. Dramatic differences in gene expression can also be found at different stages of the development of a floral organ. Figure 1 shows a comparison of proteins synthesized *in vitro* from poly-A^+ RNA from petals, seedlings, and two stages of anther development in tomato. It can be seen that there are significant quantitative and qualitative differences between the RNA populations from different organs, as well as between the two stages of development of anthers.

Identification and isolation of the developmentally regulated genes will permit a greater understanding of plant development and will provide tools for experimental manipulation of the process. The initial goal of our current work is to isolate cDNAs to mRNAs that show preferential expression in anthers and pistils. Our intention is to determine the nature of the proteins corresponding to these clones and to elucidate the role they play in floral development. The cDNA clones are being used as probes to isolate the corresponding developmentally regulated genes. Studies on the structure and function of these genes could provide information on the mechanism of developmental gene regulation and should aid in isolating and understanding the less accessible regulatory genes. In addition, if the regulatory regions of these genes can be isolated they may prove to be useful tools in studying other aspects of floral development (see discussion). In this chapter we report on our progress in using molecular techniques to isolate and characterize cDNA clones which are preferentially, or exclusively, expressed in the reproductive organs of tomato flowers.

II. Screening for Floral Specific cDNAs

Tomato *(Lycopersicon esculentum)* was chosen as a model system for our experiments because it has a number of convenient features. Flowers and fruits at all stages of development can be isolated from a single tomato plant, tomato flowers are large enough to allow isolation of individual organs, numerous floral mutants have been isolated, and a transformation system for tomato is available (McCormick *et al.,* 1986).

Several methods are available for the isolation of genes which are specific to a stage of development, or an organ. If a differentially regulated protein has been identified several approaches are available to isolate the corresponding cDNA. The protein can be purified, and used to produce specific antibodies. The antibodies can then be used to screen a cDNA library constructed in an appropriate expression vector (Young and Davis, 1983). Alternatively, the protein can be partially sequenced, and oligonucleotides can be synthesized according to the nucleic acid sequence predicted from the protein sequence. The oligonucleotides can then be used as probes on a cDNA library. This approach has been used to isolate a cDNA to the putative S_2 glycoprotein associated with self-incompatibility in *Nicotiana alata* (Anderson *et al.,* 1986, see also chapter 7). Application of this method is only possible if a specific protein product has been identified and purified.

Because our interest was in the identification of new genes with differential regulation the above methods were not applicable. Two general approaches for isolation of developmentally regulated sequences have been described. Both methods depend on obtaining different populations of messenger RNAs, one containing the sequences of interest, and one deficient in those sequences. One method, subtraction hybridization, relies on using solution, or column hybridization to the deficient population to subtract all but the differentially expressed sequences. The remaining material which is enriched for differentially expressed sequences can be used to construct a library, or can be used to synthesize a probe for an existing library. Examples of the successful use of this method can be found in chapters 2 and 10 in this volume.

In a second method, differential hybridization, duplicate plaque or colony replica filters are made from a representative population of clones from a cDNA library made with the RNA containing the desired sequences. One filter is then hybridized with a probe made from the same RNA that was used in library construction. The second filter is hybridized to a probe made from the deficient RNA. Clones which hybridize to the first probe, but not to the second can be isolated as putative organ, or time specific cDNAs. This method was used for the isolation of a self-incompatibility locus specific cDNA clone from *Brassica oleracea* (Nasaralla *et al.,* 1985, see also chapter 7). Since the second method is simpler, and requires less starting material than the subtraction method we decided to first attempt the isolation of floral specific genes by differential screening.

Because we wanted to isolate clones which showed differential expression over time within specific organs, we isolated material from tomato buds and flowers at three stages of development. Pistils, and anthers were dissected from 6—7 mm buds, 8—9 mm buds, and mature flowers near the time of anthesis. Examination of the chromosomes and structures of the developing gametophytes in each of these stages showed that meiosis was in progress in both the pistils, and anthers in the 6—7 mm buds, and was just being completed in the 8—9 mm buds. The mature flowers included those that had not yet released pollen, and those in which anthesis had just occurred. RNA was extracted from the isolated organs and the poly-A$^+$ fraction was purified by oligo-dT cellulose chromatography. The amounts of poly-A$^+$ RNA obtained from the immature organs were very small (approximately 2 µg from 500 mg of the 6—7 mm bud pistils). Since this RNA had to be sufficient for cDNA cloning and probe synthesis, we developed a very efficient cDNA cloning procedure that allowed us to begin with as little as 150 ng of poly-A$^+$ RNA (C. S. Gasser, in preparation). Using this method we were able to obtain libraries in λ-gt10 (Huynh *et al.*, 1985) containing 8 X 10^4 to 4 X 10^6 independent

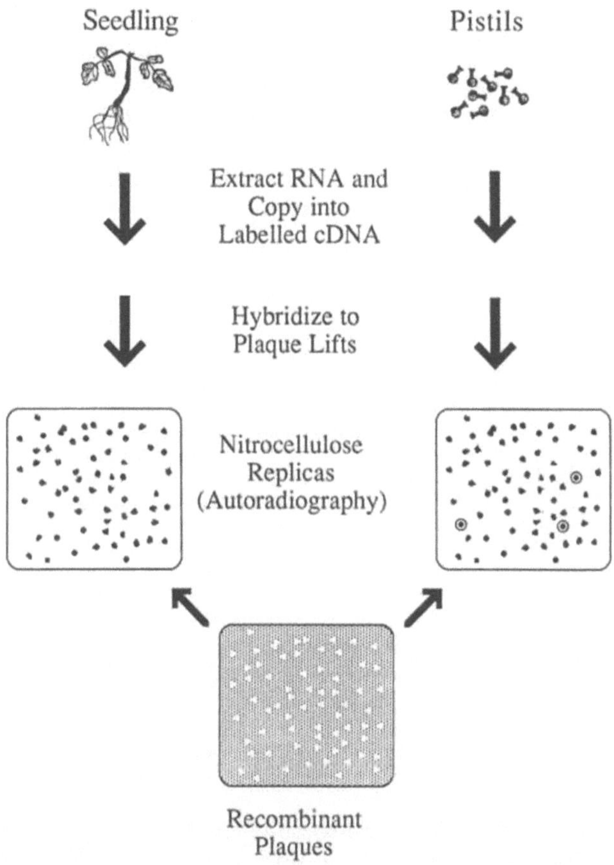

Fig. 2. Differential screening procedure. Labelled cDNA probes were prepared from poly-A$^+$ RNA from pistils and seedlings. The probes are hybridized to nitrocellulose replica plaque lifts of a pistil cDNA library. Hybridizing plaques were visualized by autoradiography, and those that showed preferential hybridization to the pistil probe (circled) were isolated as putative pistil-specific clones

clones from the anthers and pistils of 6—7 mm buds, 8—9 mm buds, and mature flowers (McCormick *et al.*, 1987).

Our initial differential screening procedure is diagrammed in Figure 2. The example shown is for screening of a pistil library. The same procedure was used to screen the anther libraries except that the second probe used was made from anther RNA. Seedling was chosen as a source for the negative screen probe because it provided a ready source of material that contained all of the major vegetative organs. The two week old seedlings that were used had not yet initiated the formation of flowers. Plaques which showed stronger hybridization to the homologous probe (from anthers or pistils) than to the seedling probe were isolated as putative floral organ-specific cDNA clones. The plaques were plated at a low density which allowed us to visualize even weakly hybridizing plaques in long exposures. This enabled us to isolate plaques which showed a differential, but weak signal, so that we would not exclusively isolate very highly expressed genes.

Since it was often difficult to conclusively isolate a single plaque from the initial screens, phage isolated in the first screen were re-plated and screened again with the same probes (Fig. 3). Phage DNA was purified from clones with a strong differential signal in both rounds of screening. As a further test of the specificity of our clones Southern blots containing digested samples of DNA from each of the clones were prepared. The blots

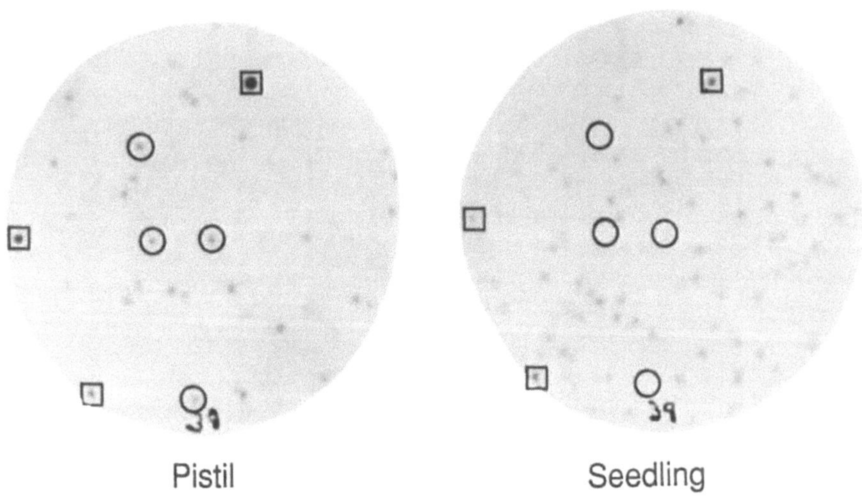

Fig. 3. Secondary screens for pistil-specific clones. An agar plug containing a putative pistil-specific clone, along with several surrounding plaques, was picked from a phage plate. Phage were eluted from the plug and replated. Duplicate nitrocellulose replicas were made from the plate and hybridized as in Figure 2 with seedling and pistil cDNA probes. Some of the plaques which show hybridization to both probes (squares) are indicated to show the orientation of the filters. Plaques which show preferential hybridization to the pistil probe can be seen (circles)

were hybridized with the same probes used in the first two rounds of screening. Since there was much more DNA on the Southern blots than there was on the plaque lifts this step was a more sensitive test for the level of expression. On the basis of this analysis we reduced the number of clones that were to be further characterized to 31 anther clones, and 23 pistil clones.

III. Organ and Temporal Specificity of Floral Clones

The Southern filters used in the final screen also proved useful in providing preliminary information concerning the expression of our genes at other stages of development, or in other floral organs. Since our only screen thus far was for differential expression in a single floral organ relative to a mixture of vegetative organs we could easily have isolated genes with an overall floral, rather than an organ-specific differential. To test this possibility the Southern blots of our cDNA clones were sequentially eluted and hybridized with probes made from either anther or pistil RNA at each stage of development. This procedure allows simultaneous characterization of a large number of clones, but is probably not as accurate a reflection of expression level as the more time consuming Northern blotting procedure (see below).

We found a broad range of patterns of hybridization among our clones from both pistils and anthers. We observed several trends in the temporal regulation and organ specificity of our clones. More than half of the pistil clones hybridized at some level to probes from all three tested stages of pistil development. In contrast, most of the anther clones were stage specific. That is, selected clones from the staged libraries did not hybridize to probes made from anthers at other stages of development (McCormick *et al.*, 1987). In addition, most of the putative pistil-specific clones had detectable hybridization to probes from some stage of anther development. Since we had not used any prior screen to eliminate clones that were expressed in more than one floral organ it is not suprising that we had selected some clones with this property. The anther clones were more specific than the pistil clones in this respect, most showing hybridization only to the homologous probe. Although our sample size of 10—30 clones from each library is small relative to the number of different messengers in anthers and pistils (Kamalya and Goldberg, 1980), our results indicate that organ and stage-specific messengers are present at higher relative concentrations in anthers than in pistils.

Those cDNAs which showed interesting patterns of expression were subcloned into plasmid vectors to allow easy isolation of DNA for probe synthesis and sequencing. The cDNAs were labelled with ^{32}P and used as probes on Northern blots of poly-A^+ RNA from vegetative and reproductive organs of tomato. Figure 4a shows an example of a Northern blot probed with a cDNA isolated from the library made from RNA from the pistils of 6—7 mm buds. The clone hybridizes to RNA from mature pistils,

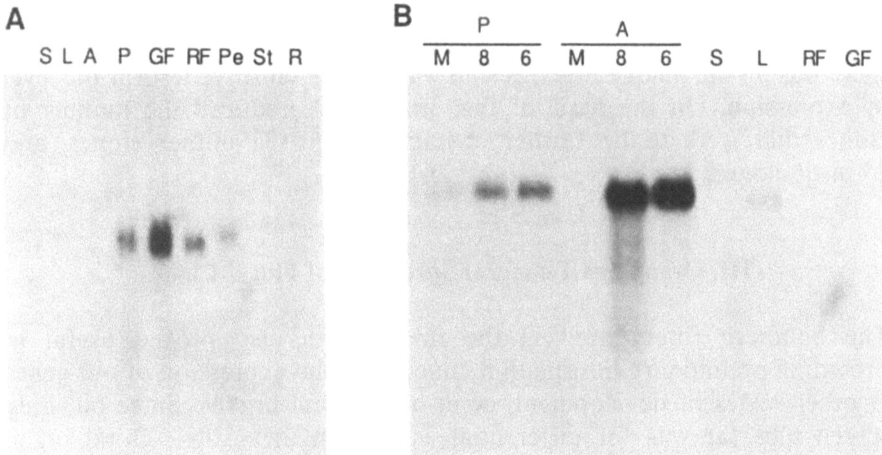

Fig. 4. Northern blots hybridized with pistil cDNA clones. One μg of poly-A^+ RNA from each of the indicated tomato organs was electrophoresed on a denaturing agarose gel and transfered to a nylon membrane. The filters were hybridized with putative pistil-specific clones. A) hybridized with pMON9606 probe; B) hybridized with pMON9601 probe. Symbols above the lanes indicate the source of the RNA in that lane: P, Pistil; A, anther; S, seedling; L, leaf; GF, green fruit; RF, ripening fruit; Pe, petal; St, stem; R, roots; M, mature tissue; 8, from 8—9 mm buds; 6, from 6—7 mm buds

petals, green fruit and maturing fruit. No hybridization is detected in the vegetative organs, or in the anthers. The intensity of hybridization is higher in the green fruit than in the mature pistil, or the ripening fruit. This demonstrates that this gene is both spatially and temporally regulated. In the pistil the level of expression increases as the ovary matures to form a fruit, and then decreases as the fruit ripens. Since the clone was isolated from the 6—7 mm bud pistil library it must also be expressing at early stages of pistil development.

Figure 4b shows a similar experiment with a clone, from the pistil cDNA library prepared from 8—9 mm buds, that has a very different pattern of expression. This clone is expressed in immature stages of both anthers and pistils, with the level of expression decreasing as the reproductive organs mature. Expression can also be detected in leaves, and weakly in seedlings. Based on differential exposures we estimate the level of expression to be 5 to 10-fold higher in 8—9 mm bud pistil than in seedlings. This demonstrates that the screening method that we used is sensitive enough to select clones having this low level of differential.

Twenty different cDNA clones have been used as probes in similar Northern blotting experiments. These clones cover a broad range of patterns of development. Clones expressed exclusively in mature pistils and green fruits, or exclusively in mature or immature anthers have been found. One clone showed increasing expression during maturation in both anthers and pistils. Another had a similar pattern of expression in pistils, but showed the opposite pattern of expression in anthers, where the expression decreased as the anthers matured. We have estimated the relative abundance of mRNA corresponding to a number of our clones by quantitation on Northern blots, and by measurements of the frequency of homologous clones in our libraries. While some of the genes we have isolated are expressed at very high levels representing approximately 1% of the mRNA in a given organ, others are expressed at a relatively low level of approximately 0.1% of the message. Thus, the differential method allowed us to isolate a wide variety of genes that show differential regulation during flower development. Even genes that are not highly expressed, and show only a 5—10-fold difference in expression between the RNA populations used to make probes can be readily detected.

IV. Tissue Specificity of Floral Clones

The methods outlined above can only be used to determine specificity at the level of organs. Anthers and pistils are complex structures consisting of a number of distinct differentiated tissues. To more precisely localize the expression of the cloned genes we have adapted an *in situ* hybridization procedure for use on floral organs (Smith *et al.*, 1987). In this procedure freshly frozen sections of whole buds or isolated flower parts are sectioned on a cryogenic microtome, the sections are fixed, treated to remove DNA and most protein, and hybridized with ^{35}S-labelled RNA probes. Autoradiography is then used to detect the location of specific hybridization between the probes and mRNA present in the tissue sections.

The results of one hybridization experiment using a cDNA clone from immature anthers is shown in Figure 5. The anti-sense probe hybridizes strongly to the layers of cells that surround the locules of the anthers. The absence of hybridization in the control experiment (Figure 5) using the sense probe shows that the hybridization pattern is not due to non-specific binding of labelled RNA. The cells which hybridize to this probe are those of the tapetum. These cells are present during pollen maturation and provide nutrients and materials for wall synthesis to the developing pollen grains. The tapetum degenerates as the pollen grains approach maturity. The pattern of hybridization of this probe to Northern blots of different stages of anther development shows that expression of this gene increases during development of the tapetum, and disappears as the tapetum degenerates. The level of resolution of this method is sufficient to determine precisely which cells in an organ are expressing a given gene.

In situ hybridization experiments have been performed with two other immature anther-specific cDNA clones. Although these cDNA clones do not cross hybridize with the clone used in Figure 5, they exhibit an identical tapetal-specific expression pattern. Since the tapetum is unique to anthers and is a very active tissue during this period it is not suprising that we were able to isolate more than one tapetal-specific clone.

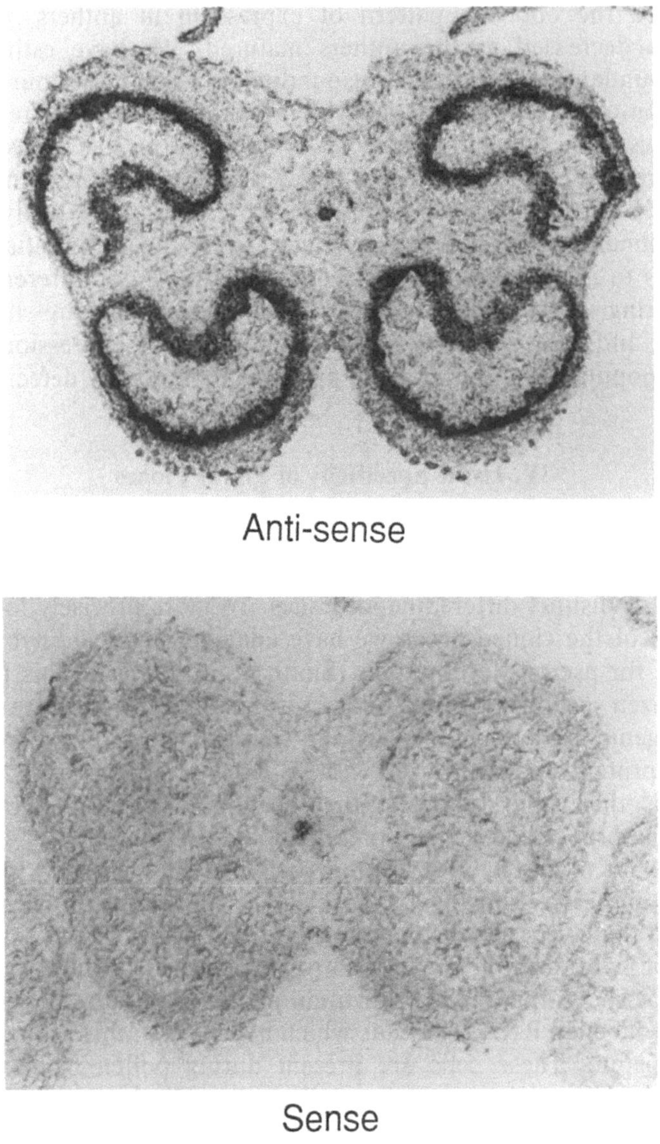

Fig. 5. *In situ* localization of gene expression. Transverse sections of immature tomato anthers were hybridized with ^{35}S-labelled RNA for either the sense, or the anti-sense strands of an anther-specific cDNA probe. Hybridization can be seen in the cells surrounding the locules of the anthers with the anti-sense probe, but not with the sense probe

Similar experiments with several of our pistil clones show that each of the clones is expressed exclusively in a single tissue of the pistil. One clone shows even greater specificity by hybridizing to a specific subset of the cells of the transmitting tissue of the style. The *in situ* hybridization experiments demonstrate that we have succeeded in isolating sequences that are tissue-specific, or specific to a subset of cells within a tissue.

V. Discussion

While we have concentrated on the isolation of floral-specific genes, several other groups have used similar approaches to isolate developmentally regulated genes from other parts of plants. Genes that are induced during the early development (Piechulla *et al.*, 1987) and ripening (Slater *et al.*, 1985; Fischer *et al.*, (this volume); Lincoln *et al.*, 1987) of tomato fruits have been isolated by differential screening or subtraction hybridization. One of the ripening-specific genes has been demonstrated to code for the enzyme polygalacturonase (Grierson *et al.*, 1986; Lincoln *et al.*, 1987). Harada *et al.* (chapter 2, this volume) have isolated sequences that are specific to germinating *Brassica napus* embryos. Mascarenhas *et al.* (chapter 6, this volume) report on the isolation of anther and pollen-specific genes. The subtraction hybridization method has been used to isolate organ specific sequences from most of the vegetative and floral organs of tobacco (R. Goldberg, personal communication). In combination, these studies promise to provide a broad range of developmentally regulated genes for future study.

One of the goals of our work is to determine the function of the sequences that we have isolated. In some cases the *in situ* localization studies give indications of the possible functions of the clones. For example, the tapetal-specific clones (Fig. 5) are likely to be involved in providing material necessary to pollen maturation. Alteration of the level or pattern of expression of the genes corresponding to our clones may produce phenotypic effects that will give clues to the functions of their gene products. In one set of experiments we have put the coding sequences of a cDNA under the control of the promoter for the 35S transcript of cauliflower mosaic virus (CaMV). This promoter directs the synthesis of large amounts of RNA in most plant organs (Sanders *et al.*, 1987). Introduction of these genes into plants by *Agrobacterium*-mediated gene transfer should lead to expression of the product of our isolated gene throughout the plant. A resulting change in phenotype would aid in assigning a function to our clone. We have also initiated construction of another set of transforming plasmids which will produce anti-sense RNA to our clones. Experiments in other systems have shown that production of anti-sense RNA can reduce or eliminate expression of the corresponding gene (Knecht and Loomis, 1987; Izant and Wientraub, 1984; McGarry and Lindquist, 1986). If a significant reduction in the level of expression of the corresponding gene can be achieved we may be able to observe a mutant phenotype in the transformed plants. Further characterization of the protein products of the cDNAs that we have isolated could be aided by having significant quantities of the protein available for biochemical characterization and antibody production. Since plant proteins can be efficiently produced in bacterial cells (Padgette *et al.*, 1987), we plan to construct appropriate vectors for high-level bacterial production of proteins encoded by our isolated cDNAs.

Genomic clones corresponding to our cDNAs have been isolated from a library of tomato DNA. We are currently isolating and characterizing the 5′-flanking regions of these genes. Work on previously isolated developmentally regulated genes from plants indicates that the control regions responsible for modulation of expression commonly reside in the 5′-flanking regions (Fraley *et al.*, 1986). We intend to use these putative regulatory regions to learn more of the role of phytohormones in flowering.

Phytohormones have been shown to have significant effects on the processes of floral induction and reproductive development. In some plants gibberellin can substitute for vernalization or appropriate photoperiod in flower induction (Lang, 1957). Auxin can induce parthenocarpic development of tomato fruits (Mapelli *et al.*, 1978), and cytokinin or cytokinin in combination with gibberellin can reduce flower abortion in stressed tomato plants (Kinet *et al.*, 1978). These are only a few examples of a very extensive literature on this subject. The complete characterization of the effects of phytohormones on reproduction is hindered by the difficulty of specific application of hormones. In most cases the hormones must be applied in a lanolin paste or as a solution in a wick that is placed on a part of the plant. Previous studies have predominantly examined the effect of chronic exposure of an entire bud or inflorescence to hormones, with the unavoidable gradients from point of contact inwards.

Isolation of organ specific clones may allow us to perform more refined experiments on the effects of phytohormones on flowering. Klee *et al.* (1987) describe a method for the production of chimeric plants with aberrant synthesis of auxin. For these experiments they used genes for auxin biosynthesis that had been isolated from the Ti plasmid of *Agrobacterium tumifaciens*. The coding sequences of these genes were put under the control of the CaMV 35S promoter or the soybean 7S seed storage protein promoter, and were introduced into petunia plants. The resulting plants had an altered phenotype consistent with the expected effects of overproduction of auxin. The Ti plasmid also includes a gene for cytokinin biosynthesis that can be used in a similar fashion (Akiyoshi *et al.*, 1983; Barry *et al.*, 1984; H. Klee, personal communication).

We are currently constructing chimeric genes containing the putative regulatory regions from our floral-specific clones, and the coding sequences of *Agrobacterium* phytohormone genes. Use of these genes in plant transformation experiments should result in plants with specifically altered patterns of hormone production. These experiments will allow us to evaluate the effect of the presence of auxin or cytokinin in particular cells of anthers or pistils at specific times in floral development. This will enable more accurate and consistent delivery of hormones than can be achieved with exogenous application. In addition cells or organs may respond differently to endogenously produced substances than to those absorbed from the outside.

The strategies described in this chapter will lead toward an understanding of genes, proteins, and their effects on floral development. It will also lead to unprecedented experimental power to intervene in the processes of development to improve the agronomic performance of fruit and seed crops.

VI. References

Akiyoshi, D. R., Morrsi, R., Hinz, R., Mischke, B., Kosuge, T., Garfinkle, D., Gordon, M., Nester, E., 1983: Cytokinin/auxin balance in crown gall tumors is regulated by specific loci in the T-DNA. Proc. Natl. Acad. Sci. U.S.A. **80,** 407—411.

Anderson, M. A., Cornish, E. C., Mau, S.-L., Williams, E. G., Hoggart, R., Atkinson, A., Bonig, I., Grego, B., Simpson, R., Roche, P. J., Haley, J. D., Penschow, J. D., Niall, H. D., Tregear, G. W., Coghlan, J. P., Crawford, R. J., Clarke, A. E., 1986: Cloning of cDNA for a stylar glycoprotein associated with expression of self incompatibility in *Nicotiana alata.* Nature **321,** 38—44.

Barry, G., Rogers, S., Fraley, R., Brand, L., 1984: Identification of a cloned cytokinin biosynthesis gene. Proc. Natl. Acad. Sci. U.S.A. **81,** 4776—4780.

Bishop, J. M., 1986: Oncogenes and hormone receptors. Nature **321,** 112—113.

Coppola, J. A., Cole, M. D., 1986: Constituitive c-myc oncogene expression blocks mouse erythrolukaemia cell differentiation but not commitment. Nature **320,** 760—763.

Fraley, R. T., Rogers, S. G., Horsch, R. B., 1986: Genetic transformation in higher plants. CRC Critical Rev. Plant Sci. **4,** 1—46.

Gehring, W. J., Hiromi, Y., 1986: Homeotic genes and the homeobox. Ann. Rev. Genet. **20,** 147—173.

Grierson, D., Tucker, G. A., Keen, J., Ray, J., Bird, J., Schuch, W., 1986: Sequencing and identification of a cDNA clone for tomato polygalacturonase. Nucleic Acids Res. **12,** 8595—8603.

Huynh, T. V., Young, R. A., Davis, R. W., 1985: Constructing and screening cDNA libraries in lambda gt10 and gt11. In: DNA Cloning: Volume I, A Practical Approach, D. Glover, ed. IRL Press (Oxford, U.K.), pp. 49—78.

Izant, J. G., Wientraub, H., 1984: Inhibition of thymidine kinase gene expression by anti-sense RNA: a molecular approach to gene analysis. Cell **36,** 1007—1015.

Kamalay, J. C., Goldberg, R. B., 1980: Regulation of structural gene expression in tobacco. Cell **19,** 935—946.

Kinet, J. M., Hurdebise, D., Parmentier, A., Stainier, R., 1978: Promotion of inflorescence development by growth substance treatments to tomato plants grown in insufficient light conditions. J. Amer. Soc. Hort. Sci. **103,** 724—729.

Klee, H. J., Horsch, R. B., Hinchee, M. A., Hein, M. B., Hoffmann, N., 1987: The effects of overproduction of two *Agrobacterium tumefaciens* T-DNA auxin biosynthetic gene products in transgenic plants. Genes Develop. **1,** 86—96.

Knecht, D. A., Loomis, W. F., 1987: Antisense RNA inactivation of myosin heavy chain gene expression in *Dictyostelium discoideum.* Science **236,** 1081—1086.

Lang, A., 1957: The effect of gibberellins upon flower formation. Proc. Natl. Acad. Sci. U.S.A. **43,** 709—717.

Lincoln, J. E., Cordes, S. P., Diekman, J., Fischer, R. L., 1987: Regulation of gene expression by ethylene in ripening tomato fruit. J. Cell. Biochem. **11B,** 59.

Mapelli, S., Frova, C., Torti, G., Soressi, G. P., 1978: Relationship between set, development, and activities of growth regulators in tomato fruits. Plant Cell Physiol. **19,** 1281—1288.

McCormick, S., Niedermeyer, J., Fry, J., Barnason, A., Horsch, R., Fraley, R., 1986: Leaf disc transformation of cultivated tomato *(L. esculentum)* using *Agrobacterium tumefaciens.* Plant Cell Rep. **5,** 81—84.

McCormick, S., Smith, A., Gasser, C., Sachs, K., Hinchee, M., Horsch, R., Fraley, R., 1987: Identification of genes specifically expressed in reproductive organs of tomato. In: Tomato Biotechnology, D. J. Niven and R. A. Jones, eds. Alan R. Liss Publ. Inc. (New York, NY), pp. 255—265.

McGarry, T. J., Lindquist, S., 1986: Inhibition of heat shock protein synthesis by heat-inducible antisense RNA. Proc. Natl. Acad. Sci. U.S.A. **83,** 399—403.

Miesfeld, R., Rusconi, S., Godowski, P. J., Maler, B. M., Okret, S., Wikström, A.-C., Gustafsson, J.-Å., Yamamoto, K. R., 1986: Genetic complementation of a glucocorticoid receptor deficiency by expression of cloned receptor cDNA. Cell **46,** 389—399.

Nasrallah, J. B., Kao, T.-H., Goldberg, M. L., Nasrallah, M. E., 1985: A cDNA clone encoding an S-locus-specific glycoprotein from *Brassica oleracea.* Nature **318,** 263—267.

Padgette, S. R., Huynh, Q. K., Borgmeyer, J., Shah, D. M., Brand, L. A., Re, D. B., Bishop, B. F., Rogers, S. G., Fraley. R. T., Kishore, G. M., 1987: Bacterial expression and isolation of *Petunia hybrida* 5-enolpyruvylshikimate-3-phosphate synthase. Arch. Biochem. Biophys. (in press).

Piechulla, B., Gruissem, W., 1987: Differential expression of several genes during fruit ripening. J. Cell. Biochem. **11B,** 42.

Sanders, P. R., Winter, J. A., Barnason, A., Fraley, R. T., 1987: Comparison of cauliflower mosaic virus 35S and nopaline synthase promoters in transgenic plants. Nucleic Acids Research **15,** 1543—1558.

Slater, A., Maunders, M. J., Edwards, K., Schuch, W., Grierson, D., 1985: Isolation and characterization of cDNA clones for polygalacturonase and other ripening-related proteins. Plant Mol. Biol. **5,** 137—147.

Smith, A. G., Hinchee, M., Horsch, R., 1987: Cell and tissue specific expression localized by *in situ* RNA hybridization in floral tissues. Plant Mol. Biol. Rep. (in press).

Verma, I. M., Mitchell, R. L., Kruijer, W., Van Beveren, C., Zokas, L., Hunter, T., Cooper, J. A., 1985: Proto-oncogene *fos:* induction and regulation during differentiation. In: Cancer Cells 3: Growth Factors and Transformation, J. Feramisco, B. Ozanne and C. Stiles, eds. Cold Spring Harbor Laboratory (Cold Spring Harbor, NY), pp. 275—287.

Wareing, P. F., Phillips, I. D. J., 1978: The Control of Growth and Differentiation in Plants. Pergamon Press (Oxford, UK).

Young, R. A., Davis, R. W., 1983: Efficient isolation of genes by using antibody probes. Proc. Natl. Acad. Sci. U.S.A. **80,** 1194—1198.

Chapter 6

Anther- and Pollen-Expressed Genes

Joseph P. Mascarenhas

Department of Biological Sciences, State University of New York at Albany, Albany, NY 12222, U. S. A.

Contents

I. Introduction

Flowering plants produce intricate, often beautiful structures, within which their reproductive processes take place. These structures are the flowers. In flowering plants, as in more primitive plants, a diploid spore-producing generation (sporophyte) alternates with a haploid, gamete-producing generation (gametophyte). The male and female gametophytes of Angiosperms are reduced to microscopic structures enclosed within the tissues of the sporophyte. The functions of the gametophytes are the production of the sperm and egg cells and their union in fertilization. The pollen grain is the male gametophyte and the embryo sac the female gametophyte. The flower consists of specialized structures, the anthers and the pistil or gynoecium, in which the sex cells are formed.

In the discussion that follows, the development of the anther and the male gametophyte has been stressed because this is the context in which gene expression is most relevant. Gene activity has been broadly defined to include the biochemistry of the cells and tissues since biochemistry is a reflection of gene expression and this information might provide a framework for molecular studies in the future.

A. Anther and Microsporangium Development

There is a wide variation among plants in the appearance and size of the stamen, the term used to describe the anther and the filament on which it is borne. An anther at maturity typically contains four elongated microsporangia attached by a thin filament to the flower axis. The microsporangia, which in essence are sacs containing the microspores or pollen grains, are arranged in pairs in the two anther lobes. At maturity just prior to anther dehiscence (the opening of the anther to release the pollen), the walls separating the two microsporangia in each anther lobe break down.

In the flower bud a very young anther consists of a mass of undifferentiated meristematic tissue surrounded by a discrete layer of epidermal cells. As this anther primordium matures, the spore-forming tissue appears in the four corners of the developing anther. In each of these regions a series of cells below the epidermis undergo periclinal divisions, i. e. divisions parallel to the outer surface. These divisions in the spore-forming regions result in two well-defined layers of cells. The inner layer consists of the primary spore producing cells which either directly or after further mitotic divisions function as pollen mother cells or microsporocytes. The outer layer of cells, the primary parietal cells, undergoes several periclinal and anticlinal (perpendicular to the surface) divisions to give rise to the cell layers of the sporangial wall. The innermost of the parietal layers is the tapetum which has special significance in pollen development. The epidermal cells of the anther primordium divide anticlinally to keep pace with the rapidly enlarging anther. The tapetum consists of densely staining cells with prominent, often polyploid nuclei. There are two major types of tapetum found in angiosperms. The glandular type of tapetal cells, such as those in *Lilium,* remain in their original position in the anther, but as pollen development progresses and reaches the stage of maximum pollen grain formation, they degenerate and finally undergo complete autolysis. In the amoeboid tapetum, as seen in *Tradescantia,* the cells lose their walls and the tapetal protoplasts intrude among the developing pollen grains and finally fuse together to form a tapetal periplasmodium which surrounds the developing pollen grains, with the tapetal cell membranes even penetrating the pollen exine during the later stages of pollen development (see review in Mascarenhas, 1975).

B. Summary of Meiosis, Pollen, and Pollen Tube Development

Pollen mother cells are produced by mitosis in the sporogenous tissue. The critical steps in the transformation of these cells into haploid microspores are two meiotic divisions which result in each pollen mother cell producing a tetrad of microspores each of which is surrounded by a layer of callose. In many species all the anthers in a flower bud are at almost the same stage of development at any given time. There is also a very good correlation between flower bud length and the stage of meiosis or pollen development. This synchrony in development makes pollen an attractive object for molecular study.

On release from the tetrad, the spores undergo a very rapid increase in size; in *Lilium* the spores increase in volume about threefold within 24 hours. With the increase in volume there is also a change in shape. This rapid growth is followed by a period of slower growth until the maximum volume of the pollen grain is reached some time before anther dehiscence. After meiosis there is a long interphase period at the end of which the microspore nucleus divides in a very unequal manner (Angold, 1968) to form two cells, the vegetative and generative cells, which inherit very different quantities and possibly qualities of cytoplasm. This division is referred to as microspore mitosis. In several plants the generative cell divides once more in the pollen grain, forming two sperm cells. In most pollens, however, the generative cell completes its division much later, during the growth of the pollen tube in the style (Brewbaker, 1967).

At maturity when the anther dehisces, the mature pollen is liberated and is carried by wind, insects, or other agents to the stigma of an appropriate flower. Here the right conditions are present for the germination of the pollen grain. The pollen grain germinates on the surface of the stigma by extruding a tube through a germ pore in the pollen wall. The tube then grows down into the style. The vegetative nucleus and the generative cell move out of the pollen grain and into the tube. In most plants germination and pollen tube growth are relatively rapid, the period from pollination to fertilization ranging from 6 to 48 hours. The rate of growth of the pollen tube varies in different species with rates as high as 35 mm/hr being reported in some cases. The pollen tube grows through the style, enters the micropyle of the ovule, and reaches the embryo sac. Here it penetrates one of the synergids, normally the one that has begun to degenerate. The tube then stops growing and the sperm cells together with some of the other tube contents are discharged into the synergid. The two sperm move by an unknown mechanism, one fusing with the egg cell to form the zygote and the other with the central cell to give rise to the primary endosperm cell, thus completing the process of double fertilization.

II. Gene Expression in the Anther

A. The Tapetum

It is generally believed, based largely on circumstantial evidence, that the tapetum is concerned with the nutrition of the developing pollen grains. The tapetum is the tissue in the anther in closest contact with them. Moreover, the developing pollen grains are completely surrounded by tapetal cells or periplasmodium. Because all materials must either pass through the tapetum or be metabolized by it (Echlin, 1971), it is assumed that the tapetum must play an essential nutritive role in the formation of the pollen grains. That the tapetum and the developing microspores interact in a developmentally meaningful way comes from studies on the dissolution of the callose walls after tetrad formation and the subsequent release of the four microspores. Several studies seem to indicate that the enzyme "callase" is synthesized transiently in the tapetum, is secreted and degrades the callose walls surrounding the microspores, and that the timing of production and release of callase by the tapetum is critical for normal pollen development (Eschrich, 1961; Mepham and Lane, 1969; Stieglitz and Stern, 1973; Frankel, Izhar and Nitsan, 1969; Izhar and Frankel, 1971). Moreover, the importance of the tapetum is supported by studies of cytoplasmic male sterility, where the initial abnormalities in anther development are not found in the sporogenous cells but in the tapetal tissue (Bino, 1985; Horner and Rogers, 1974).

The pollen grain walls of all species of plants contain proteins that are rapidly released when the pollen grains are moistened. These proteins are localized at two sites in the wall, in the inner intine layer, especially at the germinal apertures, and in the spaces and cavities of the exine and coating the exine surface. The proteins associated with the intine are thought to be incorporated during the growth of the wall and are produced by the male gametophyte. The proteins in the exine, however, appear to be transferred later and are sporophytic in origin, being derived apparently from the tapetum during its dissolution (Knox and Heslop-Harrison, 1969; Vithanage and Knox, 1980).

Pollenkitt, the lipoidal, often pigmented layer found coating the outside of pollen grains in many insect-pollinated species, is produced in the tapetum (Heslop-Harrison, 1968; Dickinson, 1973; Reznickova and Willemse, 1981). This is also true for *tryphine,* the complex mixture of hydrophobic and hydrophilic substances including proteins that coat mature pollen grains (Mepham and Lane, 1969; Albertini, Grenet-Auberger and Souvre, 1981). When dissolution of the tapetum occurs, the enclosed lipids, proteins, etc. are discharged into the loculi where they spread between the spores, filling the recesses in the surface of the exine and forming the *pollenkitt* and *tryphine.*

Flavonoids and other phenylpropanoids occur in fairly large quantities in the cavities of the exine of several plant species. In the anthers of *Tulipa* the synthesis of phenylpropanes, chalcones, flavonols, and anthocyanins occurs at different periods during post-meiotic pollen development (Wiermann, 1979). The major activity of several key enzymes of phenylpropanoid metabolism, such as phenylalanine ammonia lyase, cinnamic-acid-4-hydroxylase, etc., was found in the "tapetal plus anther loculus fluid" fraction as compared to the pollen fraction. It has been suggested that these enzymes are produced by the tapetal cells, secreted into the anther loculus, and are active in phenylpropanoid metabolism either in the loculus or in the exine cavities of the developing microspores (Herdt,

Sutfeld and Wiermann, 1978; Wiermann, 1979). The tapetum is active in the synthesis of all classes of compounds such as carbohydrates, proteins, lipids, etc., and the levels of synthesis appear to be stage specific (Sauter, 1969; Albertini *et al.*, 1981; Albertini and Souvre, 1978; Reznickova, 1978). In *Petunia hybrida,* isozyme patterns of esterase, peroxidase, alcohol dehydrogenase, and malate dehydrogenase showed changes associated with post-meiotic anther development and these enzymes exhibited a tissue- and stage-specific localization in the tapetum, anther walls, or pollen grains (Nave and Sawhney, 1986; Sawhney and Nave, 1986). A stage specific-distribution of poly(A)RNA has been observed in tapetal cells of *Hyoscyamus niger* during microsporogenesis as determined by *in situ* hybridization with ^{3}H-poly(U) (Raghavan, 1981).

B. Other Anther Tissues

Generally, studies of the role of anther tissues in pollen development have considered only the tapetum as important in this regard and have ingored the other tissues, such as the epidermis, and middle layers of cells surrounding the tapetum. Results obtained by Reznickova (1978) and Reznickova and Willemse (1980) have implicated anther tissues other than the tapetum in the nutrition of the developing pollen grains. Based on histochemical staining and ultrastructural studies, waves of starch accumulation in the developing microsporocytes, microspores, and anther tissues were found to precede the accumulation of lipids and/or carotenoids in these tissues, suggesting that lipid accumulation was preceded by starch hydrolysis. The hydrolysis of the starch that accumulated in the anther wall tissues (epidermis, endothecium, middle layers) was accompanied by lipid synthesis and accumulation in the tapetum. The subsequent decrease in the lipids in the tapetum was followed by an increase in starch in the young pollen grains. These results support the notion that anther wall layers in addition to the tapetum play an important part in the metabolism and mobilization of nutrients for the developing pollen.

Isolated microsporocytes of lily and *Trillium* at the leptotene stage completed meiosis when cultured in special media (Takegami *et al.*, 1981). Moreover, uninucleate microspores of lily and tulip at the G_1 phase of the cell cycle completed the mitotic division and produced pollen grains when successively cultured on appropriate media. When transferred to a germination medium these grains produced pollen tubes (Tanaka and Ito, 1980, 1981). These results tend to suggest that the role of the tapetum and other tissues of the anther in the nutrition of the developing pollen might be the mobilization and production of simple nutrients and the provision of the necessary physical conditions, rather than the supply of any morphogenetic components essential for critical stages of pollen development to proceed.

A large number of metabolic activities occur in the various anther tissues. From the kinetics of RNA-excess DNA-RNA hybridization experiments, Kamalay and Goldberg (1980) estimated that the total complexity of tobacco anther messenger RNA was 3.23×10^7 nucleotides, equivalent to about 26,000 diverse genes. Of these about 15,500 mRNAs were shared with those of the leaf. Approximately the same number of different mRNAs (24,000—27,000) were found in the various vegetative (leaf, stem, root) and reproductive (petal, ovary) organs. Whether the RNA isolated from the anthers included that from developing pollen grains was not specified in this study.

Goldberg (1987 and personal comunication) has obtained cDNA clones to mRNAs from tobacco anthers. *In situ* hybridization experiments utilizing some of these clones have shown that several anther mRNAs are localized within the tapetum. These mRNAs are first seen early in anther development and decay prior to anther maturation. The levels of these mRNAs are correlated with the establishment and degeneration of the tapetum. Similar cDNA clones have been transcribed and purified from mRNAs from tomato anthers (see Chapter 5 in this volume by C. Gasser *et al.*).

III. Gene Expression in the Developing Male Gametophyte

Does the differentiation of the male gametophyte depend on genes transcribed by the sporophyte, i. e. prior to meiosis, or are the genes necessary for male gametophyte differentiation transcribed from the haploid genome after meiosis? The differentiation of at least a part of the male gametophyte, that of the intricate pattern of the pollen grain wall, appears to be determined by the diploid sporophyte. The evidence is partly based on the observation that segregation of different exine types or patterns has never been recorded in one and the same tetrad. Such segregation would be expected if the exine pattern was entirely under the control of the spore nucleus. In *Acanthus spinosus* (mountain thistle), nucleus-free cytoplasmic buds are formed quite regularly after meiosis II. The buds round up within the callose wall and form an essentially normal exine. These two lines of evidence seem to indicate that the determination of the exine pattern is under the control of the genome of a sporophyte cell, probably the pollen mother cell (Heslop-Harrison, 1971). A more detailed discussion of the complex mechanisms involved in the establishment of patterns in the exine can be found in Dickinson (1976) and Rowley (1981).

A. *Specific Transcription and Translation*

Transcription of specific genes occurs in the haploid microspore after its formation and during its subsequent development. For a review of the earlier work using primarily autoradiographic and microspectrophotometric techniques see Mascarenhas (1975). The transcription of the ribosomal RNA (rRNA) genes occurs during pollen development. In lily, ribosomal RNA appears to be synthesized at a high rate after DNA synthesis is completed but prior to actual microspore mitosis, with a second lower peak of RNA synthesis following the division (Steffensen, 1966). A similar pattern of ribosome synthesis is seen in *Tradescantia paludosa* (Mascarenhas and Bell, 1970). During the last 48 hr of pollen maturation no rRNA (25S, 17S, or 5S) was synthesized. Very small amounts of labeled, newly synthesized rRNA were seen during the period 48—72 hr before anther dehiscence. Most of the rRNAs that were found in the mature pollen grain were synthesized more than 96 hr before the pollen was shed and during the period prior to microspore mitosis. Following mitosis there was a sharp decrease in the amount of rRNA synthesized (Mascarenhas and Bell, 1970; Peddada and Mascarenhas, 1975). The transcription of the transfer RNA genes also seemed to follow a similar pattern, with a greater activity prior to microspore mitosis which then decreased until the genes were transcriptionally inactive during the terminal stages of pollen maturation (Mascarenhas and Bell, 1970).

After release from the anther the pollen grain exists as a free organism for a short while until it is deposited on the stigma of a flower. Then it begins another phase of its life and development by germinating and extruding a tube within which the sperm cells are transported to the embryo sac. The discharge of the sperm cells following the rupture of the pollen tube marks the end of the life cycle of the male gametophyte.

In most plant species the pollen grain at anthesis is a dehydrated and metabolically inactive structure. The first stage in pollen germination is imbibition of water which is a rapid process. Germination is complete with the outgrowth of a tube from the grain. Pollen grains from different species vary considerably in the rapidity with which they complete germination.

No DNA synthesis occurs in the mature pollen grain and during pollen tube growth. In binucleate species where the generative cell forms two sperm cells during pollen tube growth, the DNA synthesis for this division takes place during pollen maturation prior to anther dehiscence. The generative cell in many plants is already in late prophase of mitosis at the time it is shed (Mascarenhas, 1975). Although no DNA synthesis normally occurs in the mature pollen grain and tube, both vegetative and generative nuclei contain DNA polymerase activity (Takats and Wever, 1971; Wever and Takats, 1971). DNA repair activity has also been demonstrated in germinating *Petunia* pollen (Jackson and Linskens, 1980).

In many species tested, the pollen grain at maturity contains all the proteins that are required for germination and early tube growth, since protein synthesis inhibitors do not block these events (Mascarenhas, 1975; Capkova-Balatkova *et al.*, 1980). For example, the enzymes for the synthesis of both neutral and polar lipids necessary for cell membrane formation are already present in the mature ungerminated *Tradescantia* pollen grain and are functionally stable for several hours in the pollen tube (Whipple and Mascarenhas, 1978). Similarly a large number of enzymes have been reported to be present in pollen grains of various species (Brewbaker, 1971; Bryce and Nelson, 1979; Dickinson *et al.*, 1973, 1977; Hara *et al.*, 1980; Kroh and Loewus, 1978; Lin *et al.*, 1987; Malik and Singh,

1976; Malik and Gupta, 1976; Shaykh *et al.*, 1977; Scott and Loewus, 1986).

Protein synthesis begins early in pollen germination. In *Tradescantia* protein synthesis is initiated within two minutes of placing dehydrated pollen grains in a germination medium (Mascarenhas and Bell, 1969). Similar results were found with *Petunia* (Linskens *et al.*, 1970) and tobacco (Tupy, 1977).

The ungerminated grain at anthesis contains a store of presynthesized mRNAs that are translated early during germination. The evidence for this has been obtained by studies on the effects of inhibitors of RNA and protein synthesis on germination and tube growth (Mascarenhas, 1975). In addition more direct evidence for the presence of mRNAs in the ungerminated pollen grains of *Tradescantia* and corn has been obtained by the isolation of poly(A) containing RNAs and their translation in cell-free systems into polypeptides, many of which show similarity to proteins made during germination (Frankis and Mascarenhas, 1980; N. T. Mascarenhas *et al.*, 1984). Tobacco pollen grains also contain presynthesized mRNAs (Tupy, 1982). Each corn pollen grain contains between 8.9—17.8 pg of poly(A)RNA (N. T. Mascarenhas *et al.*, 1984). The ungerminated pollen grain of *Tradescantia* has been estimated to contain approximately 6×10^6 molecules of poly(A)RNA (Mascarenhas and Mermelstein, 1981). Most of the studies in the literature seem to indicate that the mature pollen grain contains a store of stable mRNAs that are translated early during germination and that play a greater or lesser role in pollen germination and tube development depending on the plant species (Mascarenhas, 1975; Suss and Tupy, 1979).

New RNA synthesis occurs during pollen germination (Steffensen, 1966; Mascarenhas and Bell, 1970; Tupy, 1977). In *Tradescantia*, the RNA that is synthesized is not ribosomal RNA (Mascarenhas and Bell, 1970) or transfer RNA (Mascarenhas and Goralnick, 1971). A similar situation with respect to rRNA has been described in lily (Steffensen, 1966) and tobacco (Tano and Takahashi, 1964). In lily pollen tubes Steffensen (1971) has demonstrated that no ribosomal proteins are synthesized either. The observations that rRNA is not synthesized in the germinating pollen grain or pollen tube are additionally supported by the fact that in the pollen tube nuclei of most species the nucleoli are either absent or greatly reduced in size. Whereas Tupy and coworkers reported that pollen tubes of tobacco synthesize rRNA (Tupy, 1977; Tupy *et al.*, 1977), Bashe (1984) found that rRNA is not synthesized in tobacco pollen tubes. The results obtained by Tupy and coworkers could be explained by their use of nondenaturing sucrose density gradients which cause a substantial association of mRNAs with rRNAs. By utilizing sucrose gradients made up in 70% formamide this association is prevented and the radioactive profiles no longer show a coincidence with the rRNA optical density profiles (Bashe, 1984).

The mature pollen grain is equipped with a large stockpile of ribosomes and tRNA that are utilized in the protein synthesis that is required for pollen tube development and the completion of generative cell division. As discussed above, the ribosomes and tRNAs present in the mature pollen grain are synthesized during pollen maturation at a time preceding microspore mitosis.

The RNAs that are synthesized during *Tradescantia* pollen germination and tube growth appear to be mRNAs or mRNA precursors (Mascarenhas *et al.*, 1974). Based on studies of protein synthesis in the presence of actinomycin D it has been estimated that during the first hour of pollen tube growth about 50% of the translation occurs on previously existing mRNAs and the remaining 50% on newly synthesized mRNAs (Mascarenhas and Mermelstein, 1981). Are there qualitative differences between the newly synthesized and presynthesized mRNAs, or in other words, is there a different set of genes turned on during pollen germination? An analysis of protein synthesis by radioactive labeling and one- or two-dimensional polyacrylamide gel electrophoresis shows no differences in bands or spots whether or not new mRNA synthesis is blocked with actinomycin D (Mascarenhas *et al.*, 1974; Mascarenhas and Mermelstein, 1981). These results tend to support the conclusion that there are no qualitative differences at least between the more abundant types of proteins synthesized during pollen germination on previously existing and newly synthesized mRNAs.

The enzyme phytase is synthesized during germination of *Petunia* pollen on mRNAs presynthesized in the developing pollen grain prior to anthesis (Jackson and Linskens, 1982). Two phytases with different pH optima have been detected in germinating *Lilium longiflorum* pollen. One of the phytases is already present in mature ungerminated pollen whereas the other is newly synthesized during germination from preexisting mRNA (Lin *et al.*, 1987). This is the only example of an enzyme not present in the mature pollen grain but synthesized during germination and pollen tube growth on preexisting mRNA.

The results discussed above indicate that, in the pollen grain at maturity, either all the proteins that are required for germination and early tube growth are already present or, if new proteins are synthesized, the mRNAs for their synthesis already exist. Moreover, the proteins synthesized on new mRNAs during germination are similar to those synthesized on premade mRNAs. In order words, the genes for at least the prevalent and mid-prevalent mRNAs utilized during the latter part of pollen maturation and those utilized during germination and tube growth appear to be the same.

Heat shock proteins (hsps) are not synthesized by pollen tubes of *Tradescantia* at either 37° or 40°C or after a gradual increase in temperature from 29° to 40°C (Altschuler and Mascarenhas, 1982; Mascarenhas and Altschuler, 1983). The ungerminated *Tradescantia* pollen grain at the time it is released from the anther also does not appear to contain hsps that might have been synthesized during maturation (Xiao and Mascarenhas, 1985). Hsps are not synthesized in germinating maize or lily pollen either (Cooper, Ho and Hauptman, 1984), nor in pollen tubes of *Petunia hybrida* and *Lilium longiflorum* (Schrauwen *et al.*, 1986). In contrast, Zhang *et al,*

(1984) reported that heat shock proteins are synthesized in *L. longiflorum* pollen tubes that have been cultured in the absence of proline. In immature corn pollen, however, subsets of the hsps are synthesized during a heat shock depending on the stage of development (Frova *et al.*, 1987).

Within the first 30 min of germination and pollen tube growth in culture there is an almost 50% decrease in the poly(A) content of the tube. This decrease continues at a lower rate during the subsequent 30 min of culture (Mascarenhas and Mermelstein, 1981). The decrease in poly(A) content correlates with the observed rate of protein synthesis. The rate of protein synthesis is at a maximum during the first 15 min of *Tradescantia* pollen tube growth, decreases thereafter, and is only 20% of the initial rate after 60 min. Using several pollen-expressed cDNA clones (see section III C), including an actin clone, to quantify the amounts of specific mRNAs in the growing pollen tube, it was found that the rates of decline of these mRNAs during the first 60 min of germination and pollen tube growth were very similar for all the clones used (Stinson *et al.*, 1987). Since the proteins synthesized on preexisting and newly synthesized mRNAs appear to be the same, these results indicate that the rate of degradation of the mRNAs exceeds that of their synthesis. The reduction in the levels of the mRNAs correlates well with the decline in poly(A), suggesting that the entire mRNA molecules turn over, not just the poly(A) tails. The degradation of mRNAs during pollen tube growth appears to be a nonspecific degradation of all mRNAs existing in the pollen grain and there do not appear to be classes of mRNAs that differ in their rates of turnover. The significance of this rapid decrease in mRNA content and rate of protein synthesis at a time when the pollen tube continues to grow at a linear rate is not clear at the present time.

B. Estimates of Numbers of Genes Expressed in Pollen

The mature pollen grain as discussed earlier contains mRNAs that were synthesized prior to anthesis. Based on an analysis of the kinetics of hybridization of ^{3}H-cDNA with poly(A)RNA in excess, the mRNA in mature *Tradescantia* pollen was found to consist of three abundance classes with complexities of 5.2×10^{4}, 1.6×10^{6} and 2.1×10^{7} nucleotides (Willing and Mascarenhas, 1984). About 15% of the mRNA is very abundant and comprises about 40 diverse sequences each present in 26,000 copies per pollen grain. The major fraction of the mRNA (60%) is made up of about 1400 different sequences each present in about 3400 copies per grain. The least abundant fraction is a relatively small proportion in pollen (24%) compared to its presence in the vegetative shoot (64%) but consists of 18,000 different sequences each present in about 100 copies per grain. Thus in pollen about 75% of the mRNAs occur in the two more abundant frequency classes, whereas only 35% of the shoot mRNAs are abundant. The least abundant shoot mRNAs consist of 29,000 sequences each present in 5—10 copies per cell. It is interesting that the least abundant fraction in pollen contains sequences that are much more abundant (about 100 copies) than in the corresponding fraction in shoots (5—10 copies). The total complexity of pollen poly(A)RNA is 2.3×10^7 nucleotides corresponding to about 20,000 different sequences, whereas in shoots it is 3.4×10^7 nucleotides comprising about 30,000 different sequences.

Similar hybridization analyses have been carried out in maize (Willing, Bashe and Mascarenhas, 1988). In maize pollen 35% of the mRNAs are abundant and comprise about 240 sequences each present on average in about 32,000 copies per pollen grain. The middle abundance class which makes up the major fraction of the mRNAs (49%) consists of about 6000 different sequences each present in 1700 copies per grain. The least abundant fraction (15%) is made up of about 17,000 sequences each present in about 200 copies per grain. The total complexity of corn pollen poly(A)RNA is 2.4×10^7 nucleotides corresponding to about 24,000 different sequences, values very similar to those found in *Tradescantia* pollen. The complexity of pollen mRNA sequences in both *Tradescantia* and maize is about 60% that of shoot mRNAs (Willing and Mascarenhas, 1984; Willing *et al.*, 1988).

C. Cloning of Pollen-Expressed Genes and the Pattern of Transcription of Specific mRNAs

As stated above, the mature pollen grain contains a store of presynthesized mRNAs. At what stage in pollen development does the transcription for these mRNAs occur? In order to answer this question it is necessary to obtain cloned probes. Recombinant cDNA libraries have been constructed to poly(A)RNA isolated from mature pollen of *Tradescantia paludosa* and *Zea mays* (Stinson *et al.*, 1987). In these libraries there are several clones that are specific to pollen and other clones corresponding to mRNAs expressed both in pollen and vegetative tissues. Southern blot hybridizations indicate that the pollen-expressed genes are present in only one or a very few copies in the corn genome. Several of these clones have been used as probes in Northern blots to determine the stages of pollen development during which the synthesis is initiated of the mRNAs found in the mature pollen grain, and also to study the pattern of their accumulation during microsporogenesis. The mRNAs complementary to pollen-expressed sequences in both *Tradescantia* and corn are first detectable after microspore mitosis during early pollen interphase, and they continue to accumulate thereafter, reaching a maximum concentration in the mature pollen grain. Using an actin clone from soybean, actin mRNA is detectable in *Tradescantia* pollen RNA. In contrast to the pollen-specific mRNAs, actin mRNA first appears during microspore interphase prior to microspore mitosis. It accumulates thereafter and reaches a maximum at late pollen interphase before decreasing substantially in the mature pollen grain (Stinson *et al.*, 1987). The results of these analyses indicate that there are at least two different groups of mRNAs in pollen with respect to their synthesis during pollen development. The first group, represented by the pollen-specific clones, are synthesized after microspore mitosis and increase in content up to maturity. This pattern of accumulation would seem to suggest a major function for these mRNAs during the latter part of pollen maturation and during pollen germination and pollen tube growth. The second group, which includes mRNAs like that for actin, begin to accumulate soon after meiosis, reach their maximum by late pollen interphase, and then decrease substantially. Alcohol dehydrogenase (ADH) and β-galactosidase are probably other examples of proteins that are expressed in a fashion similar to that of actin. The appearance and pattern of increase of ADH enzyme activity during microsporogenesis in corn (Stinson and Mascarenhas, 1985), and of β-galactosidase in *Brassica campestris* (Singh, O'Neill and Knox, 1985), are what would be expected if their mRNAs were synthesized in a manner similar to that of the actin mRNA in *Tradescantia.*

D. Overlap of Sporophytic and Gametophytic Gene Expression

From a comparison of isozyme profiles in pollen and several vegetative tissues of tomato (nine enzyme systems coded for by 28 genes), Tanksley *et al.* (1981) found that 60% of the genes coding for these enzymes in vegetative tissues were also expressed in pollen, whereas 18 of 19 pollen enzymes or 95% were also found in one or more of the vegetative tissues. A similar study in corn showed that about 72% of the isozymes were expressed in both pollen and the sporophyte, whereas only about 6% of the isozymes studied were pollen specific (Sari-Gorla *et al.*, 1986). Extensive overlap was also found among isozyme profiles of pollen and the sporophyte tissues of three *Populus* species (Rajora and Zsuffa, 1986). Between 74% and 80% of the genes coding for these enzymes in the sporophyte were also found to be expressed in the male gametophyte.

Heterologous hybridizations, e. g. pollen cDNA to shoot poly(A)RNA or shoot cDNA to pollen RNA, indicate that in *Tradescantia* a minimum of 64% of the pollen mRNA is also found in shoot mRNA, whereas no more than 60% of the shoot mRNAs are also represented in pollen (Willing and Mascarenhas, 1985). Similarly in corn, about 65% of the sequences in pollen are similar to those in shoots (Willing, Bashe and Mascarenhas, 1988).

The pollen cDNA libraries that have been constructed include clones that are expressed in either abundance or low amounts in pollen but not at all in vegetative tissues. Other clones are expressed in abundance in both pollen and vegetative tissue, and still others are expressed in either high or low levels in both kinds of tissue (Stinson *et al.*, 1987). Only a few of the sequences appear to be unique to pollen; the great majority are expressed both in pollen and in vegetative tissues. Based on colony hybridizations to ^{32}P-cDNAs from pollen and vegetative tissues, it has been estimated that about 20% and 10% of the total sequences expressed in *Tradescantia* and corn pollen respectively are pollen specific (Stinson *et al.*, 1987).

Mulcahy (1979) proposed that genetic selection operating during male gametophyte growth in the style could have a positive effect on the sporophyte generation resulting from this selection. The speed of germination and of pollen tube growth are the primary factors governing the competition between pollen tubes in reaching and effecting fertilization in a limited number of ovules. For these variables to have selective value in the sporophyte, it is necessary that pollen germination and tube growth be regulated by a sizeable proportion of genes that are expressed and affect basic functions in both the gametophytic and sporophytic phases of development. From the studies presented earlier it is apparent that the genetic program expressed during pollen development is extensive and that there appears to be a substantial overlap between genes active in the gametophytic and sporophytic tissues. Selecting for genes in the haploid phase could thus have a positive effect on the success of the sporophyte (Mulcahy, 1979; Mulcahy and Mulcahy, 1987; Ottaviano *et al.*, 1980). Positive correlations between pollen tube growth and sporophytic traits have been reported in several plants (see review in Ottaviano and Sari-Gorla, 1979; Mulcahy and Mulcahy, 1975; Ottaviano *et al.*, 1983). Moreover, the competitive ability of the male gametophyte in the style and the selective advantages of this competition have been directly utilized in the haploid selection under selective conditions for low temperature tolerance in tomato hybrids (Zamir *et al.*, 1982; Zamir and Vallejos, 1983), for tolerance to high salt in hybrid progeny from a *Lycopersicon esculentum* (salt intolerant) × *Solanum pennellii* (salt tolerant) cross (Sacher *et al.*, 1983), and for zinc or copper tolerance in *Silene dioica* and *Mimulus guttatus* (Searcy and Mulcahy, 1985a, 1985b). The advantage of using haploid pollen to screen for desirable genes is that extremely large numbers of pollen grains can be screened within a short time on a relatively few styles in a small area, whereas a similar screening of diploid plants might require several hundred acres (Zamir, 1983).

E. Sperm Cells

There is at the present time practically no information about the molecular biology of sperm cells of flowering plants. Sperm cells in the trinucleate rye (*Secale cereale*) pollen grain synthesize RNA during germination *in vitro* as shown by ^{3}H-uridine labeling and autoradiography (Haskell and Rogers, 1985). In contrast, sperm of *Hyoscyamus niger*, in which generative cell division is completed during pollen tube growth, have not been found to synthesize RNA after short pulses of ^{3}H-uridine (Reynolds and Raghavan, 1982). These studies are apparently the extent of current understanding of the molecular biology of sperm cells.

The sperm cells of *Plumbago zeylanica* occur paired within each pollen grain. They are dimorphic, differing in size, morphology, and organelle content. The larger of the sperm cells is intimately associated with the vegetative nucleus by a long projection (Russell, 1984). The two sperm cells and the vegetative nucleus travel as a linked unit within the pollen tube. The sperm associated with the vegetative nucleus is practically devoid of plastids whereas the other sperm contains a large number of these organelles. The plastid-rich, mitochondrion-poor sperm cell preferentially fuses with the egg (Russell, 1985). This is an exciting finding because it indicates, that the two sperms have differentiated and that the process of double fertilization is not a random event. A similar dimorphism has been described for the sperm of *Brassica campestris* and *B. oleracea* (McConchie *et al.*, 1987), indicating that the preferential fusion of the two sperm cells with the female cells might be a general phenomenon.

Recently methods for the isolation of sperm cells have been described for *Plumbago* (Russell, 1986), *Brassica oleracea, Zea mays,* and *Triticum aestivum* Matthys-Rochon *et al.*, 1987). These advances now make it possible to study the biochemistry and molecular biology of flowering plant sperm, areas of knowledge which play a critical role in agriculture but which have not yet received the attention they deserve.

Acknowledgement

Work in the author's laboratory has been supported by grants from the National Science Foundation.

IV. References

Albertini, L., Grenet-Auberger, H., Souvre, A., 1981: Polysaccharides and lipids in microsporocytes and tapetum of *Rhoeo discolor* Hance. Cytochemical study. Acta Soc. Bot. Polon. **50,** 21—28.

Albertini, L., Souvre, A., 1978: Les polysaccharides des microsporocytes et du tapis chez le *Rhoeo discolor* Hance. Étude cytochimique et autoradiographie (glucose-^{3}H). Bull. Soc. Bot. Fr., Actualités botan. **125,** 45—50.

Altschuler, M., Mascarenhas, J. P., 1982: The synthesis of heat shock and normal proteins at high temperatures in plants and their possible roles in survival under heat stress. In: Heat Shock From Bacteria to Man. Schlessinger, M. J., Ashburner, M., Tissieres, A. (eds.), Cold Spring Harbor Lab. 291—297.

Angold, R. E., 1968: The formation of the generative cell in the pollen grain of *Endymion non-scriptus* (L.). J. Cell Sci. **3,** 573—578.

Bashe, D. M., 1984: Developmental control of protein and rRNA synthesis in pollen of *Tradescantia paludosa* and *Nicotiana tabacum.* M. S. Thesis, State Univ. of New York at Albany.

Bino, R. J., 1985: Histological aspects of microsporogenesis in fertile, cytoplasmic male sterile and restored fertile *Petunia hybrida.* Theor. Appl. Genet. **69,** 425—428.

Brewbaker, J. L., 1967: The distribution and phylogenetic significance of binucleate and trinucleate pollen grains in the angiosperms. Amer. J. Bot. **54,** 1069—1083.

Brewbaker, J. L., 1971: Pollen enzymes and isoenzymes. In: Pollen: Development and Physiology. Heslop-Harrison, J. (ed.), London, Butterworths. 156—170.

Bryce, W. H., Nelson, D. E., 1979: Starch-synthesizing enzymes in the endosperm and pollen of maize. Plant Physiol. **63,** 312—317.

Capkova-Balatkova, V., Hrabetova, E., Tupy, J., 1980: Effects of cycloheximide on pollen of *Nicotiana tabacum* in culture. Biochem. Physiol. Pflanzen **175,** 412—420.

Cooper, P., Ho, T.-H. D., Hauptman, R. M., 1984: Tissue specificity of the heat shock response in maize. Plant Physiol. **75,** 431—441.

Dickinson, H. G., 1973: The role of plastids in the formation of pollen grain coatings. Cytobios. **8,** 25—40.

Dickinson, H. G., 1976: Common factors in exine deposition. In: The Evolutionary Significance of the Exine, pp. 67—90. Ferguson, I. K., Muller, J. (eds.). Linnean Soc. Symp. Ser. 1.

Dickinson, D. B., Hopper, J. E., Davies, M. D., 1973: A study of pollen enzymes involved in sugar nucleotide formation. In: Biogenesis of Plant Cell Wall Polysaccharides. pp. 29—48. Loewus, F. (ed.), New York, Academic Press.

Dickinson, D. B., Hyman, D., Gonzales, J. W., 1977: Isolation of uridine 5′-pyrophosphate glucuronic acid pyrophosphorylase and its assay using ^{32}P-pyrophosphate. Plant Physiol. **59,** 1082—1084.

Echlin, P., 1971: The role of the tapetum during microsporogenesis of angiosperms. In: Pollen; Development and Physiology. pp. 41—61. Heslop-Harrison, J. (ed.). London, Butterworths.

Eschrich, W., 1961: Untersuchungen über den Ab- und Aufbau der Callose (III. Mitteilung über Callose). Z. Bot. **49,** 153—218.

Frankel, R., Izhar, S., Nitsan, J., 1969: Timing of callose activity and cytoplasmic male sterility in *Petunia.* Biochem. Genet. **3,** 451—455.

Frankis, R. C ., Mascarenhas, J. P., 1980: Messenger RNA in the ungerminated pollen grain: a direct demonstration of its presence. Ann. Bot. **45,** 595—599.

Frova, C., Binelli, G., Ottaviano, E., 1987: Isozyme and HSP gene expression during male gametophyte development in maize. In: Isozymes: Current Topics in Biological and Medical Research. Vol. 15, pp. 97—120. Scandalios, J. (ed.), New York, Alan R. Liss.

Goldberg, R. B., 1987: Emerging patterns of plant development. Cell **49,** 298—300.

Hara, A., Kawamoto, K., Funaguma, T., 1980: Inorganic pyrophosphatase from pollen of *Typha latifolia.* Plant Cell. Physiol. **21,** 1475—1482.

Haskell, D. W., Rogers, O. M., 1985: RNA synthesis by vegetative and sperm nuclei of trinucleate pollen. Cytologia **50,** 805—809.

Herdt, E., Sutfeld, R., Wiermann, R., 1978: The occurrence of enzymes involved in phenylpropanoid metabolism in the tapetum fraction of anthers. Cytobiologie **17,** 433—441.

Heslop-Harrison, J., 1968: Tapetal origin of pollen coat substances in *Lilium.* New Phytol. **67,** 779—786.

Heslop-Harrison, J., 1971: Sporopollenin in the biological context. In: Sporopollenin, pp. 1—31. Brooks, J., Grant, P. R., Muir, M., van Gjizel, P., Shaw, G. (eds.). London, Academic Press.

Horner, H. T., Rogers, M. A., 1974: A comparative light and electron microscopic study of microsporogenesis in male-fertile and cytoplasmic male-sterile pepper (*Capsicum annuum*). Canad. J. Bot. **52,** 435—441.

Izhar, S., Frankel, R., 1971: Mechanism of male sterility in *Petunia.* The relationship between pH, callose activity in the anthers, and the breakdown of microsporogenesis. Theoret. Appl. Genet. **41,** 104—108.

Jackson, J. F., Linskens, H. F., 1980: DNA repair in pollen: Range of mutagens inducing repair, effect of radiation inhibitors and changes in thymidine nucleotide metabolism during repair. Mol. Gen. Genet. **180,** 517—522.

Jackson, J. F., Linskens, H. F., 1982: Phytic acid in *Petunia hybrida* pollen is hydrolyzed during germination by a phytase. Acta Bot. Neerl. **31,** 441—447.

Kamalay, J. C., Goldberg, R. B., 1980: Regulation of structural gene expression in tobacco. Cell **19,** 934—946.

Knox, R. B., Heslop-Harrison, J., 1969: Cytochemical localization of enzymes in the wall of the pollen grain. Nature **223,** 92—94.

Kroh, M., Loewus, M. W., 1978: Pectolytic enzyme activity from *Nicotiana tabacum* pollen. Phytochem. **17,** 797—799.

Lin, J.-J., Dickinson, D. B., Ho, T.-H. D., 1987: Phytic acid metabolism in lily (*Lilium longiflorum* Thunb.) pollen. Plant Physiol. **83,** 408—413.

Linskens, H. F., Schrauwen, J. A., Konings, R. N. H., 1970: Cell-free protein synthesis with polysomes from germinating *Petunia* pollen grains. Planta **90,** 153—162.

Malik, C. P., Gupta, S. C., 1976: Changes in peroxidase isoenzymes during germination of pollen. Biochem. Physiol. Pflanzen **169,** 519—522.

Malik, C. P., Singh, M. B., 1976: Fluctuations in dehydrogenase activities during development of pollen tube of *Calotropis procera.* Biochem. Physiol. Pflanzen **169,** 583—588.

Mascarenhas, J. P., 1975: The biochemistry of angiosperm pollen development. Bot. Rev. **41,** 259—314.

Mascarenhas, J. P., Altschuler, M., 1983: The response of pollen to high temperatures and its potential applications. In: Pollen: Biology and Implications for Plant Breeding, pp. 3—8. Mulcahy, D. L., Ottaviano, E. (eds.). New York, Elsevier Biomedical.

Mascarenhas, J. P., Bell, E., 1969: Protein synthesis during germination of pollen: studies on polyribosome formation. Biochim. Biophys. Acta **179,** 199—203.

Mascarenhas, J. P., Bell, E., 1970: RNA synthesis during development of the male gametophyte of *Tradescantia.* Develop. Biol. **21,** 475—490.

Mascarenhas, J. P., Goralnick, R. D., 1971: Synthesis of small molecular weight RNA in the pollen tube of *Tradescantia paludosa.* Biochim. Biophys. Acta **240,** 56—61.

Mascarenhas, J. P., Mermelstein, M., 1981: Messenger RNAs: their utilization and degradation during pollen germination and tube growth. Acta Soc. Bot. Polon. **50,** 13—20.

Mascarenhas, J. P., Terenna, B., Mascarenhas, A. F., Rueckert, L., 1974: Protein synthesis during germination and pollen tube growth in *Tradescantia.* In: Fertilization in Higher Plants. pp. 137—143. Linskens, H. F. (ed.). North Holland.

Mascarenhas, N. T., Bashe, D., Eisenberg, A., Willing, R. P., Xiao, C. M., Mascarenhas, J. P., 1984: Messenger RNAs in corn pollen and protein synthesis during germination and pollen tube growth. Theor. Appl. Genet. **68,** 323—326.

Matthys-Rochon, E., Vergne, P., Detchepare, S., Dumas, C., 1987: Male germ unit isolation from three tricellular pollen species: *Brassica oleracea, Zea mays* and *Triticum aestivum.* Plant Physiol. **83,** 464—466

McConchie, C. A., Russell, S. D., Dumas, C., Tuohy, M., Knox, R. B., 1987: Quantitative cytology of the sperm cells of *Brassica campestris* and *B. oleracea.* Planta **170,** 446—452.

Mepham, R. H., Lane, G. R., 1969: Formation and development of the tapetal periplasmodium in *Tradescantia bracteata.* Protoplasma **68,** 446—452.

Mulcahy, D. L., 1979: The rise of the angiosperms: A genecological factor. Science **206,** 20—23.

Mulcahy, D. L., Mulcahy, G. B., 1975: The influence of gametophytic competition on sporophytic quality in *Dianthus chinensis.* Theoret. Appl. Genet. **46,** 277—280.

Mulcahy, D. L., Mulcahy, G, B., 1987: The effects of pollen competition. Amer. Sci. **75,** 44—50.

Nave, E. B., Sawhney, V. K., 1986: Enzymatic changes in post-meiotic anther development in *Petunia hybrida.* I. Anther ontogeny and isozyme analyses. J. Plant. Physiol. **125,** 451—465.

Ottaviano, E., Sari-Gorla, M., Mulcahy, D. L., 1980: Pollen tube growth rates in *Zea mays:* Implications for genetic improvement of crops. Science **210,** 437—438.

Ottaviano, E., Sari-Gorla, M., 1979: Genetic variability of male gametophyte in maize. Pollen genotype and pollen-style interaction. Monographie in Genetica Agraria. 89—106.

Ottaviano, E., Sari-Gorla, M., Arenari, I., 1983: Male gametophytic competitive ability in maize selection and implications with regard to the breeding system. In: Pollen: Biology and Implications for Plant Breeding. pp. 367—373. Mulcahy, D. L., Ottaviano, E. (eds.). New York, Elsevier Biomed.

Peddada, L., Mascarenhas, J. P., 1975: 5S ribosomal RNA synthesis during pollen development. Develop. Growth Diff. **17,** 1—8.

Raghavan, V., 1981: A transient accumulation of poly(A)-containing RNA in the tapetum of *Hyoscyamus niger* during microsporogenesis. Develop. Biol. **81,** 342—348.

Rajora, O. P., Zsuffa, L., 1986: Sporophytic and gametophytic gene expression in *Populus deltoides* Marsh, *P. nigra* L. and *P. maximowiczii* Henry. Can. J. Genet. Cytol. **28,** 476—482.

Reynolds, T. L., Raghavan, V., 1982: An autoradiographic study of RNA synthesis during maturation and germination of pollen grains of *Hyoscyamus niger.* Protoplasma **111,** 177—188.

Reznickova, S. A., 1978: Histochemical study of reserve nutrient substances in anther of *Lilium candidum.* Comp. Rend. Acad. Bulgare Sci. **31,** 1067—1070.

Reznickova, S. A., Willemse, M. T. M., 1980: Formation of pollen in the anther of *Lilium.* II. The function of the surrounding tissues in the formation of pollen and pollen wall. Acta Bot. Neerl. **29,** 141—156.

Reznickova, S. A., Willemse, M. T. M., 1981: The function of the tapetal tissue during microsporogenesis in *Lilium.* Acta Soc. Bot. Polon. **50,** 83—87.

Rowley, J. R., 1981: Pollen wall characters with emphysis upon applicability. Nord. J. Bot. **1,** 357—380.

Russell, S. D., 1984: Ultrastructure of the sperm of *Plumbago zeylanica.* II. Quantitative cytology and three-dimensional organization. Planta **162,** 385—391.

Russell, S. D., 1985: Preferential fertilization in *Plumbago:* Ultrastructural evidence for gamete-level recognition in an angiosperm. Proc. Natl. Acad. Sci. U.S.A. **82,** 6129—6132.

Russell, S. D., 1986: Isolation of sperm cells from the pollen of *Plumbago zeylanica.* Plant Physiol. **81,** 317—319.

Sacher, R., Mulcahy, D. L., Staples, R., 1983: Developmental selection for salt tolerance during self-pollination of *Lycopersicon* × *Solanum* F1 for salt tolerance of F2. In: Pollen: Biology and Implications for Plant Breeding. pp. 329—334. Mulcahy, D. L., Ottaviano, E. (eds.). New York, Elsevier.

Sari-Gorla, M., Frova, C., Binelli, G., Ottaviano, E., 1986: The extent of gametophytic-sporophytic gene expression in maize. Theoret. Appl. Genet. **72,** 42—47.

Sauter, J. J., 1969: Autoradiographische Untersuchungen zur RNS- und Proteinsynthese in Pollenmutterzellen, jungen Pollen und Tapetumzellen während der Mikrosporogenese von *Paeonia tenuifolia* L., Z. Pflanzenphysiol. **61,** 1—19.

Sawhney, V. K., Nave, E. B., 1986: Enzymatic changes in post-meiotic anther development in *Petunia hybrida*. II. Histochemical localization of esterase, peroxidase, malate- and alcohol dehydrogenase. J. Plant Physiol. **125,** 467—473.

Schrauwen, J. A. M., Reijnen, W. H., De Leeuw, H. C. G. M., van Herpen, M. M. A., 1986: Response of pollen to heat stress. Acta Bot. Neerl. **35,** 321—327.

Scott, J. J., Loewus, F. A., 1986: A calcium-activated phytase from pollen of *Lilium longiflorum*. Plant Physiol. **82,** 333—335.

Searcy, K. B., Mulcahy, D. L., 1985a: Pollen tube competition and selection for metal tolerance in *Silene diocia* (Caryophyllaceae) and *Mimulus gutatus* (Scrophulariaceae). Amer. J. Bot. **72,** 1695—1699.

Searcy, K. B., Mulcahy, D. L., 1985: Pollen selection and the gametophytic expression of metal tolerance in *Silene diocia* (Caryophyllaceae) and *Mimulus guttatus* (Scrophulariaceae). Amer. J. Bot. **72,** 1700—1706.

Shaykh, M., Kolattakudy, P. E., Davis, R., 1977: Production of a novel extracellular cutinase by the pollen and the chemical composition and ultrastructure of the stigma cuticle of *Nasturtium*. Plant Physiol. **60,** 907—915.

Singh, M. B., O'Neill, P. M., Knox, R. B., 1985: Initiation of postmeiotic β-galactosidase synthesis during microsporogenesis in oilseed rape. Plant Physiol. **77,** 229—231.

Steffensen, D. M., 1966: Synthesis of ribosomal RNA during growth and division in *Lilium*. Expt. Cell Res. **44,** 1—12.

Steffensen, D. M., 1971: Ribosome synthesis compared during pollen and pollen tube development. In: Pollen: Development and Physiology, pp. 223—229. Heslop-Harrison, J. (ed.). London, Butterworths.

Stieglitz, H., Stern, H., 1973: Regulation of β-1,3-glucanase activity in developing anthers of *Lilium*. Develop. Biol. **34,** 169—173.

Stinson, J., Mascarenhas, J. P., 1985: Onset of alcohol dehydrogenase synthesis during microsporogenesis in maize. Plant Physiol. **77,** 222—224.

Stinson, J. R., Eisenberg, A. J., Willing, R. P., Pe, M. E., Hanson, D. D., Mascarenhas, J. P., 1987: Genes expressed in the male gametophyte of flowering plants and their isolation. Plant Physiol. **83,** 442—447.

Suss, J., Tupy, J., 1979: Poly(A)RNA synthesis in germinating pollen of *Nicotiana tabacum* L., Biol. Planta (Praha) **21,** 365—371.

Takats, S. T., Wever, G. H., 1971: DNA polymerase and DNA nuclease activities in S-competent and S-incompetent nuclei from *Tradescantia* pollen grains. Expt. Cell Res. **69,** 25—28.

Takegami, M. H., Yoshioka, M., Tanaka, I., Ito, M., 1981: Characteristics of isolated microsporocytes from liliaceous plants for studies of the meiotic cell cycle *in vitro*. Plant Cell Physiol. **22,** 1—10.

Tanaka, I., Ito, M., 1980: Induction of typical cell division in isolated microspores of *Lilium longiflorum* and *Tulipa gesneriana*. Plant Sci. Lett. **17,** 279—285.

Tanaka, I., Ito, M., 1981: Studies on microspore development in liliaceous plants. II. Pollen tube development in lily pollens cultured from the uninucleate microspore stage. Plant Cell Physiol. **22,** 149—153.

Tanksley, S. D., Zamir, D., Rick, C. M., 1981: Evidence for extensive overlap of sporophytic and gametophytic gene expression in *Lycopersicon esculentum*. Science **213,** 454—455.

Tano, S., Takahashi, H., 1964: Nucleic acid synthesis in growing pollen tubes. J. Biochem. (Tokyo) **56,** 578—580.

Tupy, J., 1977: RNA synthesis and polysome formation in pollen tubes. Biol. Plantarum (Praha) **19,** 300—309.

Tupy, J., 1982: Alterations in polyadenylated RNA during pollen maturation and germination. Biol. Plantarum (Praha) **24,** 331—340.

Tupy, J., Hrabetova, E., Balatkova, V., 1977: Evidence for ribosomal RNA synthesis in pollen tubes in culture. Biol. Plantarum (Praha) **19,** 226—230.

Vithanage, H. I. M. V., Knox, R. B., 1980: Periodicity of pollen development and quantitative cytochemistry of exine and intine enzymes in the grasses *Lolium perenne* L. and *Phalaris tuberosa* L. Ann. Bot. **45,** 131—141.

Wever, G. H., Takats, S. T., 1971: Isolation and separation of S-competent and S-incompetent nuclei from *Tradescantia* pollen grains. Expl. Cell Res. **69,** 29—32.

Whipple, A. P., Mascarenhas, J. P., 1978: Lipid synthesis in germinating *Tradescantia* pollen. Phytochem. **17,** 1273—1274.

Wiermann, R., 1979: Stage-specific phenylpropanoid metabolism during pollen development. In: Regulation of Secondary Product and Plant Hormone Metabolism, pp. 231—239. Luckner, M., Schreiber, K. (eds.). Oxford, Pergamon Press.

Willing, R. P., Bashe, D., Mascarenhas, J. P., 1988: An analysis of the quantity and diversity of Messenger RNAs from pollen and shoots of *Zea mays.* Theoret. Appl. Genet. (in press).

Willing, R. P., Mascarenhas, J. P., 1984: Analysis of the complexity and diversity of mRNAs from pollen and shoots of *Tradescantia.* Plant Physiol. **75,** 865—868.

Xiao, C. M., Mascarenhas, J. P., 1985: High temperature induced thermotolerance in pollen tubes of *Tradescantia* and heat shock proteins. Plant Physiol. **78,** 887—890.

Zamir, D., 1983: Pollen gene expression and selection: applications in plant breeding. In: Isozymes in Plant Genetics and Breeding. pp. 313—330. Tanksley, S. D., Orton, T. J. (eds.). Amsterdam, Elsevier.

Zamir, D., Tanksley, S. D., Jones, R. A., 1982: Haploid selection for low temperature tolerance of tomato pollen. Genetics **101,** 129—137,

Zamir, D., Vallejos, E. C., 1983: Temperature effects on haploid selection of tomato micropores and pollen grains. In: Pollen: Biology and Implications in Plant Breeding, pp. 335—342. Mulcahy, D. L., Ottaviano, E. (eds.), New York, Elsevier.

Zhang, H., Croes, A. F., Linskens, H. F., 1984: Qualitative changes in protein synthesis in germinating pollen of *Lilium longiflorum* after a heat shock. Plant Cell Environ. **7,** 689—691.

Chapter 7

Self-Incompatibility Genes in Flowering Plants

E. C. Cornish, J. M. Pettitt, and A. E. Clarke

Plant Cell Biology Research Centre, School of Botany University of Melbourne, Parkville, Victoria 3052, Australia

Contents

I. Introduction

Within a population of flowering plants outbreeding and random mating is achieved mainly by reducing or preventing self-fertilization. While selective pollination provides for efficient pollen transfer between individuals of the same species, the principal device promoting outbreeding is self-incompatibility, a device providing for selective fertilization. Self-incompatibility operates to ensure that a plant preferentially accepts fertilization by another, genetically different, individual of the same species. In the flowering plants, unlike animal and algal systems, the incompatibility reaction is not between the gametes themselves, but between a property of the haploid genome of the male gametophyte contained in the pollen grain and the female somatic pistil tissue of the sporophyte, or between diploid factors carried by the pollen grain and the pistil tissue of the sporophyte. The former reaction is referred to as the gametophytic system of control, and the latter, the sporophytic system of control. The reaction occurs in the gametophytic system when alleles of the incompatibility gene, or genes, in the male gametophyte and pistil are matched, and in the sporophytic system when the alleles of the gene encoding the diploid factors carried by the pollen match those in the pistil. However, superimposed upon these genetical criteria of incompatibility are morphological mechanisms. The morphological classification of the incompatibility systems includes two basic types, heteromorphic and homomorphic. The terms are self-explanatory. In heteromorphic incompatibility, gross differences in floral organization distinguish the interfertile mating types, while homomorphic incompatibility shows no such differences.

II. Homomorphic Incompatibility

A. The General Features of Gametophytic Self-Incompatibility

This system of self-incompatibility is taxonomically the most widely distributed in flowering plants. It occurs, for example, in species of the Solanaceae, Liliaceae, Onagraceae, Papaveraceae and Gramineae. In most of the species studied, the grasses are an exception, self-incompatibility is governed by a single gene, the S-gene, with multiple alleles. More than 40 S-alleles have been reported in some species (Emerson, 1938; Williams, 1961; Campbell and Lawrence, 1981). The system is generally, but not invariably, correlated with bicellular pollen and a "wet" stigma, and in the majority of species the growth of an incompatible pollen tube is arrested in the transmitting tissue of the style. However, stigmatic inhibition of tube growth does occur, and is typical of the system in *Oenothera* (Onagraceae) and the grasses (Dickinson and Lawson, 1975; Heslop-Harrison and Heslop-Harrison, 1982a).

B. The General Features of Sporophytic Self-Incompatibility

The sporophytic system of self-incompatibility is well described for two families of flowering plants, the Cruciferae and Compositae. In contrast to the gametophytic system, where the S-alleles show independent action in the pollen and pistil, a feature of the sporophytic system is that the S-gene shows a complexity of dominance and codominance in allelic pairs. The system in the crucifers and composites is associated with tricellular pollen and a "dry" stigma, and the incompatibility reaction is at, or just beneath, the stigma surface. The inhibition process usually affects the initial hydration or germination of the incompatible pollen (Zuberi and Dickinson, 1985). However, if germination occurs, the incompatible tube stimulates the localized production of callose in the stigma cell at the point where penetration is attempted. A similar reaction occurs in the pollen tube and callose occludes the active region at the tip. Growth of the incompatible tube then ceases (Dickinson and Lewis, 1973).

III. Heteromorphic Incompatibility

Heteromorphic incompatibility is associated with floral polymorphism, and heterostyly is one such structural condition. Since most heterostylous species are self-incompatible (Ganders, 1979), the floral heteromorphism is seen as an adaptation to manipulate pollinator movement and pollen flow (Bawa and Beach, 1981). Distyly is the most common expression of heterostyly, and the condition has been intensively studied from Darwin (1877) to the present day. Legitimate pollen transfer is from short stamens to short styles or from long stamens to long styles, between the two floral morphs. As well as being morphologically dissimilar, the two mating types in distylous species differ genetically, in that one morph is homozygous for a recessive gene complex (supergene) and the other is heterozygous. In *Primula* (Primulaceae), for example, the long styled form, termed "pin", is the homozygote (ss) and the short styled form, termed "thrum", is the heterozygote (Ss). The genetic control of the incompatibility is from the sporophyte. This means that although half the pollen produced by the "thrum" plant carries the S-allele and half the s-allele, all the grains have the incompatibility reaction characteristic of S. The dominance of S over s is essential for the function of the outbreeding system.

IV. Nature of the Self-Incompatibility Reaction

Two different models that have been suggested to explain the phenomenon of self-incompatibility in flowering plants:

1. Oppositional models — which assume a system of inhibition of like genotypes
2. Complementary models — which assume a system of stimulation of unlike genotypes

According to the oppositional models there is active interference with pollen tube growth in the pistil. In this system, constituents of the male gametophyte and pistil interact, or combine to produce a substance, and the process of interaction, or characteristics of the product, adversely affects the development of the male gametophyte. The complementary models, on the other hand, suggest that the cessation of pollen tube growth in the self-incompatibility reaction is due to the absence of essential factors in the pollen and pistil necessary for successful penetration.

Lewis (1964) has advanced a hypothesis to explain active inhibition of growth by an oppositional reaction. On this hypothesis, called the dimer hypothesis, a substance, a polypeptide, coded by the S-allele, is present in the pollen and style, the different *S*-alleles producing different dimeric polypeptides. With self-pollination, a tetramer is formed which acts directly, or as an intermediary, to inhibit growth of the incompatible tube. However, it is not clear what would be needed to prevent the genotypically-determined products of each allele present in the dipolid somatic tissue of the pistil from forming tetramers spontaneously.

In a second hypothesis favouring an oppositional system van der Donk (1975) has suggested that identical *S*-allele polypeptides are not involved in the reaction directly. Rather, specific style components serve to activate genes coding a substance that is essential for tube growth, and the role of the pollen-specific molecules is to prevent this occurring. Since the inactivation process blocks gene activity in the style, there would be a progressive decrease in the reserve of stylar metabolites and tube growth would cease once this was exhausted. Central to this argument is the belief that the *S*-allele products in the pollen and style are different.

Among the hypotheses offered in support of the complementary system, Kroes (1973), following Bateman (1952) has suggested that incompatible tubes cease to grow because they do not have the capacity to utilize growth complexes in the style. This deficiency is genetically determined. Conversely, the growth of compatible tubes is due to their ability to hydrolyze the stylar complexes. This explanation rests on the assumption that every *S*-allele corresponds to the absence of one specific enzyme in the pollen that is required for mobilization of a nutrient held immobilized in the stylar tissue by a stylar product of the identical allele. However, as Lewis (1954) has noted, it is not clear how, on this mechanism, there is selective discrimination against incompatible tubes in mixed pollinations with grains carrying different S-alleles. In this event it would be expected that the stylar nutrient essential for tube growth would be made generally available by the enzyme secreted from the compatible tubes.

Recently, Mulcahy and Mulcahy (1983) have discussed the evidence for the oppositional models, and they have listed various observations concerning the gametophytic system of control for which none of the models can adequately account (see Mulcahy and Mulcahy, 1983). These shortcomings are explained on a new concept of complementation, termed the heterosis model. This is a multigenic model which assumes that dissimilar alleles in the pollen and style will bring about heterotic interactions, and an increase in pollen tube growth rate will result. But if the style is homozygous for a deleterious recessive allele and this allele is also carried in the pollen, the tube growth rate will be reduced. Each gene complex contains one dominant gene, and the others are taken to be deleterious recessives. The growth rate of the pollen tube, then, is the sum of the pollen-style interactions, and incompatibility is due to pollen tube growth being too slow to effect fertilization. In a compatible situation tube growth rate is promoted. This concept has been criticised by Lawrence, Marshall,

Curits and Fearon (1985) on the basis that none of the anomalies adduced by Mulcahy and Mulcahy (1983) in support of the heterosis model is inconsistent with the conventional oppositional models and that, in some instances, the nature of the evidence which they regard as conflicting has been mis-interpreted. Mulcahy and Mulcahy (1985) have restated the proposition which invokes differences in the degree of homozygosity to explain the observed variation in the strength of the incompatibility reaction and the activation of new incompatibility specificities. This is in contrast to the conventional oppositional models which accommodate the first observation with difficulty and the second observation not at all. Both observations can, they claim, can be accounted for on the heterosis model. More recently, experimental evidence has been presented in support of their contention. (Mulcahy, Mulcahy and MacMillan, 1985).

These, then, are some of the hypotheses that have been advanced to account for the self-incompatibility reaction in flowering plants. They all aim to account for the same phenomenon, the termination of growth in an incompatible pollen tube, and they differ mainly in the nature of the mechanism they postulate. All in all there is more general acceptance of the oppositional rather than the complementary scheme. This preference is based on a number of observations. First, disruptive treatments, such as gamma irradiation (Linskens *et al.*, 1960; de Nettancourt, 1969; Cresti *et al.*, 1977), increased temperature (Lewis, 1942; Ascher and Peloquin, 1966a; de Nettancourt, 1971), and treatment with inhibitors of RNA and protein synthesis (Sarfatti *et al.*, 1974; Kovaleva *et al.*, 1978) have been shown to weaken the self-incompatibility reaction. The response to these treatments is consistent with the idea of an active factor responsible for tube arrest being present in the mature style tissue. Secondly, the self-incompatibility barrier can be overcome by selfing immature flowers (de Nettancourt, 1977; Mau *et al.*, 1986) and floral immaturity is associated with low protein levels in the immature style (Cresti *et al.*, 1976). Thirdly, the strength of the incompatibility reaction declines with ageing of the flower (Ascher and Peloquin, 1966b). None of these features is as readily explained by complementation as by opposition.

This review is concerned with the recent application of recombinant DNA, peptide sequencing and carbohydrate analytical techniques that has advanced our understanding of homomorphic gametophytic and sporophytic self-incompatibility in flowering plants. For a more comprehensive discussion of the genetical and biological features of the self-incompatibility systems, the reader is referred to reviews by de Nettancourt (1977) and Heslop-Harrison (1978, 1983).

V. Nature of the S-Gene Products

The classical genetical investigations of the self-incompatibility systems would clearly suggest that there is S-gene expression in the tissues of both anther and pistil. However, in only one species, *Oenothera organensis*, which has a gametophytic system, have S-genotype differences in pollen components been found (Lewis, 1952; Lewis, Burrage and Walls, 1967). Using electrophoretic and immunological techniques these workers were able to show that the pollen contained a protein that was serologically distinct from the four alleles tested. However, two dimensional SDS polyacrylamide gel electrophoretic analysis of *Nicotiana alata* pollen and pollen tube extracts has failed to detect any component which corresponded to particular S-alleles in this species (unpublished observations). Qualitative differences related to S-genotype have been identified in the pistil proteins of six flowering plant species with the gametophytic system of control and in two with sporophytic control. The extracts used in the analyses were obtained, respectively, from the stylar and stigma tissue, and the differences were detected by electrophoretic or immunological methods. The general findings from the analyses are these:

A. Gametophytic System

1. Nicotiana alata (Solanaceae)

An association between specific proteins and different S-alleles of *N. alata* was first demonstrated by isoelectric focusing of style extracts (Bredemeijer and Blass, 1981). Following this study Anderson *et al.* (1986) isolated a style glycoprotein (Mr 32,000), which segregated with the S2 allele. In both investigations the concentration of these proteins increased with flower maturation and paralleled the ability of the style to reject self pollen. Several other S-allele associated glycoproteins have now been purified from *N. alata* style extracts (Mau *et al.*, 1986). There is a high level of homology between the N-terminal sequences of all these glycoproteins, a finding consistent with the view that they are allelic products of a single gene locus (Anderson *et al.*, 1986; Mau *et al.*, 1986).

2. Petunia hybrida (Solanaceae)

An immunological study by Linskens (1960) established the presence of antigenic style components corresponding to particular *S*-alleles. Recently, an association between basic style proteins with molecular weights varying from 27,000 to 33,000 and particular alleles has been demonstrated by Kamboj and Jackson (1986).

3. Lycopersicon peruvianum (Solanaceae)

Two dimensional gel electrophoresis of style extracts from different *S*-genotypes of *L. peruvianum* has shown an association between the S1, S2 and S3 alleles and particular proteins (Mau *et al.*, 1986). Selected two dimensional gel electrophoresis confirmed this association and allowed electroelution and sequence analysis of the proteins corresponding to the S1 and S3 alleles. Although they differ in pI, the N-terminal sequences of these proteins are very similar to each other and to the corresponding sequences of the S-associated glycoproteins of *N. alata* (Mau *et al.*, 1986).

The crosses to determine whether these proteins cosegregate with particular S-alleles have not been performed; but nevertheless, the conservation at the N-terminal region suggests a common function.

4. Prunus avium (Rosaceae)

Two components of styles which were the major antigens were identified in this species. One of these was apparently specific to the stylar tissues of the genus *Prunus* (P-antigen), and the other was specific to a particular *S*-allele group (S-antigen). The S-antigen was detected in three cultivars of S3S4 genotype, but not in cultivars of S1S2 genotype (Raff *et al.*, 1981; Mau *et al.*, 1982).

Two-dimensional SDS-polyacrylamide gel electrophoresis resolved two components which may correspond to the S3 and S4 products in the diploid stylar tissues. The proteins, when isolated under mild conditions, were found to be potent inhibitors of *in vitro* growth of pollen tubes (Williams *et al.*, 1982). If these proteins were the products of the S3 and S4 alleles, specific inhibition of pollen containing the S3 and S4 alleles might have been expected. However, the possibility was not investigated.

5. Lilium longiflorum (Liliaceae)

This species has a stylar canal from which uncontaminated samples of stylar mucilage can be readily removed for analysis. Electrofocusing showed that the mucilage contained 19 protein species. Of these, seven were found to be glycoproteins. The concentration of three of these mucilage glycoproteins was significantly reduced when the styles were incubated at 50C for 6 min (Dickinson, Moriarty and Lawson, 1982). Since this treatment has been shown to inactivate the self-incompatibility system in *Lilium* (Ascher, 1975), it is possible that the heat-labile mucilage glycoproteins are the products of the S-gene (Dickinson *et al.*, 1982).

6. Trifolium pratense (Leguminosae)

Microgradient polyacrylamide gel electrophoresis has shown that one component (Mr 24,000) is present in the style canal fluid, but not in stigma diffusates (Heslop-Harrison and Heslop-Harrison, 1982b). As the incompatibility reaction operates in the style canal and not in the stigma of this species, the protein may be implicated in tube arrest.

B. Sporophytic Systems

1. Brassica oleracea (Cruciferae)

S-specific stigma proteins were first detected in this species by immunological methods (Nasrallah and Wallace, 1967). Characteristic glycoproteins have now been identified for a number of alleles (Nasrallah *et al.*, 1970, 1972; Hinata and Nishio, 1978; Ferrari *et al.*, 1981; Nasrallah and Nasrallah, 1984). Nishio and Hinata (1982) have isolated S-specific glycoproteins from stigma extracts of plants homozygous for the alleles S39, S22 and S7. The glycoproteins have high pIs, 10.3, 11.1 and 10.6 respectively, and contain protein and carbohydrate in the ratio of 1:0.05 for S7 and S22 and 1:0.2 for S39. The molecular weight of the S7 and S39 glycoproteins is estimated to be 57,000. The S22 glycoprotein is apparently heterogeneous; two dimensional SDS-polyacrylamide gel electrophoresis resolved two components of molecular weights 60,000 and 65,000. Antisera raised to isolated S-specific glycoproteins of each genotype precipitated the homologous glycoprotein, the two other S-specific glycoproteins of *B. oleracea*, and an S7-specific glycoprotein of *B. campestris* (Hinata *et al.*, 1982). Ferrari *et al.* (1981) have isolated an S2-specific glycoprotein from a stigma extract of *B. oleracea* and shown that pre-treatment of S2 pollen with this preparation prevents germination of the pollen on a compatible stigma.

2. Brassica campestris (Cruciferae)

Genotype-related glycoproteins have been identified for a number of S-alleles of *B. campestris* using isoelectric focusing techniques (Hinata and Nishio, 1978; Hinata *et al.*, 1982). Glycoproteins associated with the S7, S8, S9 and S12 alleles have been isolated and partially characterized (Nishio and Hinata, 1982; Takayama *et al.*, 1986a, 1986b, 1987). The structural characterization of the S8, S9 and S12 related glycoproteins will be discussed in the following section.

VI. Studies of the Molecular Basis of Self-Incompatibility

Recent contributions to our understanding of the molecular basis of self-incompatibility have come from three research groups. In Japan, Hinata and co-workers at the Tohoku University in Sendai have collaborated with Takayama and co-workers at the University of Tokyo to establish protein and carbohydrate sequences of *Brassica campestris* stigma S-glycoproteins. Nasrallah and co-workers at Cornell University in the United States have determined the sequences of stigma glycoproteins encoded by three alleles of the S-locus of the related species, *Brassica oleracea*. In Australia, Clarke and co-workers at the University of Melbourne have established the sequence of a style glycoprotein corresponding to the S2-allele of *Nicotiana alata*.

In those systems examined so far it seems that different S-alleles are expressed with varying efficiency, but frequently the S-related component represents a high proportion of the total protein synthesized by the female tissue. This feature has facilitated the identification and isolation of these molecules for sequence analysis (Anderson *et al.*, 1986; Mau *et al.*, 1986). Moreover, identification of corresponding cDNA clones has been readily amenable to procedures using differential screening. This approach was adopted in the cloning strategy for isolating cDNA clones encoding an S-related protein from *Brassica oleracea* by Nasrallah *et al.* (1985) and from *Nicotiana alata* by Anderson *et al.* (1986). The first S-related cDNA to be isolated from *B. oleracea* encoded the S6 associated protein backbone. It had been previously shown that the S6 associated glycoprotein represented between six and 10 per cent of the total protein content of the stigma (Nasrallah, Doney and Nasrallah, 1985). When Nasrallah *et al.* (1985) isolated cDNA clones that hybridized strongly to mature stigma cDNA, but not to leaf or seedling cDNA, they found all the clones encoded the S6-related molecule. A similar approach was taken by Anderson *et al.* (1986) to isolate a cDNA for the S3-related glycoprotein from *N. alata,* but in addition, an S2-specific 30mer was used to screen putative S-clones. The 30mer covered part of the predicted S2 signal sequence (Anderson *et al.*, 1986). The sequence of this region was determined from a cDNA molecule which had been primed from oligonucleotides corresponding to the N-terminal sequence of the isolated S2 glycoprotein. Using the 30mer to screen clones that hybridized strongly to mature style cDNA, but not to immature style or ovary cDNA, Anderson *et al.* (1986) isolated long cDNA clones that encoded the entire sequence of mature S2 protein.

There is no homology between the S6-related sequence from *Brassica* and the S2-related sequence from *Nicotiana* (Anderson *et al.*, 1986; Nasrallah *et al.*, 1987). However, comparisons with other S-related molecules reveal sequence conservation within gametophytically and sporophytically controlled families (Anderson *et al.*, 1986; Mau *et al.*, 1986; Nasrallah *et al.*, 1987; Takayama *et al.*, 1987). Nasrallah *et al.* (1987) have recently isolated and sequenced cDNA clones encoding the S13 and S14 related glycoproteins of *B. oleracea.* Takayama *et al.* (1987) have also determined the sequence of peptides isolated from stigma glycoproteins associated with the S8, S9 and S12 alleles of *B. campestris.* As might be predicted for allelic variants, there is extensive sequence homology between these molecules, both within and between the species. The *B. oleracea* sequences have been determined in full. This analysis shows that the conserved sequences are punctuated by two variable regions where the degree of homology is less than 50 per cent. Nasrallah *et al.* (1987) have suggested that the allelic specificity of the molecules might be determined by this variation. This is an intriguing possibility because one of the variable regions is relatively hydrophilic and could therefore occupy a biologically strategic position on the surface of the molecule. Takayama *et al.* (1987) also noted a corresponding region of variation in the S-related sequences of *B. campestris.*

Takayama *et al.* (1986b) have found no major differences in the oligosaccharide side chains of the *B. campestris* glycoproteins that might confer allelic specificity. However, allelic variation may be determined by differences in the actual arrangement of carbohydrate side chains on the glycoproteins. The S6- and S14-related glycoproteins of *B. oleracea* (Nasrallah *et al.*, 1987) and the S8-related glycoprotein of *B. campestris* (Takayama *et al.*, 1987) all have nine potential N-glycosylation sites. The S13-related glycoprotein of *B. oleraceae* has eight potential sites (Nasrallah *et al.*, 1987). In the *B. oleracea* sequences, six glycosylation sites have been conserved; those that vary could conceivably confer allelic specificity. Furthermore, although eight of the nine glycosylation sites predicted for the S6-related glycoprotein from *B. oleracea* and the S8-related glycoprotein from *B. campestris* are the same, it appears that carbohydrate side chains are only attached to six of these sites in the *B. campestris* molecule (Takayama *et al.*, 1987). Characterization of the precise arrangement of carbohydrate side chains on the *B. oleracea* glycoproteins may reveal even more heterogeneity than the nucleotide sequences predict. The nucleotide sequence of the *N. alata* S2 cDNA encodes three potential N-glycosylation sites. Enzymatic deglycosylation studies indicated that a carbohydrate side chain is, in fact, attached to each site (Anderson *et al.*, 1986). Deglycosylation of the S-allele associated glycoproteins from other genotypes of *N. alata* demonstrates that there are different numbers of N-linked carbohydrate chains on each of these molecules (J. Woodward, unpublished data). This structural variation might be significant in that it denotes functional differences in the specificity of the molecule; or it might simply reflect allelic differences in the protein sequence that preclude N-glycosylation.

There appears to be significantly less conservation of S-related nucleic acid sequences in plants exhibiting gametophytic control of self-incompatibility than has been found in species with the sporophytic system of control. Although there is extensive homology between the N-terminal amino acid sequence of the S2 glycoprotein and other S-related proteins from *Nicotiana alata* and *Lycopersicon peruvianum* (Mau *et al.*, 1986) there are appreciable differences in the nucleic acid sequences encoding these molecules. This is evident from the weak hybridization of the S2 cDNA to homologous mRNA species in other genotypes of *N. alata* and *L. peruvianum* (Anderson *et al.*, 1986). This weak hybridization is not due to a lower abundance of these mRNAs (M. Anderson, pers. comm.); rather, it is attributable to sequence variation in the genes encoding them, as determined by weak cross hybridization of alleles on Southern blots (R. Bernatzy, pers. comm.).

VII. Concluding Comments

Perhaps the most convincing evidence that the *Brassica* and *Nicotiana* S-related molecules discussed here are in fact products of the S-locus comes from Southern analysis. By using the homologous cDNAs as probes, restriction fragment length polymorphisms which segregate precisely with the S-locus have been identified both for *B. oleracea* (Nasrallah *et al.*, 1985) and *N. alata* (unpublished results). An especially interesting finding to come from this analysis is that the hybridizing sequences in the *Nicotiana* genome show more variation overall than do those of *Brassica*. It is therefore tempting to speculate upon the possibility that allelic specificity in the sporophytic system is based on quite small sequence differences, whereas the gametophytic system of control requires greater sequence divergence to distinguish self from non-self. Final proof that these sequences actually do correspond to the S-locus will be obtained from gene transfer experiments. It is only then that the different models that have been suggested to explain the phenomenon of self-incompatibility in flowering plants will be amenable to investigation at the molecular level.

Lastly, it is interesting from the evolutionary point of view that there is no apparent similarity between the product of the S-loci in the gametophytic and sporophytic control systems.

VIII. References

Anderson, M. A., Cornish, E. C., Mau, S.-L., Williams, E. G., Hoggart, R., Atkinson, A., Bonig, I., Grego, B., Simpsom, R., Roche, P. J., Haley, J. D., Penschow, J. D., Niall, H. D., Tregear, G. W., Coghlan, J. P., Crawford, R. J., Clarke, A. E., 1986: Cloning of cDNA for a stylar glycoprotein associated with expression of self-incompatibility in *Nicotiana alata*. Nature, Lond. **321,** 38—44.

Ascher, P. D., 1975: Special stylar property required for compatible pollen tube growth in *Lilium longiflorum* Thunb. Bot. Gaz. **136,** 317—321.

Ascher, P. D., Peloquin, S. J., 1966a: Influence of temperature on incompatible and compatible pollen tube growth in *Lilium longiflorum*. Can. J. Genet. Cytol. **8,** 661—664.

Ascher, P. D., Peloquin, S. J., 1966b: Effect of floral ageing on the growth of compatible and incompatible pollen tube growth in *Lilium longiflorum*. Am. J. Bot. **53,** 99—102.

Bateman, A. J., 1952: Self-incompatibility systems in angiosperms. I. Theory. Heredity **6,** 285—310.

Bawa, K. S., Beach, J. H., 1981: Evolution of sexual systems in flowering plants. Ann. Mo. Bot. Gard. **68,** 254—274.

Bredemeijer, G. M. M., Blass, J., 1981: S-specific proteins in styles of self-incompatible *Nicotiana alata*. Theor. appl. Genet. **59,** 185—190.

Campbell, J. M., Lawrence, M. J., 1981: The population genetics of self-incompatibility polymorphism in *Papaver rhoeas*. 2. Number and frequency of S-alleles in a natural population (R 106). Heredity **46,** 81—90.

Cresti, M., Ciampolini, F., Pacini, E., 1977: Ultrastructurall aspects of pollen tube growth inhibition after gamma irradiation in *Lycopersicon peruvianum*. Theor. appl. Genet. **49,** 297—303.

Cresti, M., van Went, J. L., Pacini, E., Willemse, M. T. M., 1976: Ultrastructure of transmitting tissue of *Lycopersicon peruvianum* style: development and histochemistry. Planta **132,** 305—312.

Darwin, C., 1877: The different forms of flowers on plants of the same species. John Murray, London.

Dickinson, H. G., Lawson, J., 1975: Pollen tube growth in the stigma of *Oenothera organensis* following compatible and incompatible intraspecific pollinations. Proc. Roy. Soc. Lond. B. **188,** 327—344.

Dickinson, H. G., Lewis, D., 1973: Cytochemical and ultrastructural differences between intraspecific compatible and incompatible pollinations in *Raphanus*. Proc. Roy. Soc. Lond. B. **183,** 21—28.

Dickinson, H. G., Moriarty, J., Lawson, J., 1982: Pollen-pistil interaction in *Lilium longiflorum:* the role of the pistil in controlling pollen tube growth following cross- and self-pollinations. Proc. Roy. Soc. Lond. B. **215,** 45—62.

Donk, J. A. W. van der, 1975: Recognition and gene expression during the incompatibility reaction in *Petunia hybrida* L. Mol. Gen. Genet. **141,** 305—316.

Emerson, S., 1938: The genetics of self-incompatibility in *Oenothera organensis.* Genetics **23,** 190—202.

Ferrari, T. E., Bruns, D., Wallace, D. H., 1981: Isolation of a plant glycoprotein involved with control of intercellular recognition. Plant Physiol. **67,** 270—277.

Ganders, F. R., 1979: The biology of heterostyly. N. Z. J. Bot. **17,** 607—635.

Heslop-Harrison, J., 1978: Genetics and physiology of angiosperm incompatibility systems. Proc. Roy. Soc. Lond. B. **202,** 73—92.

Heslop-Harrison, J., 1983: Self-incompatibility: phenomenology and physiology. Proc. Roy. Soc. Lond. B. **218,** 371—395.

Heslop-Harrison, J., Heslop-Harrison, Y., 1982a: The pollen-stigma interaction in the grasses. 4. An interpretation of the self-incompatibility response. Acta Bot. Neerl. **31,** 429—439.

Heslop-Harrison, J., Heslop-Harrison, Y., 1982b: Pollen-stigma interaction in the Leguminosae: constituents of the stylar fluid and the stigma secretion of *Trifolium pratense* L. Ann. Bot. **49,** 729—735.

Hinata, K., Nishio, T., 1978: S-allele specificity of stigma proteins in *Brassica oleracea* and *B. campestris.* Heredity **41,** 93—100.

Hinata, K., Nishio, T., Kimura, J., 1982: Comparative studies on S-glycoproteins purified from different S-genotypes in self-incompatible *Brassica* species. II. Immunological specificities. Genetics **100,** 649—657.

Kamboj, R. K, Jackson, J. F., 1986: Self-incompatibility alleles control of a low molecular weight, basic protein in pistils of *Petunia hybrida.* Theor. appl. Genet. **71,** 815—819.

Kovaleva, L. V., Milyaeva, E. L., Chuilukhyan, M. K. H., 1978: Overcoming self-incompatibility by inhibitors of nucleic acid and protein metabolism. Phytomorphology **28,** 445—449.

Kroes, H. W., 1973: An enzyme theory of self-incompatibility. Incompat. Newsletter Assoc. EURATOM-ITAL., Wageningen **2,** 5—14.

Lawrence, M. J., Marshall, D. F., Curtis, V. E., Fearon, C. H., 1985: Gametophytic self-incompatibility re-examined: a reply. Heredity **54,** 131—138.

Lewis, D., 1942: The physiology of incompatibility in plants. I. The effect of temperature. Proc. Roy. Soc. Lond. B. **131,** 13.

Lewis, D., 1952: Serological reactions of pollen incompatibility substances. Proc. Roy. Soc. Lond. B. **140,** 127—135.

Lewis, D., 1954: Comparative incompatibility in angiosperms and fungi. Advan. Genet. **6,** 235—285.

Lewis, D., 1964: A protein dimer hypothesis on incompatibility. Proc. 11th Internat. Congr. Genet. The Hague, 1963. In: Genetics Today, S. J. Geerts (ed.) **3,** 656—663.

Lewis, D., Burrage, S., Walls, D., 1967: Immunological reactions of single pollen grains, electrophoresis and enzymology of pollen protein exudates. J. Exper. Bot. **18,** 371—378.

Linskens, H. F., 1960: Zur Frage der Entstehung von Abwehrkörpern bei der Inkompatibilitätsreaktion von *Petunia.* III. Mitteilung: Serologische Teste mit Leitgewebs- und Pollenextrakten. Z. Bot. **48,** 126—135.

Linskens, J. F., Schrauwen, J. A. M., Donk, M. van der, 1960: Überwindung der Selbstinkompatibiliter durch Röntgenbestrahlung des Griffels. Naturwissenschaften **47,** 547.

Mau, S.-L., Raff, J., Clarke, A. E., 1982: Isolation and partial characterization of components of *Prunus avium* L. styles, including an antigenic glycoprotein associated with a self-incompatibility genotype. Planta **156,** 505—516.

Mau, S.-L., Williams, E. G., Atkinson, A., Anderson, M. A., Cornish, E. C., Grego, B., Simpson, R. J., Kheyer-Pour, A., Clarke, A. E., 1986: Style proteins of a wild tomato (*Lycopersicon peruvianum*) associated with expression of self-incompatibility. Planta **169,** 184—191.

Mulcahy, D. L., Mulcahy, G. B., 1983: Gametophytic self-incompatibility reexamined. Science N. Y. **220,** 1247—1251.

Mulcahy, D. L., Mulcahy, G. B., 1985: Gametophytic self-incompatibility or the more things change ... Heredity **54,** 139—144.

Mulcahy, D. L., Mulcahy, G. B., MacMillan, D., 1985: The Heterosis Model: a progress report. pp. 245—250. In: Biotechnology and Ecology of Pollen, D. L. Mulcahy, G. B. Mulcahy, E. Ottaviano (eds.), Springer-Verlag, New York.

Nasrallah, J. B., Doney, R. C., Nasrallah, M. E., 1985: Biosynthesis of glycoproteins involved in the pollen-stigma interaction of incompatibility in developing flowers of *Brassica oleracea* L. Planta **165,** 100—107.

Nasrallah, J. B., Kao, T.-H., Goldberg, M. L., Nasrallah, M. E., 1985: A cDNA clone encoding an S-locus-specific glycoprotein from *Brassica oleracea.* Nature, Lond. **318,** 263—267.

Nasrallah, J. B., Kao, T.-H., Chen, C.-H., Goldberg, M. L., Nasrallah, M. E., 1987: Amino-acid sequence of glycoproteins encoded by three alleles of the S locus of *Brassica oleracea.* Nature, Lond. **326,** 617—619.

Nasrallah, M. E., Barber, J. T., Wallace, D. H., 1970: Self-incompatibility proteins in plants: detection, genetics, and possible mode of action. Heredity **25,** 23—27.

Nasrallah, M. E., Nasrallah, J. B., 1984: Electrophoretic heterogeneity exhibited by the S-allele specific glycoproteins of *Brassica.* Experientia **40,** 279—281.

Nasrallah, M. E., Wallace, D. H., 1967: Immunogenetics of self-incompatibility in *Brassica oleracea* L. Heredity **22,** 519—527.

Nasrallah, M. E., Wallace, D. H., Savo, R. M., 1972: Genotype, protein, phenotype relationships in self-incompatibility of *Brassica.* Genet. Res. Camb. **20,** 151—160.

Nettancourt, D. de, 1969: Radiation effects on the one locus-gametophytic system of self-incompatibility in higher plants. Theor. appl. Genet. **39,** 187—196.

Nettancourt, D. de, 1971: The generation of new S alleles at the incompatibility locus of *Lycopersicon peruvianum* Mill. Theor. appl. Genet. **41,** 120—129.

Nettancourt, D. de, 1977: Incompatibility in Angiosperms. Springer-Verlag, Berlin.

Nishio, T., Hinata, K., 1982: Comparative studies on S-glycoproteins purified from different self-incompatible *Brassica* species. I. Purification and chemical properties. Genetics **100,** 641—647.

Raff, J. W., Knox, R. B., Clarke, A. E., 1981: Style antigens of *Prunus avium* L. Planta **153,** 125—129.

Sarfatti, G., Ciampolini, F., Pacini, E., Cresti, M., 1974: Effects of actinomycin on *Lycopersicon peruvianum* pollen tube growth and the self-incompatibility reaction. In: Fertilization in Higher Plants, pp. 293—300, H. F. Linskens (ed.), North Holland, Amsterdam.

Takayama, S., Isogai, A., Tsukamoto, C., Ueda, Y., Hinata, K., Okazaki, K., Suzuki, A., 1986 a: Isolation and some characterization of S-locus specific glycoproteins associated with self-incompatibility in *Brassica campestris.* Agric. Biol. Chem. **50,** 1365—1367.

Takayama, S., Isogai, A., Tsukamoto, C., Ueda, Y., Hinata, K., Okazaki, K., Koseki, K., Suzuki, A., 1986 b: Structure of carbohydrate chains of S-glycoproteins in *Brassica campestris* associated with self-incompatibility. Agric. Biol. Chem. **50,** 1673—1676.

Takayama, S., Isogai, A., Tsukamoto, Y. U., Hinata, K., Okazaki, K., Suzuki, A., 1987: Sequence of the S-glycoproteins, products of the *Brassica campestris* self-incompatibility locus. Nature, Lond. **326,** 102—105.

Williams, E. G., Ramm-Anderson, S., Dumas, C., Mau, S.-L., Clarke, A. E., 1982: The effect of isolated components of *Prunus avium* L. styles on an *in vitro* growth of pollen tubes. Planta **156,** 517—519.

Williams, W., 1961: Genetics of red clover (*Trifolium pratense* L.) compatibility. III. The frequency of incompatibility S-alleles in two non-pedigree populations of red clover. J. Genet. **48,** 69—79.

Zuberi, M. I., Dickinson, H. G., 1985: Pollen-stigma interaction in *Brassica.* III. Hydration of the pollen grains. J. Cell Sci. **76,** 321—336.

Chapter 8

Regulatory Circuits of Light-Responsive Genes

Maria Cuozzo, Steve A. Kay, and Nam-Hai Chua

Laboratory of Plant Molecular Biology, The Rockefeller University, 1230 York Avenue, New York, NY 10021-6399, U.S.A.

With 1 Figure

Contents

I. Introduction

Light is utilized by plants not only as an energy source to drive photosynthesis, but also as a trigger for a series of developmental events, from seed germination to flowering. Light also acts as a modulator of gene activity in response to changing light conditions, such as quantity, quality, duration and direction of the light source. This chapter will present a brief description of light-regulated processes in plants. We will then review recent findings on the proteins and DNA sequences involved in light responses. Finally, we will speculate on mechanisms of signal transduction, if only to point out areas which require more intensive investigation. For a comprehensive review of light regulation of gene expression in higher plants, the reader is referred to Tobin and Silverthorne (1985), as well as to more general recent reviews (Thompson *et al.*, 1985; Ellis, 1986; Kuhlemeier *et al.*, 1987b).

There are at least three photoreceptors in higher plants which mediate responses to light: protochlorophyllide, one or more blue light receptors, and phytochrome. Phytochrome has been the most extensively characterized photoreceptor, and we will focus on its physical properties. The process through which the reception of light is converted to a change in gene expression is, for the most part, unknown. Experimental findings using kinetic approaches have been extensively reviewed by Schafer and Briggs (1986). We have included in this chapter a discussion of possible mediators and mechanisms of signal transduction in plants, mainly by comparison with animal systems. Recent advances in the technology of plant cell transformation and subsequent regeneration to transgenic plants have provided an excellent system in which to identify the DNA sequences which confer light responsiveness to genes. The latest findings in this area will be treated here, as well as recent efforts to isolate protein factors which bind to these regulatory sequences. The reader is also referred to a review by Kuhlemeier *et al.* (1987b), for a further treatment on this topic.

II. Multiplicity of Light Effects

A. Photomorphogenesis

Light affects virtually every stage of the plant life cycle, including germination, leaf initiation, plastid differentiation, vascular differentiation, biosynthesis of chlorophyll and other pigments, and flowering. Normal plant development, photomorphogenesis, requires light. The requirement for light is independent of and precedes the requirement of light for photosynthesis. This is illustrated dramatically by comparing seedlings grown in light to those grown in darkness. Dark grown seedlings have elongated stems, no expanded leaves, and lack chlorophyll, a condition called "etiolation". If etiolated sedlings are then exposed to light, normal development can proceed: leaves are expanded, and greening occurs.

The changes in morphology that occur following treatment of dark grown plants with light are accompanied by changes in the amount of extractable activity of dozens of enzymes (for reviews, see Lamb and Lawton, 1983; Tobin and Silverthorne, 1985). These enzymes are involved in many metabolic pathways, including photosynthesis, nucleic acid and protein biosynthesis, photorespiration, and others. Regulation of the amount of enzyme activity occurs at several levels. Transcriptional regulation was implicated by the finding that *in vitro* translation of poly(A)RNA from light- and dark-grown plants yielded qualitatively different sets of proteins (Tobin and Klein, 1975; Apel and Kloppstech, 1978; de Vries *et al.*, 1982). Later, the availability of specific hybridization probes showed, in the case of certain proteins involved in photosynthesis (rbcS and Cab) that increase in translatable mRNA resulted from increases in mRNA abundance rather than conversion from an untranslatable form (Smith and Ellis, 1978; Gollmer and Apel, 1983; Coruzzi *et al.*, 1984; Nelson *et al.*, 1984; Timko *et al.*, 1985). Nuclear runoff transcription experiments have further shown in a few cases that increased mRNA abundance results from increased initiation events (Gallagher and Ellis, 1982; Silverthorne and Tobin, 1984; Berry-Lowe and Meagher, 1985; Gallagher *et al.*, 1985; Mosinger *et al.*, 1985).

A list of genes expressed differentially in light- and dark-grown plants is given in Tobin and Silverthorne (1985) and includes method of mRNA measurement and plant species examined. It must be pointed out that regulation of the same enzyme in different species can be different, so that it is prudent not to generalize. Note that although light causes dramatic increases in rbcS mRNA in many species, including pea, tobacco, maize and soybean, the stimulatory effect is much reduced in mung bean and barley.

Not all light effects are due to an increase in mRNA abundance, particularly in the case of chloroplast genes. Post-transcriptional effects of light have been documented in the case of rbcL from pea (Inaminc *et al.*, 1985) and amaranthus (Berry *et al.*, 1985), and for nuclear-encoded Cab from pea (Bennett, 1981) and Lemna (Slovin and Tobin, 1982).

B. Effects on Gene Expression

The first step of any photoresponse is the absorption of light by a photoreceptor. There are at least three types of photoreceptors in higher plants, and more than one photoreceptor type can participate in any given process. Indeed, all three types of photoreceptors are involved in chloroplast development (Sengar, 1982; Harpster and Apel, 1985). Blue light/UV photoreceptors participate in a variety of responses. Blue light effects include phototropic curvature, carotenogenesis, and stomatal opening (for reviews, see Sengar, 1982; Briggs and Iino, 1983). Irradiation by UV light causes induction of enzymes involved in flavonoid biosynthesis (Kreuzaler *et al.*, 1983; Kuhn *et al.*, 1984). A second type of photoreceptor absorbs red light, which results in the reduction of protochlorophyllide to chlorophyllide, an essential step in the biosynthesis of chlorophyll.

By far, the best characterized photoreceptor is phytochrome (for review, see Pratt, 1982). The structure of phytochrome will be considered in detail in section III A, below. The protein exists in two forms which are interconvertible by light of the appropriate wavelength: P_r which absorbs red light, and P_{fr} which absorbs far-red light. P_{fr} is the active form of the molecule in most systems, in that absorption of red light, which causes conversion of P_r to P_{fr}, induces physiological responses. Subsequent irradiation by far-red light can usually cancel the inductive effect, if given within a characteristic "escape time". Thus, induction by red light, and reversibility by far-red light, is diagnostic of phytochrome involvement in a process. The fact that the two forms of phytochrome have different absorption spectra provides a mechanism by which plants can sense changes in light quality. Shading by a canopy of green leaves absorbs much of the red light, and the plant uses its P_r to P_{fr} ratio as a sensor to adjust its growth strategy accordingly (Smith, 1982).

Tobin and Silverthorne (1985) listed the following specific products whose genes have been demonstrated to be regulated by phytochrome: rbcL and rbcS polypeptides, chlorophyll a/b apoprotein, 32 kd thylakoid membrane protein, rRNA, NADPH-protochlorophyllide reductase, and phytochrome itself. Since that time, there have been reports of phytochrome control of other genes, including: maize plastid genes encoding the alpha, beta and epsilon subunits of coupling factor 1, subunit III of CFo, the reaction center proteins A1 and A2 of photosystem I, two membrane proteins of photosystem II (Zhu *et al.*, 1985); and pea ferrodoxin I (Dobres *et al.*, 1987). In the cases of protochlorophyllide reductase and phytochrome, the effect of illumination by red light is to reduce the expression of these genes rather than to induce expression (Apel, 1981; Colbert *et al.*, 1983, 1985; Otto *et al.*, 1984; Hershey *et al.*, 1984; Mosinger *et al.*, 1985).

The amount of red light, or fluence, required for induction of individual transcripts falls into two categories: low fluence (LF) and very low fluence (VLF) (Kaufman *et al.*, 1984). The majority of responses are low fluence responses, which can be reversed if far-red illumination occurs before the escape time elapses. VLF responses have also been reported for the induction of Cab genes in pea and barley (Kaufman *et al.*, 1984; Briggs *et al.*, 1986) and for attenuation of phytochrome transcription in Avena seedlings (Colbert *et al.*, 1983). VLF responses are not reversible by far-red illumination. In fact, far-red illumination alone is enough to initiate these responses. Although P_r has an absorption maximum in the red range, far-red light of 730 nm is about 10% as efficient as red light of 667 nm in the conversion of P_r to P_{fr} (Thompson *et al.*, 1985). Thompson *et al.*, (1985) caution that green "safe" lights can induce VLF responses: the threshold for the VLF response is equivalent to the amount of light generated by several discharges of static electricity.

Different genes induced by red light often respond to the same light regimen in different ways. Kaufman *et al.* (1986) followed the kinetics of accumulation of twelve phytochrome regulated transcripts in developing pea buds following a red light pulse: four different patterns of accumulation were noted. The kinetics of far-red reversibility were also found to vary. The diversity of responses for different genes implies several intermediate steps or branch points in the pathway between the photoreceptor and gene expression. Schafer and Briggs (1986) have recently discussed the implication of kinetic data for the understanding of signal transduction.

A gene can respond in a different way to the same stimulus depending on the physiological state of the plant. For example, two members of the rbcS gene family in pea, rbcS-3A and 3C, show classical phytochrome induction with a short pulse of red light in etiolated seedlings of pea and transgenic petunia, whereas the same illumination has no effect in mature green leaves (Fluhr and Chua, 1986): In this tissue only prolonged red or blue illumination induces these genes, with blue light having a much greater quantum efficiency. This has led to the proposal of simple phytochrome induction of the rbcS-3A and 3C genes in etiolated tissue, with a more complex regulation in mature leaves involving co-action between phytochrome and a putative blue light photoreceptor. A similar co-action has been proposed for anthocyanin induction in sorghum (Oelmuller and Mohr, 1985).

The kinetics of induction of the rbcS gene members also changes with the developmental stage of the tissue (Fluhr *et al.*, 1986b). Two members of the family (3A and 3C) are induced much more rapidly in etiolated tissue than two others (E9 and 8.0), while in dark-adapted mature tissue these four gene members are induced with similar kinetics. These differences could be due to the presence of functionally distinct forms of phytochrome at different developmental stages (see section III A), or a change in the activity or presence of transduction chain intermediates, or different affinities of *cis*-acting DNA sequences for limiting transcription factors.

Characterization of light effects on gene expression in plants has clearly relied upon the greening of the etiolated plant as a model to approximate normal photomorphogenesis. This has provided an attractive system because dramatic changes can be observed upon illumination, and the perception of light by photoreceptors is not complicated by the presence of chlorophyll. However, etiolation is usually only a short phase in the normal growth and development of a plant, thus providing a limited model for light regulation under more natural physiological conditions.

C. *Rhythms*

A wide range of plant responses have been characterized in terms of diurnal and circadian rhythms (Mansfield and Snaith, 1984). Diurnal rhythms are usually interpreted as fluctuations that occur in phase with exogenous stimuli, such as light/dark cycles, whereas circadian rhythms occur independently of external signals and are thought to represent the activity of an endogenous biological clock. Such diverse phenomena as leaf movements (Satter and Galston, 1981), CO_2 metabolism (Wilkins, 1959), stomatal movements (Martin and Meidner, 1971) and many floral induction patterns (Vince-Prue, 1983) have been shown to be under the control of endogenous circadian rhythms.

While such complex physiological processes are difficult to define in molecular terms, many workers have now demonstrated that individual enzymes are under diurnal or circadian control. Many enzymes involved in nitrogen metabolism, including glutamine synthetase (Knight and Weissman, 1982), asparaginase (Sieciehowicz *et al.*, 1985) and nitrate reductase (Duke *et al.*, 1978; Lillo, 1984) have enzyme activities that vary diurnally. Griffiths *et al.* (1985), were able to show that both enzyme activity and protein level of protochlorophyllide reductase fluctuate diurnally in oat and barley seedlings. These authors have suggested that the fluctuation may be required for daily chloroplast maintenance and repair.

The activity of RUBISCO in several species undergoes diurnal changes by a mechanism that controls the level of an endogenous inhibitor of the enzyme. This nocturnal inhibitor has recently been identified (Gutteridge *et al.*, 1986; Berry *et al.*, 1987) and is a molecule that closely resembles an intermediate in the carboxylation reaction.

Kloppstech (1985) has demonstrated that in mature pea leaves mRNA levels for the rbcS, Cab and "early light" induced polypeptides exhibit circadian rhythmicity, either in light/dark cycles or extended continuous illumination. Work in our own laboratory (Nagy, F., Kay, S. A. and Chua, N.-H., manuscript in press) has revealed a dramatic circadian oscillation in the mRNA level of the wheat Cab-1 gene. Furthermore, this rhythmicity is conserved when the wheat gene is transfered to tobacco, and it appears to be influenced by phytochrome. The discovery of the control of gene expression via endogenous rhythms in mature tissue should lead to new insights into the subtle requirement of the plant to maintain its growth and development under normal light/dark cycles. Genes that are responsive to circadian rhythms will also prove useful in the molecular analysis of the biological clock itself, perhaps in a similar way to the dissection of the period locus of Drosophila (Reddy *et al.*, 1986; Jackson *et al.*, 1986).

III. Effectors of Photoreception

A. Phytochrome

To begin to understand the mechanisms by which phytochrome transduces a spectral signal into a cellular response, it is necessary to have a detailed knowledge of the structure of the photoreceptor. This will allow identification of regions of the molecule which are important for *in vivo* function. We attempt here to summarize the wealth of recent data obtained on phytochrome, and as space does not permit an exhaustive description of the literature, we refer to reviews by Pratt (1982) and Smith (1983) for a historical perspective and to Lagarias (1985) for an excellent treatise of both biophysical and biochemical aspects.

The difficulties in purification of phytochrome are noteworthy. Most early workers attempted to purify P_r, hence all steps were carried out under dim green safelights. The purification is monitored spectrally by measuring A_{666}/A_{280}. Early attempts to characterize phytochrome were complicated by reports of differing molecular weights. For example, a 118 kDa P_r was first assumed to be the undegraded form in oats (Hunt and Pratt, 1980). However, it was later shown (Vierstra and Quail, 1982a; Vierstra and Quail, 1983) that extraction of oat phytochrome in either the P_{fr} form or with boiling SDS buffer containing protease inhibitors, yielded a molecule of 124 kDa. Most interestingly this undegraded form of phytochrome demonstrated spectral properties different from those of the "large" phytochrome of 118 kDa described previously (Vierstra and Quail, 1982b). The cleavage by endogenous proteases during isolation provided the first evidence that a 6 kDa fragment was preferentially susceptible to cleavage when phytochrome was in the P_r form, and therefore this fragment is involved in a light-induced conformational change. The 6 kDa fragment was assigned to the extreme N-terminal end of the molecule by use of monoclonal antibodies. These antibodies defined 3 domains of oat phytochrome: the 6 kDa N-terminal region, a larger N-terminal domain containing the chromophore (an open chain tetrapyrrole) and a C-terminal domain. This domain map was related to an endogenous protease cleavage map (Vierstra *et al.*, 1984) which confirmed that the P_r hypersensitive site was at the N-terminus and was conserved among plant species, emphasizing that a conserved domain assumed a different conformation in P_r versus P_{fr}. The 6 kDa region also demonstrated differential sensitivity to exogenous proteases added to purified oat phytochrome (Lagarias and Mercurio, 1985). The chromophore binding domain is dependent upon the 6 kDa N-terminal fragment for spectral integrity (Jones *et al.*, 1985), whereas the C-terminal domain does not appear to interact with the chromophore directly.

In conjunction with protease mapping, immunological comparisons of phytochrome from different species have also helped identify conserved domains which are therefore thought to be important in terms of function. Initial purification and immunological analysis of phytochrome from dicot species (Cordonnier and Pratt, 1982a and b) revealed that monocot phytochromes were overall poorly related antigenically to dicot phytochromes, although both groups exhibited similarities in spectral properties and molecular weights (Vierstra *et al.*, 1984). The use of monoclonal antibodies to both classes of plant phytochromes revealed some common epitopes between the two (Cordonnier *et al.*, 1984). Purified Cucurbita phytochrome (Vierstra and Quail, 1985) has a protease sensitive N-terminal domain similar to that found in oats as well as a P_{fr} sensitive site that appears to be a hinge between the major N- and C-terminal domains. Recent immunological studies have revealed that an oat monoclonal which preferentially binds P_r (Cordonnier *et al.*, 1985) recognizes an epitope within the extreme N-terminal domain that is only exposed in P_r, thus corroborating the protease data. Another set of oat monoclonal antibodies has been identified which recognize a different epitope nearer the middle of the N-terminal chromophore domain, but is exposed preferentially in P_{fr} (Shimazaki *et al.*, 1986). One particular epitope recognized by a pea monoclonal antibody appeas to be highly conserved, as it is present in phytochromes of mosses, algae and higher plants (Cordonnier *et al.*, 1986). The identification of these conserved domains now awaits more rigorous biochemical determination of their importance *in vivo*.

The deduction of the primary structure of phytochrome of oat (Hershey *et al.*, 1985) and Cucurbita (Sharrock *et al.*, 1986) has of course allowed a more direct comparison of the two molecules and thus a better idea of important conserved regions. Oat appears to contain a small multigene family encoding phytochrome, with the cDNA sequences of two members being highly homologous in the coding and 5′ untranslated regions.

Localized regions of homology exist at the amino acid level between oat and Cucurbita (Sharrock *et al.*, 1986). As predicted, the major N-terminal chromophore binding region is highly conserved, whereas the C-terminal portion has only interspersed regions of homology. Surprisingly, the extreme N-terminal fragment has poor homology at the amino acid level, although the two have a similar hydrophobicity profile. The highest locally conserved regions are the twenty amino acids around the point of attachment of the chromophore, and a hydrophobic pocket just N-terminal to this which is thought to harbor the chromophore. The overall hydrophobicity of these two apoproteins is interesting in view of the classical "pelletability" of phytochrome (Marme, 1977) and recent evidence demonstrating aggregation and sequestration upon photoconversion *in vivo* (McCurdy and Pratt, 1986; Speth *et al.*, 1986).

The oat phytochrome molecule has recently been subjected to direct biochemical characterization (Jones and Quail, 1986; Lagarias and Mercurio, 1985). The molecule appears to exist as an elongated dimer of two elliptical monomers. The dimerization site resides 42 kDa from the C-terminal end and the interaction is ionic in nature. The large N-terminal domain is globular whereas the C-terminal domain is more extended. A simplistic interpretation of this data gives rise to a "dumbell" shape for phytochrome in solution.

Finer light-induced conformational mapping has been elegantly provided by Wong *et al.* (1986), who have probed the accessibility of sites within purified phytochrome to phosphorylation by exogenously added kinases. P_r is more susceptible to phosphorylation that P_{fr} and tryptic mapping of the phosphopeptides has revealed that these phosphorylation sites reside mostly near the N-terminus. Upon photoconversion to P_{fr} these sites are masked and a new C-terminal domain site becomes available for phosphorylation. The N-terminal domain becomes more compacted and the central hinge between the two domains is more exposed. Interestingly, an endogenous kinase activity is associated with highly purified 124 kDa phytochrome, which preferentially phosphorylates P_r. The possibility therefore exists for phytochrome auto-kinase activity. Further work will be required to ascertain whether this is an activity of phytochrome itself or is due to a co-purifying kinase.

Another area which should prove interesting for phytochrome structure/function relationships is the presence of distinct forms of phytochrome in etiolated vs. mature green leaves. Tokuhisa *et al.* (1985) have detected two species of phytochrome in green oats, a major species of 118 kDa and a minor one of 124 kDa (Tokuhisa *et al.*, 1985). The major species is immunologically and physically distinct from etiolated oat phytochrome. Shimazaki and Pratt (1985, 1986) have similarly demonstrated distinct forms of phytochrome in green oat, or in norflurazon bleached mature leaf tissue. Interestingly, if mature leaf tissue is darkened for 48 h the etiolated form accumulates preferentially. Abe *et al.* (1985) have also detected two types of phytochrome in mature pea leaves, designated forms I and II. Form I is identical to the phytochrome found in etiolated tissue whereas form II differs by immunological and physical criteria. This raises the possibility of a different functional pool of phytochrome in mature tissue, which may be responsible for processes such as interaction with circadian rhythms of gene expression observed for some phytochrome regulated genes. It will be of great interest to clone a phytochrome apoprotein gene that is specific for the green form, to allow both structural and regulatory comparisons to be made.

B. Signal Transduction

Although the list of genes which are regulated by phytochrome is growing rapidly, and despite the detailed knowledge of the structure of the photoreceptor, very little is known about the events occurring between P_{fr} formation and the corresponding change in mRNA level. A simplistic model would invoke a direct effect of phytochrome (P_{fr}) upon the gene itself, perhaps in a similar manner to the activation of the glucocorticoid receptor and the binding to its target gene (Yamamoto, 1985). Despite the large number of gene promoter sequences now available and the rapid purification procedures for native phytochrome, no report to date has demonstrated direct sequence-specific binding of the photoreceptor to DNA. Thus it appears unlikely that phytochrome activates gene transcription directly. Considering the diversity of phytochrome responses discussed earlier, we need to invoke a model which predicts common initial steps in the pathway that would branch out into different classes of responses, thus requiring different intracellular messengers.

Recent advances in animal systems may shed some light on possible plant mediators of photoperception. Animal cells have at least two major signal transducing systems in response to extracellular stimuli: one is via the formation of cAMP leading to the activation of protein kinase A (Flockhart and Corbin, 1982); the other involves transduction of receptor occupancy via GTP binding proteins (G-proteins) which activates phospholipase C, causing rapid turnover of inositol phospholipids, Ca^{2+} mobilization and the activation of the phospholipid dependent protein kinase C (for reviews see Berridge and Irvine, 1984; Bell, 1986; Majerus *et al.*, 1986; Nishizuka, 1986; Houslay, 1987; Sibley *et al.*, 1987). It is therefore pertinent to review the literature to establish whether plants contain analogous components of these transduction chains.

Clearly, plants contain many constituents of a functional cAMP regulated system including cAMP, adenylate cyclase, cAMP phosphodiesterase and calmodulin (reviewed by Brown and Newton, 1981). However, no unequivocal demonstration of a protein kinase A-like activity has been given. Evidence exists for Ca^{2+} fluxes in plant cells and organelles, some of which have been related to phytochrome photoconversion (see Roux, 1984 and refs. therein). Oat protoplasts show a net efflux of Ca^{2+} upon P_{fr} formation. This is postulated to work via calmodulin activation of a Ca^{2+} ATPase (Hale and Roux, 1980). Similarly P_{fr} is thought to mobilize Ca^{2+} from the mitochondrial matrix leading to activation of an outer membrane

ATPase (Serlin *et al.*, 1984). Roux (1984) has proposed a model in which the photoconversion of phytochrome leads to an increase in cytosolic Ca^{2+} and subsequent activation of enzymes via calmodulin. In an effort to relate this hypothesis to the observed activation of gene expression by phytochrome, Datta *et al.* (1985) demonstrated a slight increase in phosphorylation of three unidentified nuclear proteins following red illumination *in vitro*. They have also observed the activation of a chromatin associated NTPase in isolated nuclei (Roux *et al.*, 1986), confirming the earlier *in vitro* finding in pea seedlings (Wagle and Jaffe, 1980). Both *in vitro* activations are prevented somewhat by EGTA and calmodulin antagonists. These experiments imply that P_{fr} may activate genes via calmodulin/Ca^{2+} stimulation of enzymatic activities. However, two major criteria need to be satisfied for the Ca^{2+} model. Firstly, a demonstration of calmodulin stimulated kinase and phosphatase activities is required. Secondly, one would expect only transient increases in Ca^{2+} concentration if it were a second messenger of phytochrome action. Rapid turnover of P_{fr} *in vivo* could be a mechanism by which this is achieved. Better measurements of whole cell Ca^{2+} fluxes using non-disruptive fluorescent chelators (Tsien, 1981) are required.

Another route for Ca^{2+} involvement in phytochrome-regulated events is via protein kinase C and phospholipid metabolism. A Ca^{2+} phospholipid dependent kinase activity has been detected in extracts from zucchini cotyledons (Schafer *et al.*, 1985), and a similar enzyme activity is present in wheat cells (Olah and Kiss, 1986), which is activated by phorbol esters. However, it is not clear which components of a protein kinase C transduction chain are present in plant cells. For example, is phospholipase C activity present which would result in the production of inositol triphosphate, the Ca^{2+} mobilization signal? Does diacylglycerol stimulate plant protein kinase C *in vivo*? Do plant cells contain regulatory G proteins? These questions will be answered by the use of activators and inhibitors of the pathway at the whole cell level. One model we would like to propose involves the binding of P_{fr} to an intracellular or membrane bound receptor (or G-protein) following phototransformation. This would in turn activate phospholipase C, producing inositol triphosphate and diacylglycerol. The former compound releases Ca^{2+} into the cytoplasm and the latter activates protein kinase C. Phosphorylated intermediates would then activate gene transcription (or mRNA degradation) directly.

Finally, we would like to make a tentative comparison between plant transduction systems and that of the vertebrate visual perception system. The recent isolation of the genes for the cone pigment apoproteins (Nathans *et al.*, 1986a and 1986b) revealed strong homology between the red and green pigment genes, and much lower homology between these and the blue pigment gene. All three apoproteins bear some homology with opsin, the monochromatic rod cell receptor. In many ways, phytochrome and the other plant photoreceptors can be considered as the "eyes" or visual system of the plant, covering an even broader spectral range than that of animals. Thus the small gene families found for phytochrome in oats, and more recently rice (S. A. Kay *et al.*, unpublished data), coupled with the presence of different pools of phytochrome with varying degrees of homology, can be considered the plant counterparts of the cone pigment gene family. The transduction of light by rhodopsin is well characterized (Altman, 1985), involving interaction with a GTP binding protein, transducin, and the subsequent activation of a cGMP phosphodiesterase. The identification of a plant version of "transducin", which would be a putative phytochrome intracellular receptor, will be one of the major steps towards understanding the molecular mechanism underlying phytochrome action on gene expression.

C. Cis-Acting DNA Sequences

The ability to reintroduce defined DNA segments into plant cells has allowed the identification of sequences which mediate light-regulated transcription. The most frequently used method of gene transfer to higher plants exploits the natural ability of *Agrobacterium tumefaciens* to transfer DNA into the genomes of many dicotyledonous species (for reviews, see Bevan and Chilton, 1982; Nester *et al.*, 1984). With the development of "disarmed" strains, gene transfer occurs without disturbing the endogenous hormone balance (Zambryski *et al.*, 1983; Bevan, 1984; Horsch *et al.*, 1984, 1985; Klee *et al.*, 1985). Transgenic plants can be regenerated which differ from untransformed plants only by the presence of the transfered genes of interest. Although regeneration of transgenic plants takes at least 12 weeks, there are many advantages to studying gene expression in plants rather than calli. While white light regulation of rbcS can be demonstrated in calli, results obtained with that assay system are not always consistent with results obtained with transgenic plants. Furthermore, phytochrome regulation of transferred genes at different developmental stages cannot be investigated with the calli assay system. The following section will focus on expression studies of three light-responsive genes, including one from a monocot species, assayed in dicot transgenic plants. Light-regulated gene expression has been most recently reviewed by Kuhlemeier *et al.* (1987b).

The regulatory sequences, or light-responsive elements (LREs), of the rbcS genes have been the most extensively characterized. Early gene transfer experiments with calli demonstrated that DNA sequences important for light regulation were located in the 5′ flanking region of these genes, and that it was possible to obtain correct transcription initiation of the transferred gene in calli. Broglie *et al.* (1984) demonstrated that a pea gene, rbcS-E9, could be accurately transcribed and light regulated in transformed petunia calli. Herrera-Estrella *et al.* (1984) constructed a chimeric gene having 1 kb of the pea rbcS-SS3.6 5′ flanking region fused to the bacterial CAT gene coding region, and showed light-inducible expression in tobacco calli.

The rbcS genes in pea constitute a small multigene family (Bedbrook *et al.*, 1980; Coruzzi *et al.*, 1983), each member of which has a characteristic expression pattern which can be distinguished by use of gene specific probes (Fluhr *et al.*, 1986b). When three members of this gene family, rbcS-E9, -3A, and -3C were introduced into transgenic plants, rather than calli, it was found that the subtle differences of expression were reproduced in transgenic tobacco or petunia (Nagy *et al.*, 1985; Fluhr and Chua, 1986). The use of transgenic F1 seedlings (Fluhr and Chua, 1986) demonstrated that light regulation was under phytochrome control at that developmental stage.

The light regulation and organ specificity of the transferred rbcS genes led to efforts to define more precisely the sequences in the 5′ flanking region of these genes that conferred these regulatory properties. Construction of chimeric rbcS genes *in vitro*, and expression in transgenic plants, has revealed the presence of enhancer-like regions upstream of rbcS genes, i. e. DNA fragments which activate transcription in either orientation relative to the structural gene. For the rbcS-3A gene, a 280 bp fragment (from −330 to −50) could bestow light regulation and organ specificity to a heterologous gene (Fluhr *et al.*, 1986a). An even smaller fragment (−327 to −82) from the pea rbcS-E9 gene was found sufficient for the same regulatory responses. Both fragments also conferred phytochrome regulated expression in transgenic F1 seedlings.

Comparison of sequences of the 5′ flanking regions of the rbcS genes has revealed certain conserved regions, designated boxes I—V (Fluhr *et al.*,

```
-410                                                        -352
 GATCCAAAAGCTTGGACAGGAACAAATGTTACCCATACATAAAAGATATTTGTGAAGTAA

                      -330                                  -292
CAGTCACAAAATTCCATGAGGCCAACATACTACAATTGAATTTTCATGGATACAATTCTT

    -285                                  III*              -232
ACAAAATAAAAATATCGACATAACCACCATCACACATTTACACTCTTCACATGAAAAGAT

        II*                                   -189          -172
AAGATCAGTGAGGTAATATCCACATGGCACTGTCCTATTGGTGGCTTATGATAAGGCTAG

-169         I        -149   II      -136              III  -112
CACACAAAATTTCAAATCTTGTGTGGTTAATATGGCTGCAAACTTTATCATTTTCACTAT

            -100                                             -52
CTAACAAGATTGGTACTAGGCAGTAGCTAAGTACCACAATATTAAGACCATAATATTGGA

-50                    IV                          V  -1       9
 AATAGATAAATAAAAACATTATATATAGCAAGTTTTAGCAGAAGCTTTGCAATTCATACA

             22
GAAGTGAGAAAAATG
```

Fig. 1. Reprinted from Kuhlemeier *et al.* (1987a). The upstream sequence of the pea rbcS-3A gene is shown. Sequences (I—V) which are conserved between members of the pea rbcS gene family are boxed. Also boxed are sequences which are similar to the sequences of box II and box III, designated II* and III*

1986a; Kuhlemeier *et al.*, 1987a, b). As shown in Figure 1, in the pea rbsS-3A gene, boxes II and III (near −150) have homologous but non-identical copies further upstream near −220. Boxes II and III, and their homologous copies II* and III*, compete for the same protein factors in *in vitro* binding studies. This will be discussed further in section III D. Experiments with mutated upstream fragments of the pea rbcS-3A gene in transgenic plants has revealed that functional redundancy of regulatory elements exists (Kuhlemeier *et al.*, 1987a).

Kuhlemeier *et al.* (1987a) have demonstrated that more than one segment of the 5′ flanking region of rbcS-3A, which contain copies of boxes II and III, can act to turn off transcription in the dark, when these segments are placed between a constitutive viral enhancer and a reporter gene. This indicates the presence of transcriptional "silencer" elements as have been found in yeast (Brand *et al.*, 1985) and mammalian systems (Goodbourne *et al.*, 1986; Laimins *et al.*, 1986; Nir *et al.*, 1986) as well as in the pea Cab upstream region (Simpson *et al.*, 1986a). An interesting aspect of the box II and box III sequences is their homology to mammalian constitutive enhancer elements. As detailed in Kuhlemeier *et al.* (1987a), these sequences share homology with the SV40 core enhancer GT motif, the adenovirus 5 E1A −200 enhancer, and the constitutive enhancer in the human beta-interferon upstream region.

The sequences in the −150 region which turn off rbcS-3A transcription in the dark may overlap enhancer or positive regulatory sequences. Deletion analysis has revealed that the 5′ endpoint of a positive element is located between −149 and −166 (Kuhlemeier *et al.*, 1987a). Recently, a construct in which just two nucleotides of box II were substituted, was found to be inactive, further indicating the presence of a positive element in the −150 region (Kuhlemeier *et al.*, submitted). Experiments in progress include testing of all the conserved boxes in single or multiple copies for positive and negative regulatory activity, and the definition of the critical nucleotides within these sequences.

Regulatory sequences in the 5′ flanking region of a pea Cab gene have been identified by Simpson *et al.*, (1985, 1986a, 1986b). Results with transgenic plants have shown that a 400 bp fragment is sufficient to direct white-light regulation and organ-specific expression of a chimeric gene. A Cab gene from wheat, a monocot, has been expressed in a light-regulated and organ-specific manner in transgenic tobacco (Lamppa *et al.*, 1985b). Nagy *et al.* (1986) demonstrated a characteristic VLF phytochrome response of a chimeric gene with approximately 2 kb of 5′ flanking Cab sequences present. More recent work has identified a 268 bp fragment (from −357 to −89) which can function bidirectionally to confer enhancement, organ specificity and phytochrome response (Nagy *et al.*, 1987). Work is continuing to define the regulatory capabilities of this sequence and to search for other possible light-responsive sequences in the Cab upstream region. It should be noted that not all monocot genes tested thus far can function in dicot plants. Expression of a wheat rbcS gene could not be detected after transfer to tobacco plants (Keith and Chua, 1986).

Upstream sequences determining the UV-light induced expression of chalcone synthase were examined by Kaulen *et al.* (1986). *Cis*-acting elements located on a 2.3 kb promoter fragment are sufficient for UV-dependent induction of a chimeric gene in transgenic tobacco plants.

D. Trans-Acting Factors

In vitro experimental approaches provide complementary data to the identification of *cis*-acting DNA sequences assayed *in vivo.* As noted above, work in this laboratory has identified sequences in the 5′ upstream region of the pea rbcS-3 A gene which bind protein factors present in nuclei isolated from pea leaves. DNaseI footprinting experiments have revealed protection of conserved boxes II and III (P. Green *et al.*, 1987), which as noted above display regulatory activity in transgenic plants. An additional protection site has been found further upstream, in a region which contains homologous boxes II* and III*.

To address the issue of whether these regions bind the same factor, the technique of gel retardation was used (for review see Hendrickson, 1985). Using this approach with nuclear extracts of pea leaves, it was found that a DNA fragment containing sequences of boxes II and III competed for factors in the binding assay with a fragment containing boxes II* and III* (P. Green *et al.*, 1987). The binding activity, designated GT-1, was insensitive to nonspecific competitor DNA, and had little affinity for a DNA fragment from the constitutive CaMV 35 S promoter, which contains three sequence motifs homologous to the SV40 core enhancer. Fragments of the rbcS-3 A gene containing either a 12 bp substitution of box II or box III were also unable to compete for binding.

Nuclei prepared from dark-grown plants also contain GT-1. Therefore, a regulatory role for GT-1 does not depend on its *de novo* synthesis in the light. Use of the technqiue of *in vivo* footprinting (Ephrussi *et al.*, 1985; Becker *et al.*, 1986) will determine whether GT-1 is bound constitutively *in vivo.* In that case, light induction of the rbcS-3 A gene would require another factor to bind to or covalently modify the constitutively bound GT-1. Either of these changes could potentiate transcription. A recent precedent for factor modification was reported by Sen and Baltimore (1986). They found that the activity of a DNA binding protein, NF-KB, was induced by a posttranslational modification mediated directly or indirectly by protein kinase C. Alternatively, GT-1 could act as a receptor for the binding of other factors in the light which are required for rbcS-3 A transcription. It will be of great interest to purify and characterize the binding activity from both light and dark-grown pea leaf extracts, as a step in unravelling the black box of the signal transduction pathway.

IV. Conclusion

Light, via photoreceptors, elicits an impressive array of responses on the activity of plant genes and their products. These effects are exerted at the levels of transcription, mRNA stability and translation, to produce either up- or down-regulation. The response of individual genes to light varies according to the spectral quality and intensity of the signal, as well as the developmental stage of the cells. Thus, many reports now document light effects on genes whose products are as diverse as light-harvesting to flavonoid biosynthesis.

The most highly characterized photoreceptor is phytochrome. Much is now known about its structure in both monocots and dicots, and exciting new evidence suggests the presence of different forms in mature plant tissues. The isolation of the apoprotein genes will make possible the functional mutation analysis of the molecule *in vivo*. For example, modification of the ubiquitination site (Shanklin *et al.*, 1987) or identification of mRNA sequences critical for stability can now be analyzed. A potentially powerful approach might combine both molecular analysis and genetics via complementation of phytochrome deficient (or reduced) mutants (Koorneef *et al.*, 1980) with different gene constructs.

The power of the transgenic plant system in studying photoregulation cannot be overemphasized. In this respect, plant molecular biologists can expect to challenge the wealth of knowledge on *in vivo cis*-acting elements in animals, whose transgenic models are, at present, more limited. The identification of small active sequences, both negative and positive, has opened the door to identification of the protein factors that specifically recognize them, and which therefore potentially modulate gene activity *in vivo*. The analysis of plant transcription factors is impeded, however, by the lack of a soluble cell free transcription system. Despite the many gaps that exist in our knowledge of the transduction chain, we are confident that a combination of molecular biology, protein biochemistry and genetics, will eventually lead to the elucidation of the full sequence of events between a photon entering a cell, and the corresponding gene response.

Acknowledgements

This work was supported in part by a grant from Monsanto Company. M. C. is the recipient of a postdoctoral fellowship from the NIH; S. A. K. was supported by postdoctoral fellowships from SERC/NATO and The Winston Foundation. We thank our colleagues in the laboratory for fruitful discussions and contributions to the ideas described in this article

V. References

Abe, H., Yamamoto, K. T., Nagatani, A., Furuya, M., 1985: Characterization of green tissue-specific phytochrome isolated immunochemically from pea seedlings. Plant Cell Physiol. **26,** 1387—1399.

Altman, J., 1985: New visions in photoreception. Nature **313,** 264—265.

Apel, K., 1981: The protochlorophyllide holochrome of barley (Hordeum vulgare L.). Phytochrome-induced decrease of translatable mRNA coding for the NADPH protochlorophyllide oxido-reductase. Eur. J. Biochem. **120,** 89—93.

Apel, K., Kloppstech, K., 1978: The plastid membranes of barley (Hordeum vulgare): Light-induced appearance of mRNA coding for the apoprotein of the light-harvesting chlorophyll a/b protein. Eur. J. Biochem. **85,** 581—588.

Becker, P. B., Gloss, B., Schmid, W., Strahle, U., Schutz, G., 1986: In vivo protein-DNA interactions in a glucocorticoid response element required the presence of the hormone. Nature **324,** 686—688.

Bedbrook, J. R., Smith, S. M., Ellis, R. J., 1980: Molecular cloning and sequencing of cDNA encoding the precursor to the small subunit of chloroplast ribulose-1,5-bisphosphate carboxylase. Nature **287,** 692—697.

Bell, R. M., 1986: Protein kinase C activation by diacylglycerol second messengers. Cell **45,** 631—632.

Bennett, J., 1981: Biosynthesis of the light-harvesting chlorophyll a/b protein; polypeptide turnover in darkness. Eur. J. Biochem. **118,** 61—70.

Berridge, M. J., Irvine, R. F., 1984: Inositol triphosphate, a novel second messenger in cellular signal transduction. Nature **312,** 315—321.

Berry, J. A., Lorimer, G. H., Pierce, J., Seemann, J. R., Meek, J., Freas, S., 1987: Isolation, identification, and synthesis of 2-carboxyarabinitol 1-phosphate, a diurnal regulator of ribulose-bisphosphate carboxylase activity. Proc. Natl. Acad. Sci. U.S.A. **84,** 734—738.

Berry, J. O., Nikolau, B. J., Carr, J. P., Klessig, D. F., 1985: Transcriptional and post-transcriptional regulation of ribulose-1,5-bisphosphate carboxylase gene expression in light- and dark-grown amaranth cotyledons. Mol. Cell. Biol. **5,** 2238—2246.

Berry-Lowe, S. L., Meagher, R. B., 1985: Transcriptional regulation of a gene encoding the small subunit of ribulose-1,5-bisphosphate carboxylase in soybean tissue is linked to phytochrome response. Mol. Cell. Biol. **5,** 1910—1917.

Bevan, M., 1984: Binary Agrobacterium vectors for plant transformation. Nuc. Acids Res. **12,** 8711—8721.

Bevan, M. W., Chilton, M.-D., 1982: T-DNA of the agrobacterium Ti- and Ri-plasmids. Ann. Rev. Genetics **16,** 357—384.

Brand, A. H., Breeden, L., Abraham, J., Sternglaz, R., Nasmyth, K., 1985: Characterization of a "silencer" in yeast: a DNA sequence with properties opposite those of a transcriptional enhancer. Cell **41,** 41—48.

Briggs, W. R., Iino, M., 1983: Blue light absorbing photoreceptors in plants. Phil. Trans. R. Soc. London, Ser. B **303,** 347—359.

Broglie, R., Coruzzi, G., Fraley, R. T., Rogers, S. G., Horsch, R. B., Niedermeyer, J. G., Fink, C. L., Flick, J. S., Chua, N.-H., 1984: Light-regulated expression of a pea ribulose-1,5-bisphosphate carboxylase small subunit gene in transformed plant cells. Science **224,** 838—843.

Brown, E. G., Newton, R. P., 1981: Cyclic AMP and higher plants. Phytochemistry **20,** 2453—2463.

Castelfranco, P. A., Beale, S. J., 1983: Chlorophyll biosynthesis: Recent advances and areas of current interest. Ann. Rev. Plant Physiol. **34,** 241—278.

Colbert, J. T., Hershey, H. P., Quail, P. H., 1983: Autoregulatory control of translatable phytochrome mRNA levels. Proc. Natl. Acad. Sci. U.S.A. **80,** 2248—2252.

Colbert, J. T., Hershey, H. P., Quail, P. H., 1985: Phytochrome regulation of phytochrome mRNA abundance. Plant Mol. Biol. **5,** 91—101.

Cordonnier, M.-M., Pratt, L. H., 1982a: Immunopurification and initial characterization of dicotyledonous phytochrome. Plant Physiol. **69,** 360—365.

Cordonnier, M.-M., Pratt, L. H., 1982b: Comparative phytochrome immunochemistry as assayed by antisera against both monocot and dicot phytochrome. Plant Physiol. **70,** 912—916.

Cordonnier, M.-M., Greppin, M., Pratt, L. H., 1984: Characterization by enzyme linked immunosorbent assay of monoclonal antibodies to Pisum and Avena phytochrome. Plant Physiol. **74,** 123—127.

Cordonnier, M.-M., Greppin, H., Pratt, L. H., 1985: Monoclonal antibodies with differing affinities to the red-absorbing and far-red absorbing forms of phytochrome. Biochemistry **24,** 3246—3252.

Cordonnier, M.-M., Greppin, H., Pratt, L. H., 1986: Identification of a highly conserved domain on phytochrome from angiosperms to algae. Plant Physiol. **80,** 982—987.

Coruzzi, G., Broglie, R., Cashmore, A. R., Chua, N.-H., 1983: Nucleotide sequences of two pea cDNA clones encoding the small subunit of ribulose 1,5-bisphosphate carboxylase and the major chlorophyll a/b-binding thylakoid polypeptide. J. Biol. Chem. **258,** 1399—1402.

Coruzzi, G., Broglie, R., Edwards, C., Chua, N.-H., 1984: Tissue-specific and light-regulated expression of a pea nuclear gene encoding the small subunit of ribulose-1,5-bisphosphate carboxylase. EMBO J. **3,** 1671—1679.

Daniels, S. M., Quail, P. H., 1984: Monoclonal antibodies to three separate domains on 124 kilodalton phytochrome from Avena. Plant Physiol. **76,** 622—626.

Datta, N., Chen, Y.-R., Roux, S. J., 1985: Phytochrome and calcium stimulation of protein phosphorylation in isolated pea nuclei. Biochem. Biophys. Res. Comm. **128,** 1403—1408.

De Vries, S. C., Springer, J., Wessels, J. G. H., 1982: Diversity of abundant mRNA sequences and patterns of protein synthesis in etiolated and greened pea seedlings. Planta **156,** 129—135.

Dobres, M. S., Elliot, R. C., Watson, J. C., Thompson, W. F., 1987: A phytochrome regulated pea transcript encodes ferredoxin I. Plant Mol. Biol. **8,** 53—59.

Duke, S. H., Friedrich, J. W., Schrader, L. E., Koukkari, W. L., 1978: Oscillations in the activities of enzymes of nitrate reduction and ammonia assimilation in Glycine max and Zea mays. Physiol. Plant. **42,** 269—276.

Ellis, R. J., 1986: Photoregulation of plant gene expression. Bioscience Reports **6,** 127—136.

Ephrussi, A., Church, G. M., Tonegawa, S., Gilbert, W., 1985: B lineage-specific interactions of an immunoglobulin enhancer with cellular factors in vivo, Science **227,** 134—140.

Faciotti, D., O'Neal, J. K., Lee, S., Shewmaker, C. K., 1985: Light-inducible expression of a chimeric gene in soybean tissue transformed with Agrobacterium. Biotechnology **3,** 241—246.

Flockhart, D. A., Corbin, J. D., 1982: Regulatory mechanisms in the control of protein kinases. Crit. Rev. Biochem. **12,** 133—186.

Fluhr, R., Chua, N.-H., 1986: Developmental regulation of two genes encoding ribulose-bisphosphate carboxylase small subunit in pea and transgenic petunia: phytochrome responses and blue light induction. Proc. Natl. Acad. Sci. U.S.A. **83,** 2358—2362.

Fluhr, R., Kuhlemeier, C., Nagy, F., Chua, N.-H., 1986a: Organ-specific and light induced expression of plant genes. Science **232,** 1106—1112.

Fluhr, R., Moses, P., Morelli, G., Coruzzi, G., Chua, N.-H., 1986b: Expression dynamics of the pea rbcS multigene family and organ distribution of the transcripts. EMBO J. **5,** 2063—2071.

Gallagher, T. F., Ellis, R. J., 1982: Light-stimulated transcription of genes for two chloroplast polypeptides in isolated pea leaf nuclei. EMBO J. **1,** 1493—1498.

Gallagher, T. F., Jenkins, G. I., Ellis, R. J., 1985: Rapid modulation of transcription of nuclear genes encoding chloroplast proteins by light. FEBS Lett. **186,** 241—245.

Gollmer, I., Apel, K., 1983: The phytochrome controlled accumulation of mRNA sequences encoding the light harvesting chlorophyll a/b-protein of barley (Hordeum vulgare L.). Eur. J. Biochem. **133,** 309—313.

Goodbourn, S., Burstein, H., Maniatis, T., 1986: The human beta-interferon gene enhancer is under negative control. Cell **45,** 601—610.

Green, P. J., Kay, S. A., Chua, N.-H., 1987: Sequence-specific interactions of a pea nuclear factor with light-responsive elements upstream of the rbcS-3A gene. EMBO J. **6,** 2543—2549.

Griffiths, W. T., Kay, S. A., Oliver, R. P., 1985: The presence and photoregulation of protochlorophyllide reductase in green tissues. Plant Mol. Biol. **4,** 13—22.

Gutteridge, S., Parry, M. A. J., Burton, S., Keys, A. J., Mudd, A., Feeney, F., Servaites, J. C., Pierce, J., 1986: A nocturnal inhibitor of carboxylation in leaves. Nature **324,** 274—276.

Hale, C. C. H., Roux, S. J., 1980: Photoreversible calcium fluxes induced by phytochrome in oat coleoptile cells. Plant Physiol. **65,** 658—662.

Harpster, M., Apel, K., 1985: The light-dependent regulation of gene expression during plastid development in higher plants. Physiol. Plant **64,** 147—152.

Hendrickson, W., 1985: Protein-DNA interactions studied by the gel electrophoresis-DNA binding assay. Biotechniques **3,** 198—207.

Herrera-Estrella, L., Van den Broeck, G., Maenhaut, R., Van Montagu, M., Schell, J., Timko, M., Cashmore, A., 1984: Light-inducible and chloroplast-associated expression of a chimeric gene introduced into Nicotiana tabacum using a Ti plasmid vector. Nature **310,** 115—120.

Hershey, H. P., Colbert, J. T., Lissemore, J. L., Barker, R. F., Quail, P. H., 1984: Molecular cloning of cDNA for Avena phytochrome. Proc. Natl. Acad. Sci. U.S.A. **81,** 2332—2336.

Hershey, H. P., Barker, R. F., Idler, K. B., Lissemore, J. L., Quail, P. H., 1985: Analysis of cloned cDNA and genomic sequences for phytochrome: complete amino acid sequences for two gene products expressed in etiolated Avena. Nuc. Acids Res. **13,** 8543—8559.

Horsch, R. B., Fraley, R. T., Rogers, S. G., Sanders, P. R., Lloyd, A., Hoffmann, N., 1984: Inheritance of functional foreign genes in plants. Science **223,** 496—498.

Horsch, R. B., Fry, J. E., Hoffmann, N. L., Eichholtz, D., Rogers, S. G., Fraley, R. T., 1985: A simple and general method for transferring genes into plants. Science **227,** 1229—1231.

Houslay, M. D., 1987: Egg activation unscrambles a potential role for IP4. Trends in Biochem. Sci. **12,** 1—2.

Hunt, R. E., Pratt, L. H., 1980: Partial characterization of undegraded oat phytochrome. Biochemistry **19,** 390—394.

Inamine, G., Nash, B., Weissbach, H., Brot, N., 1985: Light regulation of the synthesis of the large subunit of ribulose-1,5-bisphosphate carboxylase in peas: evidence for translation control. Proc. Natl. Acad. Sci. U.S.A. **82,** 5690—5694.

Jackson, F. R., Bargiello, T. A., Yun, S.-H., Young, M. W., 1986: Product of per locus of Drosophila shares homology with proteoglycans. Nature **320,** 185—188.

Jones, A. M., Quail, P. H., 1986: Quaternary structure of 124-kilodalton phytochrome from Avena sativa L. Biochemistry **25,** 2987—2995.

Jones, A. M., Vierstra, R. D., Daniels, S. M., Quail, P., 1985: The role of separate molecular domains in the structure of phytochrome from etiolated Avena sativa L. Planta **164,** 501—506.

Kaufman, L. S., Thompson, W. F., Briggs, W. R., 1984: Different red light requirements for phytochrome-induced accumulation of Cab RNA and rbcS RNA. Science **226,** 1447—1449.

Kaufman, L. S., Roberts, L. L., Briggs, W. R., Thompson, W. F., 1986: Phytochrome control of specific mRNA levels in developing pea buds. Plant Physiol. **81,** 1033—1038.

Kaulen, H., Schell, J., Kreuzaler, F., 1986: Light-induced expression of the chimeric chalcone synthase — NPTII gene in tobacco cells. EMBO J. **5,** 1—8.

Keith, B., Chua, N.-H., 1986: Monocot and dicot pre-mRNAs are processed with different efficiencies in transgenic tobacco. EMBO J. **5,** 2419—2426.

Klee, H. J., Yanofsky, M. F., Nester, E. W., 1985: Vectors for transformation of higher plants. Biotechnology **3,** 637—642.

Kloppstech, K., 1985: Diurnal and circadian rhythmicity in the expression of light-induced plant nuclear mRNAs. Planta **165,** 502—506.

Knight, T. J., Weissman, G. S., 1982: Rhythms in glutamine synthetase activity energy charge and glutamine in sunflower roots. Plant Physiol. **70,** 1683—1688.

Koorneef, M., Rolff, E., Spruit, C. J. P., 1980: Genetic control of light-inhibited hypocotyl elongation in Arabidopsis thaliana (L.) Heynh. Z. Pflanzenphysiol. **100,** 147—160.

Kreuzaler, F., Ragg, H., Fautz, E., Kuhn, D. N., Hahlbrock, K., 1983: UV induction of chalcone synthase mRNA in cell suspension cultures of Petroselinum hortense. Proc. Natl. Acad. Sci. U.S.A. **80,** 2591—2598.

Kuhlemeier, C., Fluhr, R., Green, P. J., Chua, N.-H., 1987a: Sequences in the pea rbcS-3A gene have homology to constitutive mammalian enhancers but function as negative regulatory elements. Genes and Develop. **1,** 247—255.

Kuhlemeier, C., Green P. J., Chua, N.-H., 1987b: Regulation of gene expression in higher plants. Ann. Rev. Plant Physiol. **38,** 221—257.

Kuhn, D. N., Chappel, J., Boudet, A., Hahlbrock, K., 1984: Induction of phenylalanine ammonia-lyase and 4-coumarate: CoA ligase mRNAs in cultured plant cells by UV light or fungal elicitor. Proc. Natl. Acad. Sci. U.S.A. **81,** 1102—1106.

Lagarias, C., 1985: Progress in the molecular analysis of phytochrome. Photochem. Photobiol. **42,** 811—820.

Lagarias, C., Mercurio, F. M., 1985: Structure function studies on phytochrome. Identification of light induced conformational changes in 124 kDa Avena phytochrome in vitro. J. Biol. Chem. **260,** 2415—2423.

Laimins, L., Holmgren-Konig, M., Khoury, G., 1986: Transcriptional "silencer" element in rat repetitive sequences associated with the rat insulin 1 gene locus. Proc. Natl. Acad. Sci. U. S. A. **83,** 3151—3155.

Lamb, C. J., Lawton, M. A., 1983: Photocontrol of gene expression. In: Encyclopedia of Plant Physiology: Photomorphogenesis (NS 16 A). Shropshire, W., Mohr, H. (eds.), pp. 213—257. Berlin: Springer-Verlag.

Lamppa, G., Morelli, G., Chua, N.-H., 1985 a: Structure and developmental regulation of a wheat gene encoding the major chlorophyl a/b-binding polypeptide. Mol. Cell. Biol. **5,** 1370—1378.

Lamppa, G., Nagy, F., Chua, N.-H., 1985 b: Light-regulated and organ-specific expression of a wheat Cab gene in transgenic tobacco. Nature **316,** 750—752.

Lillo, C., 1984: Diurnal variations of nitrate reductase, glutamine synthetase, glutamate synthetase, alanine aminotransferase and aspartate aminotransferase in barley leaves. Physiol. Plant. **61,** 214— 218.

Majerus, P. W., Connolly, T. M., Deckmyn, H., Ross, T. S., Bross, T. E., Ishii, H., Bansal, V. S., Wilson, D. B., 1986: The metabolism of phosphoinostitide-derived messenger molecules. Science **234,** 1519—1526.

Mansfield, T. A., Snaith, P. J., 1984: Circadian rhythms. In: Advanced Plant Physiology. Wilkins, M. B. (ed.), pp. 201—218. Pitman, London.

Marme, D., 1977: Phytochrome: membranes as possible sites of primary action. Annual Rev. Plant Physiol. **28,** 173—198.

Martin, E. S., Meidner, H., 1971: Endogenous stomatal rhythm in Tradesiantia virginiana. New Phytol. **71,** 1045—1054.

McCurdy, D. W., Pratt, L. H., 1986: Kinetics of intracellular redistribution of phytochrome in Avena coleoptiles after its photoconversion to the active, far red absorbing form. Planta **167,** 330—336.

Morelli, G., Nagy, F., Fraley, R. T., Rogers, S. G., Chua, N.-H., 1985: A short conserved sequence is involved in the light-inducibility of a gene encoding ribulose 1,5-bisphosphate carboxylase small subunit of pea. Nature **315,** 200—204.

Mosinger, E., Batschauer, A., Schafer, E., Apel, K., 1985: Phytochrome control of in vitro transcription of specific genes in isolated nuclei from barley (Hordeum vulgare). Eur. J. Biochem. **147,** 137—142.

Nagy, F., Morelli, G., Fraley, R. T., Rogers, S. G., Chua, N.-H., 1985: Photoregulated expression of a pea rbcS gene in leaves of transgenic petunia plants. EMBO J. **4,** 3063—3068.

Nagy, F., Kay, S. A., Boutry, M., Hsu, M.-Y., Chua, N.-H., 1986: Phytochrome controlled expression of a wheat Cab gene in transgenic tobacco seedlings. EMBO J. **5,** 1119—1124.

Nagy, F., Boutry, M., Hsu, M.-Y., Wong, M., Chua, N.-H., 1987: 5′ proximal region of the wheat Cab-1 gene contains a 268 bp enhancer-like sequence for phytochrome response. EMBO J. **9,** 2537—2542.

Nagy, F., Kay, S. A., Chua, N.-H., 1988 (in press): The circadian expression of the wheat Cab-1 gene is regulated at the transcription level in transgenic tobacco. Genes and Develop.

Nathans, J., Thomas, D., Hogness, D. S., 1986 a: Molecular genetics of human color vision: the genes encoding blue, green and red pigments. Science **232,** 193—202.

Nathans, J., Piantanida, T. P., Eddy, R. L., Shows, T. B., Hogness, D. S., 1986 b: Molecular genetics of inherited variation in human color vision. Science **232,** 203—210.

Nelson, T., Harpster, M. H., Mayfield, S. P., Taylor, W. C., 1984: Light-regulated gene expression during maize leaf development. J. Cell Biol. **98,** 558—564.

Nester, E. W., Gordon, M. P., Amasino, R. M., Yanofsky, M. F., 1984: Crown gall: a molecular and physiological analysis. Ann. Rev. Plant Physiol. **35,** 387—413.

Nir, U., Walker, M. C., Rutter, W. J., 1986: Regulation of rat insulin 1 gene expression: evidence for negative regulation in nonpancreatic cells. Proc. Natl. Acad. Sci. U.S.A. **83,** 3180—3184.

Nishizuka, Y., 1986: Studies and perspectives of protein kinase C. Science **233,** 305—312.

Oelmuller, R., Mohr, H., 1985: Mode of coaction between blue/UV light and light absorbed by phytochrome in light-mediated anthocyanin formation in the milo seedling. Proc. Natl. Acad. Sci. U.S.A. **82,** 6124—6128.

Olah, Z., Kiss, Z., 1986: Occurrence of lipid and phorbol ester activated protein kinase in wheat cells. FEBS Lett. **195,** 33—37.

Otto, V., Schafer, E., Nagatani, A., Yamamoto, K. T., Furuya, M., 1984: Phytochrome control of its own synthesis in Pisum sativum. Plant Cell Physiol. **25,** 1579—1584.

Pratt, L. H., 1982: Phytochrome: the protein moiety. Annual Rev. Plant Physiol. **33,** 557—582.

Reddy, P., Jacquier, A. C., Abovich, N., Petersen, G., Rosbash, M., 1986: The period clock locus of D. melanogaster codes for a proteoglycan. Cell **46,** 53—61.

Roux, S. J., 1984: Ca^{2+} and phytochrome action in plants. Bioscience **34,** 25—29.

Roux, S. J., Datta, N., Chen, Y.-R., Kim, S.-H., 1986: Light, calcium and calmodulin regulation of enzyme activities in isolated nuclei. J. Cell Biochem. Supplement 10 B, 14.

Satter, R. L., Galston, A. W., 1981: Mechanisms of control of leaf movements. Ann. Rev. Plant Physiol. **32,** 83—110.

Schafer, E., Briggs, W. R., 1986: Photomorphogenesis from signal perception to gene expression. Photobiochem. Photobiophys. **12,** 305—320.

Schafer, A., Bygrave, F., Matzenauer, S., Marme, D., 1985: Identification of a calcium- and phospholipid-dependent protein kinase in plant tissue. FEBS Lett. **187,** 25—28.

Schopfer, P., 1977: Phytochrome control of enzymes. Annual Rev. Plant Physiol. **28,** 223—252.

Sen, R., Baltimore, D., 1986: Inducibility of K immunoglobulin enhancer-binding protein NF-KB by a posttranslational mechanism. Cell **47,** 921—928.

Sengar, H., 1982: The effect of blue light on plants and microorganisms. Photochem. Photobiol. **35,** 911—920.

Serlin, B. S., Sopory, S. K., Roux, S. J., 1984: Modulation of oat mitochondrial ATPase activity by Ca^{2+} and phytochrome. Plant Physiol. **74,** 827—833.

Shanklin, J., Jabben, M., Vierstra, R. D., 1987: Red light-induced formation of ubiquitin-phytochrome conjugates: identification of possible intermediates of phytochrome degradation. Proc. Natl. Acad. Sci. U.S.A. **84,** 359—363.

Sharma, R., 1985: Phytochrome regulation of enzyme activity in higher plants. Photochem. Photobiol. **41,** 747—755.

Sharrock, R. A., Lissemore, J. L., Quail, P. H., 1986: Nucleotide and amino acid sequence of a Cucurbita phytochrome cDNA clone: identification of conserved features by comparison with Avena phytochrome. Gene **47,** 287—295.

Shimazaki, Y., Pratt, L. H., 1985: Immunochemical detection with rabbit polyclonal and mouse monoclonal antibodies of different pools of phytochrome from etiolated and green Avena shoots. Planta **164,** 333—344.

Shimazaki, Y., Pratt, L. H., 1986: Immunoprecipitation of phytochrome from green Avena by rabbit antisera to phytochrome from etiolated Avena. Planta **168,** 512—515.

Shimazaki, Y., Cordonnier, M.-M., Pratt, L. H., 1986: Identification with monoclonal antibodies of a second antigenic domain on Avena phytochrome that changes upon its photoconversion. Plant Physiol. **82,** 109—113.

Sibley, D. R., Benovic, J. L., Caron, M. G., Lefkowitz, R. J., 1987: Regulation of transmembrane signaling by receptor phosphorylation. Cell **48,** 913—922.

Sieciehowicz, K., Ireland, R. J., Joy, K. W., 1985: Diurnal variation of asparaginase in developing pea leaves. Plant Physiol. **77,** 506—508.

Silverthorne, J., Tobin, E. M., 1984: Demonstration of transcriptional regulation of specific genes by phytochrome action. Proc. Natl. Acad. Sci. U.S.A. **81,** 1112—1116.

Simpson, J., Timko, M. P., Cashmore, A. R., Schell, J., Van Montagu, M., Herrera-Estrella, L., 1985: Light-inducible and tissue specific expression of a chimeric gene under control of the 5′ flanking sequence of a pea chlorophyll a/b binding protein gene. EMBO J. **4,** 2723—2729.

Simpson, J., Schell, J., Van Montagu, M., Herrera-Estrella, L., 1986a: The light-inducible and tissue specific expression of a pea LHCP gene involves an upstream element combining enhancer and silencer-like properties. Nature **323,** 551—553.

Simpson, J., Van Montagu, M., Herrera-Estrella, L., 1986b: Photosynthesis associated gene families: differences in response to tissue-specific and environmental factors. Science **233,** 34—38.

Slovin, J. P., Tobin, E. M., 1982: Synthesis and turnover of the light-harvesting chlorophyll a/b protein in Lemna gibba grown with intermittent red light: possible translational control. Planta **154,** 465—472.

Smith, H., 1982: Light quality, photoperception, and plant strategy. Annual Rev. Plant Physiol. **33,** 481—518.

Smith, W. O., 1983: Phytochrome as a molecule. In: Encyclopedia of Plant Physiology (NS) 16A: Photomorphogenesis. Shropshire, W., Mohr, H. (eds.), pp. 96—1118. Springer-Verlag, Berlin.

Smith, S. M., Ellis, R. J., 1978: Light-stimulated accumulation of transcripts of nuclear and chloroplast genes for ribulose-bisphosphate carboxylase. J. Mol. Appl. Genet. **1,** 127—137.

Speth, V., Otto, V., Schafer, E., 1986: Intracellular localization of phytochrome in oat coleoptiles by electron microscopy. Planta **168,** 299—304.

Thompson, W. F., Kaufman, L. S., Watson, J. C., 1985: Induction of plant gene expression by light. Bioessays **3,** 153—159.

Timko, M. P., Kausch, A. P., Castresana, C., Fassler, J., Herrera-Estrella, L., Van den Broeck, G., Van Montagu, M., Schell, J., Cashmore, A. R., 1985: Light regulation of plant gene expression by an upstream enhancer-like element. Nature **318,** 579—582.

Tobin, E. M., 1978: Light regulation of specific mRNA species in Lemna gibba L. G-3. Proc. Natl. Acad. Sci. U.S.A. **75,** 4749—4753.

Tobin, E. M., Klein, A. O., 1975: Isolation and translation of plant messenger RNA. Plant Physiol. **56,** 88—92.

Tobin, E. M., Silverthorne, J., 1985: Light regulation of gene expression in higher plants. Ann. Rev. Plant. Physiol. **36,** 569—593.

Tokuhisa, J. G., Daniels, S. M., Quail, P. H., 1985: Phytochrome in green tissue: spectral and immunochemical evidence for distinct molecular species of phytochrome in light grown Avena sativa L. Planta **164,** 321—332.

Tsien, R. Y., 1981: A non-disruptive technique for loading calcium buffers and indicators into cells. Nature **290,** 527—528.

Vierstra, R. D., Quail, P. H., 1982a: Native phytochrome: inhibition of proteolysis yields a homogenous monomer of 124 kilodaltons from Avena. Proc. Natl. Acad. Sci. **79,** 5272—5276,

Vierstra, R. D., Quail, P. H., 1982b: Proteolysis alters the spectral properties of 124 kdalton phytochrome from Avena. Planta **156,** 158—165.

Vierstra, R. D., Quail, P. H., 1983: Purification and initial characterization of 124 kdalton phytochrome from Avena. Biochemistry **22,** 2498—2505.

Vierstra, R. D., Quail, P. H., 1985: Spectral characterization and proteolytic mapping of native 120-kilodalton phytochrome from Cucurbita pepo L. Plant Physiol. **77,** 990—998.

Vierstra, R. D., Cordonnier, M.-M., Pratt, L. H., Quail, P. H., 1984: Native phytochrome: immunoblot analysis of relative molecular mass and in vitro proteolytic degradation for several plant species. Planta **160,** 521—528.

Vince-Prue, D., 1983: Photomorphogenesis and flowering. Encyclopedia of Plant Physiology (NS) 16B: Photomorphogenesis. Shropshire, W., Mohr, H. pp. 457—490. Springer-Verlag, Berlin.

Wagle, J., Jaffe, M. J., 1980: Plant Physiol. Supplement **65,** 3.

Wilkins, M. B., 1959: An endogenous rhythm in the rate of CO_2 output of Bryophyllum. I. Some preliminary experiments. J. Exp. Bot. **10,** 377—390.

Wong, Y.-S., Cheng, H.-C., Walsh, D. A., Lagarias, J. C., 1986: Phosphorylation of Avena phytochrome in vitro as a probe of light-induced conformational changes. J. Biol. Chem. **261,** 12089—12097.

Yamamoto, K. R., 1985: Steroid receptor regulated transcription of specific genes and gene networks. Ann. Rev. Genet. **19,** 209—252.

Zambryski, P., Joos, H., Genetello, C., Leemans, J., Van Montagu, M., Schell, J., 1983: Ti-plasmid vector for the introduction of DNA into plant cells without alteration of their normal regeneration capacity. EMBO J. **2,** 2143—2150.

Zhu, Y. S., Kung, S. D., Bogorad, L., 1985: Phytochrome control of levels of mRNA complementary to plastid and nuclear genes of maize. Plant Physiol. **79,** 371—376.

Chapter 9

Regulation of Gene Expression by Ethylene

James E. Lincoln and Robert L. Fischer

Division of Molecular Plant Biology, 313 Hilgard Hall, University of California at Berkeley, Berkeley, CA 94720, U. S. A.

With 6 Figures

Contents

I. Introduction

A. Plant Hormones

In plants, many processes are regulated by a small number of hormones: auxin, abscisic acid, cytokinin, gibberellin, and ethylene (for review, see Wareing and Phillips, 1981). Plant hormones, like animal hormones, are active in extremely small quantities and influence the growth and differentiation of tissues and organs. However, certain important aspects about plant hormone action are unique. That is, each plant hormone is synthesized in many parts of the plant and each regulates a wide variety of different processes depending on the organ or tissue. Furthermore, plants regulate many processes by modulating both hormone concentration and sensitivity to the hormone (Trewavas, 1982; Trewavas *et al.*, 1983). Thus, understanding the mechanism of plant hormone action involves knowing how both hormone concentration and tissue sensitivity regulate cellular differentiation.

B. Ethylene and the Control of Tomato Fruit Ripening

The plant hormone ethylene has a profound influence on plant development. Active in trace amounts, it affects seed germination, seedling growth, root and leaf growth, many stress phenomena, plant senescence, and fruit development (Abeles, 1973; Lieberman, 1979). In particular, ethylene has been shown to be intimately involved in the initiation of ripening of many climacteric fruits, such as tomato. This is because the onset of fruit ripening is associated with an increase in ethylene biosynthesis, the onset of ripening is hastened when unripe fruit are exposed to exogenous ethylene, and removal of ethylene from fruit or exposure of fruit to specific inhibitors of ethylene biosynthesis greatly retards ripening (Lyons and Pratt, 1964; Rhodes, 1980; Biale and Young, 1981; Yang, 1985). Associated with tomato fruit ripening is an accumulation of carotenoid pigments, conversion of chloroplasts to chromoplasts, breakdown of cell wall components, and changes in gene expression (Rhodes, 1980; Biale and Young, 1981; Su *et al.*, 1984; Grierson, 1985; Tucker *et al.*, 1985; Biggs *et al.*, 1986; Lincoln *et al.*, 1987). Thus, ethylene plays a critical role in regulating the complex fruit ripening developmental program.

Tomato fruit ripening, like other ethylene mediated processes (e. g., leaf abscission, flower senescence) involves both changes in ethylene concentration and changes in sensitivity of tissue to ethylene (Abeles, 1967; Kende and Hanson, 1976; McGlasson *et al.*, 1978). As fruits approach the onset of ripening, they become increasingly sensitive to ethylene. That is, lower concentrations of exogenous ethylene are required to induce fruit ripening. These experiments suggest that when fruit tissue becomes sufficiently responsive to basal ethylene levels, the ripening process is initiated, resulting in an autocatalytic burst of ethylene production (McGlasson *et al.*, 1978; McGlasson, 1985; Yang, 1985). Continued high level ethylene biosynthesis is required for successful completion of ripening. Thus, the onset of ripening may be controlled by an increasing sensitivity to ethylene, while subsequent processes are most likely controlled by the large increase in ethylene concentration (McGlasson, 1985; Yang *et al.*, 1986). Furthermore, as described below, the induction of gene expression associated with tomato fruit ripening is likewise regulated by both changes in ethylene sensitivity and concentration (Lincoln *et al.*, 1987).

C. Induction of Gene Expression by Exposure to Exogenous Ethylene

Exposing plant tissue to exogenous ethylene results in the induction of gene expression (Christoffersen and Laties, 1982; Zurfluh and Guilfoyle,

1982; Boller *et al.*, 1983; Grierson, 1985). mRNAs that accumulate in different tissues in response to ethylene have been cloned (Tucker *et al.*, 1985; Broglie *et al.*, 1986; Lincoln *et al.*, 1987) and the increase in mRNA concentration has been shown to reflect increased gene transcription (see below and Nichols *et al.*, 1984). Recent efforts to understand the regulatory role of ethylene during fruit ripening have focused on the activation of gene expression in unripe fruit exposed to elevated levels of exogenous ethylene (Grierson, 1985; Tucker *et al.*, 1985; Lincoln *et al.*, 1987). As described below, studies in my laboratory indicate that when unripe tomato fruits are exposed to exogenous ethylene the genetic response is quite rapid. Within 2 hours there is a significant increase in the mRNA level and relative transcription rates of specific genes.

D. A Model System for Studying Hormonal Regulation of Gene Expression During Plant Development

The regulation of gene expression by ethylene during tomato fruit ripening is an excellent model system for the following reasons: (1) Dramatic changes in both ethylene hormone concentration and sensitivity play a critical role in regulating fruit ripening. (2) Mutants defective in ethylene biosynthesis and fruit ripening are available (Smith and Ritchie, 1983). (3) Tomato plants and fruit tissue are amenable to molecular, biochemical, and genetic experimentation. That is, DNA regulatory sequences can be identified by gene transfer experiments using efficient plant transformation procedures (McCormick *et al.*, 1986; Zambryski *et al.*, 1983), and techniques for isolating proteins that specifically bind DNA regulatory sequences can be used to identify cellular factors that may regulate gene expression (Deikman and Fischer, unpublished results). For these reasons, tomato fruit ripening is a system where one can begin to define the molecular and cellular processes which regulate ethylene hormone-inducible gene expression.

II. Analysis of Ethylene-inducible Gene Expression

A. Isolation of cDNA Clones

To develop a strategy to clone ethylene-responsive mRNAs, it was necessary to know when ethylene biosynthesis rates increased during fruit development. Unripe fruits produce low basal levels of ethylene, while ripe fruits produce elevated levels of ethylene (Abeles, 1973). However, a more detailed analysis (Fig. 1) indicated that immature and early mature green stage (MG1 and MG2) fruit evolved ethylene at a low, basal rate. At the MG3 stage a small increase in ethylene evolution rate was detected. At the MG4 stage when overt signs of ripening (e. g., carotenoid pigment accumulation) were observed, the ethylene evolution rate sharply increased.

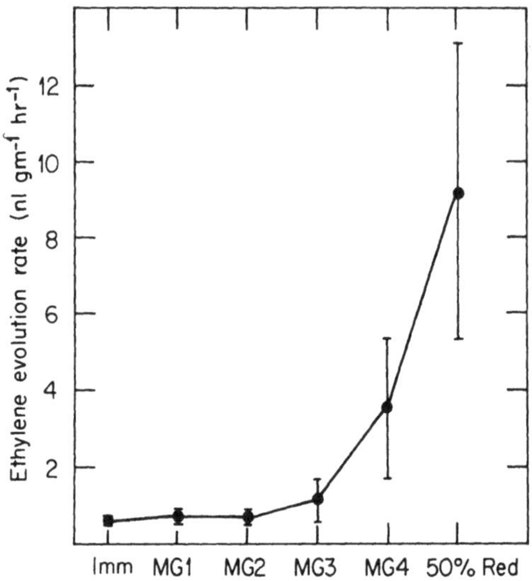

Fig. 1. Ethylene production rate during fruit development. Individual fruits were placed in 50 ml containers that were sealed and incubated 1 hr. A 1 ml sample from the closed atmosphere was removed and the ethylene content determined by gas chromatography (Varian 500). Fruit maturity stage was determined as follows. Immature fruits were 50% full size. Full size mature green (MG) stages were identified by the extent of locular tissue breakdown resulting in the formation of a viscous gel. In MG 1 fruit the locular tissue was firm, in MG 2 fruit a small amount of gel was present, and in MG 3 fruit the formation of the gel was complete. In MG 4 fruit pigmentation was just detectable in the interior of the fruit, while later stages of fruit development were defined by further pigment accumulation. Results represent the mean ±SD. Sample size for each group was as follows: immature (Imm), 5; MG 1, 39; MG 2, 57; MG 3, 15; MG 4, 26; 50% red, 20

Based on the measurements of ethylene biosynthesis, we reasoned that ethylene-inducible mRNAs were more abundant in ripening fruit than in unripe fruit. Thus, a cDNA library enriched for ripening-specific sequences was constructed using subtraction hybridization techniques (Davis, *et al.*, 1984) as shown in Figure 2. The library was then screened with a probe enriched for sequences that accumulate when MG 1 fruit are exposed to ethylene for 8 hr as described in Lincoln *et al.* (1987). Using these procedures four classes of ethylene-induced cDNA clones were isolated represented by E 4, E 8, E 17, and J 49. Northern blot analysis indicated that the E 4, E 8, E 17, and J 49 cDNA clones hybridized to 0.9, 1.6, 0.6, and 1.0 kb mRNAs, respectively (Lincoln *et al.*, 1987).

To identify the proteins encoded by the cloned mRNAs, we compared their respective DNA sequences and predicted amino acid sequences to those stored in the NIH DNA Sequence Library, EMBL DNA Sequence Library, and the NBRF Protein Sequence Library. This analysis revealed that E 17 was related (67% homologous) to the DNA sequence of a proteinase inhibitor I gene (Lincoln *et al.*, 1987). Proteinase inhibitors are a class of low molecular weight polypeptides that accumulate in wounded

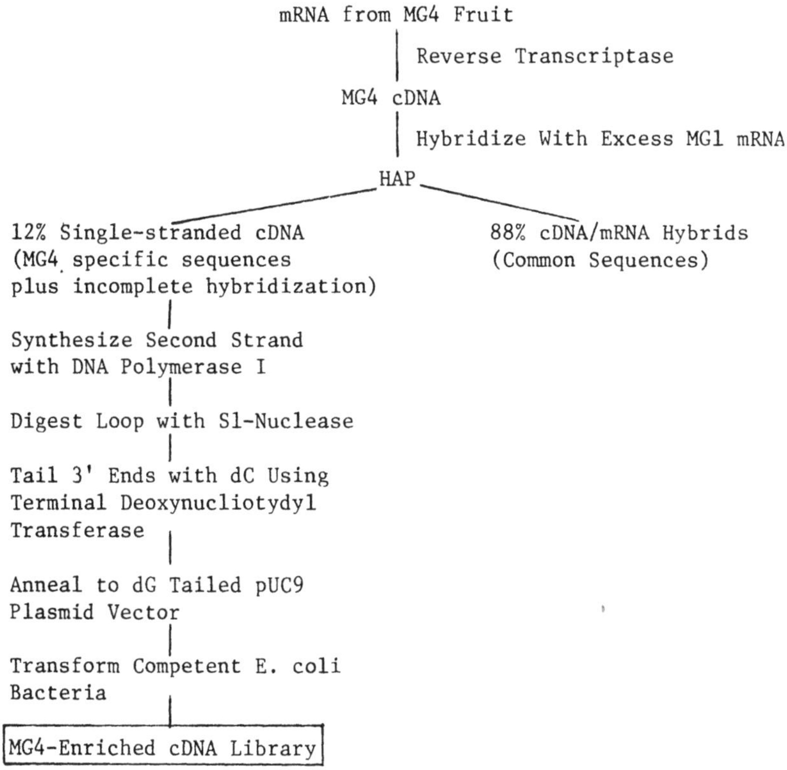

Fig. 2. Constructing a cDNA library enriched for ripening-specific sequences using subtraction hybridization procedures. Seven μg of ^{32}P-labeled MG4 cDNA and 70 μg MG1 mRNA were hybridized for 16.5 hr at 70°C (equivalent R_0t 4,000) in a 12 μl reaction containing 33 mM PIPES (pH 6.9), 1 M NaCl, 0.2 mM EDTA, 0.2% $NaDodSO_4$. Unreacted cDNA (12% of the total cDNA mass) was isolated by passing the reaction over a 1 ml hydroxyapatite column at 68°C equilibrated in 0.12 M NaH_2PO_4, 0.12 M Na_2HPO_4, 0.2% $NaDodSO_4$. Enriched for MG4-specific sequences, the unreacted cDNA was used to construct a cDNA library as indicated

leaves (Graham *et al.*, 1985). The DNA sequence homology between *E17* and a proteinase inhibitor gene indicates they may have evolved from a common ancestral gene, and suggests that the regulation of gene expression by wounding and by ethylene may be related processes.

B. Induction of Gene Expression by Exogenous Ethylene in Unripe Tomato Fruit

To determine how rapidly cloned mRNAs accumulate in ethylene treated fruit, the labeled cDNA clones were hybridized to mRNA isolated from MG1 fruit exposed to either ethylene or air for 0.5 to 8 hr. As shown in Figure 3, within 2 hr the concentration of each cloned mRNA had increased significantly over that of the air control. Thus, ethylene rapidly induces the expression of specific genes.

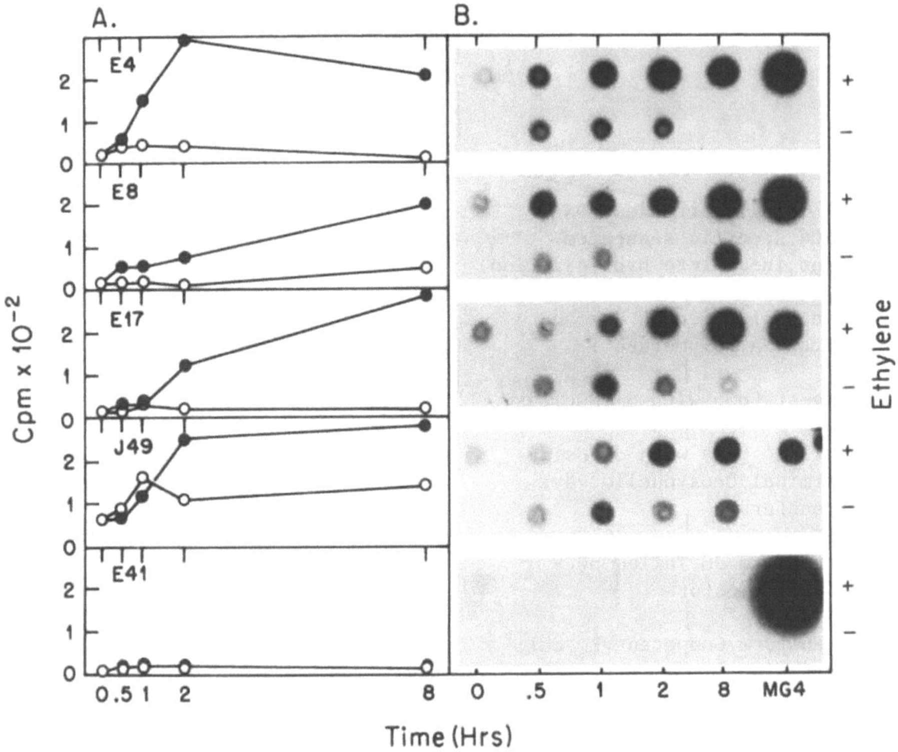

Fig. 3. Accumulation of specific mRNAs in fruit exposed to ethylene. mRNA was isolated from MG 1 fruit treated with either 10 µl/Liter ethylene (●) or air (○) for the indicated period of time, or from MG 4 fruit. One µg of each mRNA was dotted onto nitrocellulose and hybridized to the indicated ^{32}P-labeled DNA probes. Following autoradiography (B), each dot was cut out and the extent of hybridization was determined by liquid scintillation spectrometry (A). *E41* is a polygalacturonase cDNA clone (Lincoln *et al.*, 1987)

To determine whether the rapid ethylene-induced accumulation of the cloned mRNAs was an organ-specific process, the labeled cDNA clones were hybridized to mRNA isolated from leaves treated with ethylene for 2 hours. As shown in Figure 4, ethylene does not rapidly induce *E8* gene expression in leaves. Hence, E8 represents a gene whose response to ethylene is fruit specific. In contrast, ethylene rapidly induces *E4, E17,* and *J49* gene expression in leaves. Thus, they represent genes whose response to exogenous ethylene is not organ-specific. However, the capacity to produce elevated levels of ethylene during plant development is not limited to fruit. Leaves increase ethylene biosynthesis rates as they senesce (McGlasson *et al.*, 1975), and in response to a variety of stresses (Yang and Pratt, 1978; Yang, 1985). We hope to determine whether activation of *E4, E17,* and *J49* gene expression represents a common response to increase in ethylene concentration that occur in different plant organs during development.

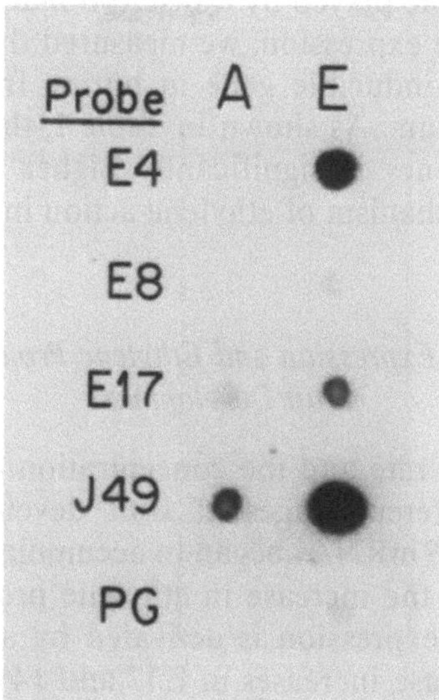

Fig. 4. Accumulation of specific mRNAs in leaves exposed to ethylene. Intact plants were exposed to either 10 µl/liter ethylene (E) or to air (A) for 2 hr. Polysomal, poly A^+ mRNA was isolated from leaves. One µg of each mRNA was dotted onto nitrocellulose and hybridized to the indicated ^{32}P-labeled probes

Table 1. Relative transcription rates of ethylene-inducible genes in unripe fruit treated with ethylene or air

Gene	% [^{32}P] nuclear RNA × 10^3	
	Ethylene	Air
E4	3.5	<0.2
E8	2.8	<0.2
E17	<0.2	<0.2
J49	5.0	<0.2
D21	2.5	2.1

MG1 fruit were treated with 10 µl/Liter ethylene or air for 2 hr. Nuclei were isolated and ^{32}P-labeled nuclear RNAs were synthesized, purified, and hybridized to plasmid DNAs bound to filters as described by Walling *et al.* (1986). Each hybridization reaction contained 1×10^7 cpm of ^{32}P-labeled nuclear RNA, plasmid DNA filters, and a pUC18 background hybridization control. D21 is a constitutively expressed control clone.

$$\%\,[^{32}P]\text{ nuclear RNA} = \frac{(\text{cpm hybridized} - \text{pUC 18 cpm})\,(R)\,(100)}{(\text{input cpm})\,(H)}$$

Where R is the ratio of mRNA to plasmid insert lengths. H is the hybridization efficiency (0.2) which was estimated by reacting 25 S ^{32}P-labeled ribosomal RNA with a filter containing excess 25 S ribosomal DNA.

To investigate the role played by transcriptional processes in regulating ethylene-inducible gene expression, we measured the relative transcription rates of each ethylene-inducible gene in unripe fruit treated with either ethylene or air for 2 hours. As shown in Table 1, the relative transcription rates of the cloned genes is significantly higher following exposure to ethylene. Thus, the mechanism of ethylene action involves the induction of gene transcription.

C. Activation of Gene Expression and Ethylene Production During Tomato Fruit Development

The ethylene evolution rate and the concentration of each cloned mRNA were measured at different stages of fruit development. As shown in Figure 5, the *E4* and *E8* mRNAs began to accumulate relatively late (MG 3 stage), coincident with the increase in ethylene production. This suggests that *E4* and *E8* gene expression is activated by an increase in ethylene concentration. In contrast, increases in E 17 and J 49 mRNA concentration were detected relatively early (MG 1 and MG 2 stages), when low constant basal levels of ethylene production were detected. This indicates that *E17* and *J49* gene expression is not activated by an increase in ethylene concentration during fruit ripening.

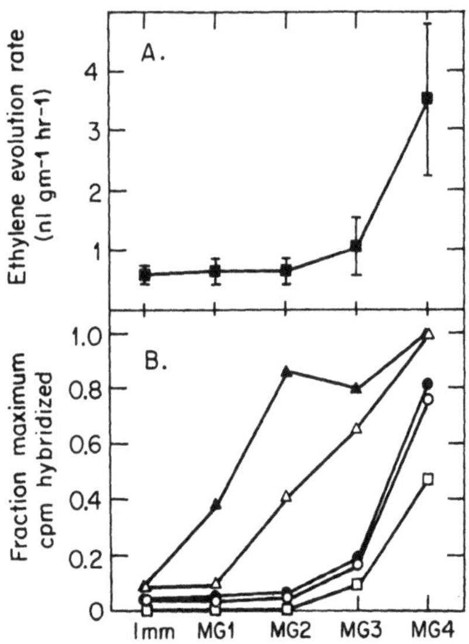

Fig. 5. Ethylene production rate and the accumulation of specific mRNAs during fruit development. (A) Ethylene production rate was determined as described in Figure 1. (B) One µg mRNA isolated from fruit at the indicated stages was dotted onto nitrocellulose filters, hybridized to the indicated ^{32}P-labeled DNA probes. Following autoradiography, each dot was cut out and the extent of hybridization was determined by liquid scintillation spectrometry. The maximum cpm hybridized for each gene: *E4* (○), 6625 cpm at red stage; *E8* (●), 2876 cpm at 50% red stage; *E17* (△), 559 cpm at MG4 stage; *J49* (▲), 3848 cpm at MG4 stage, *E41* (□), 12,352 cpm at red stage

D. Repression of Gene Expression by a Competitive Inhibitor of Ethylene Action

We reasoned that early (*E17* and *J49*) gene expression might be activated by an increased capacity of fruit tissue to respond to low basal levels of ethylene. Alternatively, even though their expression was inducible when MG1 stage fruit were exposed to exogenous ethylene, it was possible that *E17* and *J49* gene expression was not regulated by ethylene during fruit development. To investigate the role played by ethylene in regulating cloned gene expression, fruit were exposed to norbornadiene, a highly specific competitive inhibitor of ethylene action (Yang, 1985). As shown in Figure 6, treatment with norbornadiene repressed the expression of all the cloned ethylene-inducible genes when compared to the control fruit exposed to air. The norbornadiene did not affect the concentration of an additional cloned mRNA encoded by a gene expressed constitutively throughout fruit development (Lincoln and Fischer, unpublished results). These results show that ethylene is needed for the maximal accumulation of both early and late cloned mRNAs. In particular, they suggest that low basal levels of ethylene are required for *E17* and *J49* gene expression.

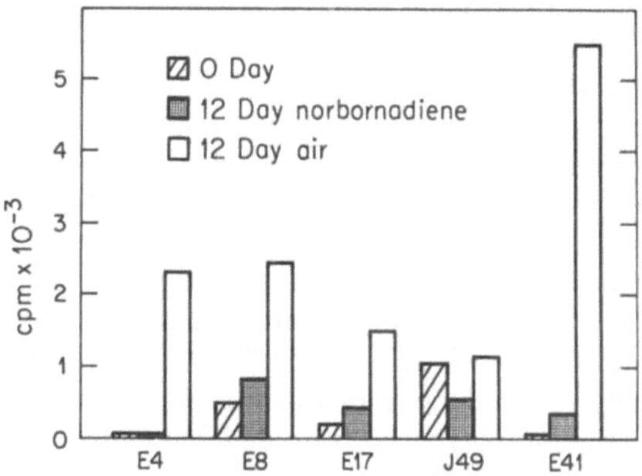

Fig. 6. Expression of cloned genes in the presence of an inhibitor of ethylene action, norbornadiene. One hundred and forty mature green fruit were divided into 3 groups. One group was immediately harvested (0 day control). The other two groups were put in 60 liter chambers for 12 days. One chamber contained air plus 1000 µl/liter norbornadiene. The other chamber contained only air. One µg of mRNA isolated from each group was dotted and hybridized to the indicated ^{32}P-labeled probes. Because mature green substages could only be determined by examining internal morphological markers, initially the population was a mixture of MG1 and MG2 fruit. As a result, *J49*, a gene activated at the MG1 stage (Fig. 6) was expresed in the untreated (0 day) control fruit. As expected, expression of all the other cloned genes was low in the untreated control fruit. Ripening (accumulation of carotenoid pigments) was observed in the air controls, but not in the norbornadiene treated fruit

III. Discussion

Ethylene has been shown to be intimately involved in the initiation and maintenance of ripening of tomatoes and many other fruits. In order to better understand the relationship between ethylene, gene expression, and the onset and maintenance of fruit ripening, we have studied a set of genetic responses to this important plant hormone.

We have found that treating unripe fruit with exogenous ethylene rapidly induces gene expression. A similar induction has been observed when soybean hypocotyls were exposed to ethylene (Zurfluh and Guilfoyle, 1982). Adjusting for the approximately 15 minutes it takes for ethylene gas to diffuse into fruit (Cameron and Yang, 1982), the induced gene expression we observe is nearly as rapid as auxin-induced gene expression demonstrated in pea epicotyls and soybean hypocotyls (Hagen *et al.*, 1984; Theologis *et al.*, 1985; Key *et al.*, 1986), and may represent a primary genetic response to ethylene. These results suggest that plant hormones such as auxin and ethylene are capable of inducing changes in the physiology of plants by rapidly altering patterns of gene expression. In addition, we have found that the induction of cloned gene expression by ethylene involves changes in the relative rates of gene transcription. Thus, the mechanism of ethylene action can be investigated in part by studying proteins that regulate the transcription of ethylene-inducible genes.

Our results suggest that gene expression can be regulated by changes in ethylene concentration and by changes in sensitivity to ethylene. The importance of ethylene concentration is demonstrated by the fact that exposure of fruits to exogenous ethylene rapidly activates cloned gene expression (Fig. 3), and the onset of *E4* and *E8* gene expression is approximately coincident with the increase in ethylene evolution rate during fruit ripening (Fig. 5). These results are consistent with the idea that increasing ethylene concentration leads to the onset of gene expression. Similar results have been obtained in studies of the regulation of cellulase gene expression during avocado fruit ripening (Tucker *et al.*, 1985). The importance of sensitivity to ethylene is demonstrated by the fact that E17 and J49 mRNA accumulate early in fruit development when the ethylene concentration is basal (Fig. 5), and these low levels of ethylene appear to be required for maximal gene expression (Fig. 6). These results suggest that during fruit development *E17* and *J49* gene expression is activated by an increase in sensitivity to basal, low levels of ethylene. This is consistent with physiological evidence that most fruits become more sensitive to ethylene as they mature and approach the onset of ripening (McGlasson *et al.*, 1978; Yang, 1985). In addition to fruit ripening, other important ethylene-mediated processes such as leaf abscission (Abeles, 1968) and flower senescence (Kende and Hanson, 1976) involve both changes in ethylene concentration and changes in the capacity of tissues to respond to ethylene. Thus, the activation of gene expression by respective changes in ethylene sensitivity and concentration during fruit ripening might be an important general phenomenon during plant development.

Trewavas (1982) described the interaction of a growth substance such as ethylene with plant tissue by the relationship:

growth substance + receptor $\leftrightarrow$ active complex $\rightarrow$ biological response.

Depending on whether ethylene or the putative receptor(s) are limiting, the relationship predicts that a biological response (e. g., activation of gene expression) may result from increasing ethylene concentration, receptor concentration, and/or receptor activity. Thus, activation of *E4* and *E8* gene expression may result from increased ethylene concentration, while *E17* and *J49* gene expression may result from increased ethylene receptor concentration or activity. In the future it will be important to identify the cellular factors that regulate ethylene-inducible gene expression during plant development.

IV. References

Abeles, F. B., 1967: Mechanism of action of abscission accelerators. Physiol. Plantarum **20,** 442—454.

Abeles, F. B., 1973: Ethylene in plant biology. pp. 1—302. Academic Press, New York.

Biale, J. B., Young, R. E., 1981: Respiration and ripening in fruits — retrospect and prospect. In: Recent advances in the biochemistry of fruit and vegetables. pp. 1—39. Friend, J., Rhodes, M. J. C. (eds.). Academic Press, London.

Biggs, M. S., Harriman, R. W., Handa, A. K., 1986: Changes in gene expression during tomato fruit ripening. Plant Physiol. **81,** 395—403.

Boller, T., Gehri, A., Mauch, F., Vogeli, U., 1983: Chitinase in bean leaves: induction by ethylene, purification, properties, and possible function. Planta **157,** 22—31.

Broglie, K. E., Gaynor, J. J., Broglie, R. M., 1986: Ethylene-regulated gene expression: molecular cloning of the genes encoding an endochitinase from *Phaseolus vulgaris.* Proc. Natl. Acad. Sci. U. S. A. **83,** 6820—6824.

Cameron, C. C., Yang, S. F., 1982: A simple method for the determination of resistance to gas diffusion in plant organs. Plant Physiol. **70,** 21—33.

Christoffersen, R. E., Laties, G. G., 1982: Ethylene regulation of gene expression in carrots. Proc. Natl. Acad. Sci. U. S. A. **79,** 4060—4063.

Davis, M. M., Cohen, D. I., Nielsen, E. A., Steinmetz, M., Paul, W. E., Hood, L., 1984: Cell-type-specific cDNA probes and the murine I region: the localization and orientation of A^d. Proc. Natl. Acad. Sci. U. S. A. **81,** 2194—2198.

Graham, J. S., Pearce, G., Merryweather, J., Titani, K., Ericsson, L., Ryan, C. A., 1985: Wound-induced proteinase inhibitors from tomato leaves. J. Biol. Chem. **260,** 6555—6560.

Grierson, D., 1985: Gene expression ripening tomato fruit. CRC Critical Rev. in Plant Sci. **3,** 113—132.

Hagen, G. H., Kleinschmidt, A., Guilfoyle, T., 1984: Auxin-regulated gene expression in intact soybean hypocotyl and excised hypocotyl sections. Planta **162,** 147—153.

Kende, H., Hanson, A. D., 1976: Relationship between ethylene evolution and senescence in morning-glory flower tissue. Plant Physiol. **57,** 523—527.

Key, J. L., Kroner, P., Walker, J., Hong, J. C., Ulrich, T. H., Ainley, W. M., Nagao, T. H., 1986: Auxin-regulated gene expression. Phil. Trans. R. Soc. Lond. B. **314,** 427—440.

Lieberman, M., 1979: Biosynthesis and action of ethylene. Ann. Rev. Plant Physiol. **30,** 533—591.

Lincoln, J. E., Cordes, S., Read, E., Fischer, R. L., 1987: Regulation of gene expression by ethylene during *Lycopersicon esculentum* (tomato) fruit development. Proc. Natl. Acad. Sci. U.S.A. **84,** 2793—2797.

Lyons, J. M., Pratt, H. K., 1964: Effect of stage of maturity and ethylene treatment on respiration and ripening of tomato fruits. Proc. Amer. Soc. Hort. Sci. **84,** 491—500.

McCormick, S., Niedermeyer, J., Fry, J., Barnason, A., Horsch, R., Fraley, R., 1986: Leaf disc transformation of cultivated tomato (*L. esculentum*) using *Agrobacterium tumefaciens.* Plant Cell Reports **5,** 81—84.

McGlassen, W. B., 1985: Ethylene and fruit ripening. Hort. Sci. **20,** 51—54.

McGlassen, W. B., Wade, N. L., Adato, I., 1978: Phytohormones and fruit ripening. In: Phytohormones and related Compounds — a comprehensive treatise. Vol. 2, pp. 447—493. Letham, D. S., Goodwin, P. B., Higgins, T. J. V. (eds.). Elsevier, Amsterdam.

Nichols, S. E., Laties, G. G., 1984: Ethylene-regulated gene transcription in carrot roots. Plant Mol. Biol. **3,** 393—401.

Rhodes, M. J. C., 1980: The maturation and ripening of fruits: In: Senescene in plants, pp. 157—205. Thimann, K. V. (ed.) CRC Press, Boca Raton.

Smith, J. W. M., Ritchie, D. B., 1983: A collection of near-isogenic lines of tomato: research tool of the future. Plant Mol. Biol. Reporter **1,** 41—45.

Speirs, J., Brady, C. J., Grierson, D., Lee, E., 1984: Changes in ribosome organization and messenger RNA abundance in ripening tomato fruits. Aust. J. Plant Physiol. **11,** 225—233.

Su, L.-Y., McKeon, T., Grierson, D., Cantwell, M., Yang, S. F., 1984: Development of 1-aminocyclopropane-1-carboxylic acid synthase and polygalacturonase activities during the maturation and ripening of tomato fruit. Hort. Science **19,** 576—578.

Theologis, A., Huynh, T., Davis, R. W., 1985: Rapid induction of specific mRNAs by auxin in pea epicotyl tissue. J. Mol. Biol. **153,** 53—68.

Trewavas, A. J., 1982: Growth substance sensitivity: the limiting factor in plant development. Physiol. Plant. **55,** 60—72.

Trewavas, A. J., 1983: Is plant development regulated by changes in the concentration of growth substances or by changes in the sensitivity to growth substances? Sensitivity is the regulating factor. Trends in Biochemical Sciences **8,** 354—357.

Tucker, M. L., Christoffersen, R. E., Woll, L., Laties, G. G., 1985: Induction of cellulase by ethylene in avocado fruit. In: Ethylene and plant development. pp. 163—175. Roberts, J. A., Tucker, G. A. (eds.). Butterworths, London.

Walling, L., Drews, G. N., Goldberg, R. B., 1986: Transcriptional and post-transcriptional regulation of soybean seed protein mRNA levels. Proc. Natl. Acad. Sci. U.S.A. **83,** 2123—2127.

Wareing, P. G., Phillips, I. D. J., 1981: Mechanisms of action of plant growth hormones. In: Growth and differentiation in plants. pp. 75—104. Pergamon Press, Oxford.

Yang, S. F., 1985: Biosynthesis and action of ethylene. Hort. Sci. **20,** 41—45.

Yang, S. F., Pratt, H. K., 1978: The physiology of ethylene in wounded plant

tissues. In: The biochemistry of wounded plant tissues. pp. 595—622. Kahl, G. (ed.). Walter de Gruyter & Co., Berlin.
Yang, S. F., Liu, Y., Lau, O. L., 1986: Regulation of ethylene biosynthesis in ripening apple fruits. Acta Horticulturae **179,** 711—720.
Zambryski, P., Joos, H., Genetello, C., Leemans, J., Van Montagu, M., Schell, J., 1983: Ti plasmid vector for the introduction of DNA into plant cells without alteration of their normal regeneration capacity. EMBO J. **2,** 2143—2150.
Zurfluh, L. L., Guilfoyle, T. J., 1982: Auxin- and ethylene-induced changes in the population of translatable messenger RNA in basal sections and intact soybean hypocotyl. Plant Physiol. **69,** 338—340.

Chapter 10

Root Nodule Symbiosis: Nodulins and Nodulin Genes

Desh Pal S. Verma[1] and Ashton J. Delauney

Center for Plant Molecular Biology, McGill University, Montreal, Canada

With 5 Figures

Contents

---

[1] *Present address:* Biotechnology Center and the Department of Molecular Genetics, Ohio State University, Columbus, OH 43210-1002, U.S.A.

I. Introduction

The soils in which plants grow are heavily populated with a variety of microorganisms, and a wide range of plant-microbe interactions have evolved, driven in the main by the microbes' requirements for organic nutrients. Many of these interactions involve potent pathogens that are detrimental to the plant; however, some microorganisms such as diazotrophic bacteria and mycorrhizal fungi enter into beneficial, symbiotic associations with plants.

Nitrogen fixation by bacteria in symbiosis with plants is the single most important contributor to the nitrogen cycle (Postgate, 1978) and is of great agricultural importance since it allows certain plants to grow in nitrogen-deficient soils. The complexity of the symbiotic interactions between plants and diazotrophs ranges from "loose" associations exemplified by the prevalence of *Beijerinckia* in the phyllosphere of sugar cane and cocoa (Postgate, 1978) to associative symbioses as that of *Azospirillum* with the roots of monocots (Dobereiner and Boddey, 1981) and the intricate intracellular interactions exhibited by *Rhizobium* with legume plants (Verma and Long, 1983).

The formation of an effective association between *Rhizobium* and its legume host involves a complex series of events including mutual recognition of the symbiotic partners, infection of plant cells by the bacteria, development of the nodule structure, differentiation of the bacteriods, nitrogen fixation and assimilation, and ultimately senescence of the nodule (reviewed by Verma and Long, 1983). During these processes, changes in the gene expression of both organisms occur resulting in the synthesis of nodule-specific bacteria-encoded proteins (bacteroidins) and host-encoded proteins (nodulins). It is the expression of nodulin genes which forms the focus of this chapter. Other aspects of the symbiotic process have been extensively covered in recent reviews (Bauer, 1981; Verma and Long 1983; Sutton, 1983; Dazzo and Gardiol, 1984; Verma and Nadler, 1984; Rolfe and Shine, 1984; Long, 1984; Kondorosi and Kondorosi, 1986) and will be discussed only briefly here. We shall review in detail the known structure and function of nodulins and their genes, as well as the emerging picture of how these genes are regulated during the differentiation of the nodule. Evidence will be discussed which suggests that the legume-*Rhizobium* symbiosis is in a process of dynamic evolution and consequently may provide some insight as to how an endosymbiont becomes a fully-integrated cellular organelle.

II. An Overview of Legume Nodulation

To ward off attack by potential pathogens inhabiting the rhizosphere and phyllosphere, plants have involved a myriad of defensive mechanisms. The establishment of an effective legume-*Rhizobium* symbiosis requires that the plant selectively lowers these protective barriers to accommodate a controlled invasion by *Rhizobium*. This little-understood process is highly specific and a given plant host is generally nodulated by a single *Rhizobium* species (see Lim and Burton, 1982). It has been demonstrated that rhizobia in the soil are chemotactically attracted to their specific hosts (Currier and Strobel, 1976) and subsequently become attached to the root surface via a specific receptor molecule, probably a lectin (Dazzo and Gardiol, 1984).

Attachment of the rhizobia is followed by penetration into the plant cell, occurring either via a root hair or through openings which arise due to lateral root emergence. Entry via the root hair is achieved by partial degradation of the host cell wall with concomitant deposition of new cell-wall-like material to form an "infection thread" along the path of invasion. Eventually, the bacteria are released from the infection thread into cells derived from the root cortex. Only about half of the nodule cells become infected; uninfected cells, however, play an active role in the functioning of nodules as they are involved in the carbon and nitrogen assimilatory processes. As the bacteria are released they are compartmentalized within a membrane, the peribacteroid membrane (pbm), originating from the plasma membrane of the host cell. By this stage, the bacteroid form of the rhizobia has changed significantly from the free-living form and has a thinner, less rigid cell wall, as well as an altered outer membrane. This transition is accompanied by the expression of several new genes, including the genes involved in nitrogen fixation.

Meristematic activity in the root cortex gives rise to nodules which fall into two main morphological categories: determinate (spherical) or cylindrical with apical meristems. In most legumes, cell division ceases upon rhizobial infection. The nodule does not, however, merely represent the proliferation of root cells to house the invading rhizobia. Rather, it is a unique, highly specialized organ geared towards the effective exploitation of bacteroidal nitrogen fixation. This process requires that an abundant supply of energy be delivered to the bacteroids to fuel nitrogen fixation and that the nitrogenase enzyme be protected from free oxygen; also, the reduced nitrogen must be efficiently assimilated by the host and transported away from the nodules. The nodulins play a key role in satisfying the structural, metabolic and transport requirements of symbiotic nitrogen fixation, and the specific functions of individual nodulins have recently begun to be elucidated.

III. Induction of Plant Genes Coding for Nodulins

A. Nodulin Structure and Function

Nodulins were first identified in soybean nodules. *In vitro* translation of soybean nodule polysomal mRNA followed by immunoprecipitation with nodule-specific antibodies detected about 20 nodulins (Legocki and Verma, 1980), and analysis of mRNAs from nodules and uninfected roots demonstrated that approximately 40 moderately abundant transcripts are

Table 1. Characterised nodule-specific genes and proteins[a]

Nodulin	Apparent MW (kDa)[b]	Actual MW (kDa)[c]	Subcellular location	Function	References
Lba[d]	13	15,7	cytosol	oxygen carrier	Hyldig-Nielsen *et al.*, 1982
Lbc_1	13	15,7	cytosol	oxygen carrier	Hyldig-Nielsen *et al.*, 1982
Lbc_2	13	15,7	cytosol	oxygen carrier	
Lbc_3	13,5	15,8	cytosol	oxygen carrier	Brisson and Verma, 1982
N-35	33	35.1	peroxisome (uninfected cell)	uricase II	Bergman *et al.*, 1983; Nguyen *et al.*, 1985
N-23 (C 51)[e]	25	24.3	pbm	—	Mauro *et al.*, 1985; Jacobs *et al.*, 1987
N-26 b	25.5	23.5	—	—	Jacobs *et al.*, 1987
N-27	27	22.4	cytosol	—	Jacobs *et al.*, 1987
N-44 (E 27)[f]	42	39.0	—	—	Sengupta-Gopalan *et al.*, 1986
N-20	—	20.0	—	—	Sandal *et al.*, 1987
N-22	—	22.7	—	—	Sandal *et al.*, 1987
N-24	24	15.1	pbm	—	Katinakis and Verma, 1985; Fortin *et al.*, 1985
N-26	21.5	22.5	pbm	—	Fortin *et al.*, 1987
N-75	75	≤45.0	—	—	Franssen *et al.*, 1987
N-100[g]	90	—	—	Sucrose synthase	Thummler and Verma, 1987
N-25 (alfalfa)	25	—	—	—	Kiss *et al.*, 1987; G. Kiss (pers. comm.)
GS[h]	38,40	—	—	Glutamine synthetase	Sengupta-Gopalan and Pitas, 1986; Hirel *et al.*, 1987
XDH[g]	141	—	—	Xanthine dehydrogenase	Triplett, 1985
Purine nucleosidase II		—	—	Purine nucleosidase	Larsen and Jochimsen, 1987

a. The data pertain to soybean nodulins except nod-25 of alfalfa.
b. Estimated by SDS-PAGE.
c. Calculated from deduced amino-acid sequences.
d. Leghemoglobins have been identified and characterized to varying degrees in several legumes other than soybean (e. g., see Appleby, 1984).
e. The C 51 sequence (Sengupta-Gopalan *et al.*, 1986) is identical to that of nodulin-23 (Mauro *et al.*, 1985) except for 3 nucleotides.
f. The E 27 sequence (Sengupta-Gopalan *et al.*, 1986) appears to be identical to the partially determined nodulin-44 sequence (Verma *et al.*, unpublished).
g. These proteins have not been demonstrated to be absolutely nodule-specific.
h. Nodule GS has also been characterized in *P. vulgaris* (Gebhardt *et al.*, 1986) and pea (Tingey et al., 1987).

specific to the nodule (Auger and Verma, 1981). cDNA cloning subsequently revealed that approximately 7% of nodule mRNA consisted of four RNA species (designated A to D) other than the super abundant leghemoglobin mRNAs which comprised 12—15% of total mRNA (Fuller *et al.*, 1983; Fuller and Verma, 1984). Nodulins have since been identified in several other legumes including pea (Bisseling *et al.*, 1983), alfalfa (Lang-Unnasch and Ausubel, 1985; Vance *et al.*, 1985; Kiss *et al.*, 1987) *Phaseolus vulgaris* (Campos *et al.*, 1987), *Sesbania rostrata* (de Lajudie and Huguet, 1987) and *Vicia faba* (Mohapatra *et al.*, 1987).

Initially, due to the dearth of information regarding the precise functions of nodulins, they were broadly classified as C-nodulins (nodulins which play a central role in supporting nitrogen fixation and might therefore be common to all nodules), and S-nodulins (species-specific nodulins) (Verma and Nadler, 1984). More recently, as the functions of increasing numbers of nodulins have been ascertained, it has become possible to categorize these proteins on a functional basis. However, the functions of many nodulins still remain to be elucidated, and possible functions in some instances may be gleaned from structural analyses of the nodulin genes. The current status of knowledge of characterized nodulins is summarized in Table 1.

1. Leghemoglobins

The most abundant and historically most extensively studied nodulins are the leghemoglobins (Lbs). They are products of a multigene family. Four major Lb species have been characterized in soybean: Lba, Lbc_1, Lbc_2 and Lbc_3, differing only slightly in amino acid sequence (Fuchsman and Appleby, 1979). Nucleotide sequences have been determined for the major Lb genes (Brisson and Verma, 1982; Hyldig-Nielsen *et al.*, 1982; Wiborg *et al.*, 1982), as well as for 2 Lb pseudogenes (Brisson and Verma, 1982; Wiborg *et al.*, 1983) and 2 truncated genes (Brisson and Verma, 1982). Analyses of gene organization (Lee *et al.*, 1983; Bojsen *et al.*, 1983) have revealed that the Lb genes are clustered into two major loci in soybean: one contains the Lba, Lbc_1, Lbc_3 genes and a pseudogene, while the other contains the Lbc_2 gene and the second pseudogene. The truncated genes are found at two other loci which are not closely linked.

Leghemoglobins are homologous to the animal oxygen-binding hemoglobins and myoglobins (Hunt *et al.*, 1978). They comprise an approximately 16,000 M_r apoprotein moiety linked to a heme prosthetic group. In the nodule, Lbs facilitate the delivery of oxygen to the respiring bacteroids, whilst limiting the concentration of free oxygen to the oxygen-labile nitrogenase enzyme (Appleby, 1984). It has also been suggested (Verma *et al.*, 1978; Robertson *et al.*, 1984a) that Lbs may play a role in supplying oxygen to the host cell mitochondria; however, Appleby (1984) has argued that mitochondria are unlikely to respire efficiently at a free oxygen levels of 5—10 nM which correspond to the concentration at which oxygen dissociates from oxygenated Lb. Thus, further biochemical studies are required to resolve this issue.

It has been generally accepted that the heme moiety is supplied by the bacteroids (see Verma and Nadler, 1984). Consistent with this view, *R. meliloti* mutants deficient in heme biosynthesis (due to a transposon insertion in the gene for δ-aminolevulinic acid synthase) induce Fix^- nodules on alfalfa (Leong *et al.*, 1982). However, similar *Bradyrhizobium japonicum* mutants form effective nodules on soybean (Guerinot and Chelm, 1985), suggesting that certain legumes may be able to synthesize adequate supplies of heme for Lb synthesis, independently of the bacteroids.

While it is unclear whether the different Lbs (including a number of minor variants resulting from post-translational modifications) have different roles in the nodule, evidence that the relative levels of the Lbs with higher oxygen affinities increase after the onset of nitrogen fixation has led to the suggestion that the evolution of Lb heterogeneity may be geared towards more effective nitrogen fixation (Bisseling *et al.*, 1986). Mutations accumulated in different Lbs may account for the different oxygen-binding efficiencies of these molecules, and since this protein is essential for symbiotic nitrogen fixation, the pressure to keep multiple forms remains on the host plant. No legume plant has yet been found to contain only a single gene for Lb.

The intracellular location of Lbs has, until recently, been a subject of considerable controversy. Based on studies using subcellular fractionation techniques, some investigators have claimed that Lb is located both in the nodule cytoplasm and within the pbm (Bergersen and Appleby, 1981; Appleby, 1984). However, using immunocytological procedures to localize the protein directly, Verma and co-workers have shown that Lb is present exclusively in the cytoplasm of soybean nodules (Verma and Bal, 1976; Verma *et al.*, 1978; Nguyen *et al.*, 1985). Similar techniques have been utilized in nodules of lupin (Robertson *et al.*, 1978) and pea (Robertson *et al.*, 1984b) to establish that Lb is localized in the host cytoplasm. These findings are entirely consistent with the absence of a signal peptide on newly-synthesized Lb proteins (Verma *et al.*, 1979), suggesting that these proteins are unable to traverse the pbm. The localization of Lb exclusively in the host cytoplasm may require the presence of an auxiliary oxygen carrier in the pbm to transport oxygen to the bacteroid surface.

Not only have Lbs been detected in all legume nodules screened to date, but it has recently been demonstrated that a number of non-leguminous plants which form symbiotic associations with diazotrophic organisms also possess hemoglobins (Hbs). Rhizobia are known to nodulate only one non-leguminous plant, *Parasponia* (a member of the family Ulmaceae). *Parasponia* nodules have been shown to contain a dimeric Hb which shares considerable amino acid sequence homology with soybean Lba (Appleby *et al.*, 1984). Knowledge of Hb in non-legumes has been further extended by the discovery of Hb proteins and genomic sequences homologous to soybean Lb probes in several unrelated actinorrhizal plants which form nitrogen-fixing associations with *Frankia spp.* (Tjepkema, 1983; Roberts *et al.*, 1985; Hattori and Johnson, 1985).

Moreover, two species (carob, a legume; birch, a non-legume) which do not enter into nitrogen-fixing symbioses also contain genomic sequences homologous to those for soybean Lb (Hattori and Johnson, 1985). It should be emphasized, however, that in none of these cases where Hb sequences have been detected by hybridization to an Lb probe is it known whether the identified genes are functional.

Following the characterization of soybean Lb genes which revealed a close similarity to animal Hb genes, it was proposed that the latter might have been transferred from an animal to an ancestral legume relatively recently in evolutionary history (Lewin, 1981; Hyldig-Nielsen *et al.*, 1982). However, the demonstrated presence of Hb genes in a wide variety of unrelated plant species in addition to legumes argues against a recent transfer of these genes from animals to plants. This view is consistent with the data of Brown *et al.* (1984) who dated the divergence of plant and animal globin genes to about 0.9 to 1.4 billion years ago, prior to the metazoan radiations which gave rise to the plant and animal kingdoms.

All nitrogen-fixing organisms have evolved strategies to protect the nitrogenase enzyme from oxygen which is often required in the same cell for oxidative phosphorylation (Postgate, 1978). In plant-diazotroph symbioses, the oxygen-buffering capacity of Lb appears to be universally utilized to make oxygen consumption and nitrogen fixation physiologically compatible.

2. *Nitrogen-Assimilatory Enzymes*

The ammonia synthesized by the bacteroids is secreted into the host cell cytoplasm where it is assimilated into either amides (asparagine and glutamine) or ureides (allantoin and allantoic acid) for transport to other parts of the plant (reviewed by Miflin and Cullimore, 1984; Schubert, 1986). The synthesis of these nitrogenous intermediates occurs via complex pathways, and a substantial proportion of the enzymatic machinery of the nodule is devoted to the assimilation of fixed nitrogen. Several enzymes involved in amide and ureide biogenesis have been identified in nodules. Some of these enzymes appear to be nodulins, i. e., unique to the nodule whereas others are common to several plant organs but show enhanced activity in nodule tissue. The rationale for the existence of nodule-specific forms of nitrogen assimilatory enzymes is that this tissue functions at a high pH and under low pO_2, thus necessitating the adaptation of these enzymes to the unique nodule environment.

a. *Glutamine Synthetase:* The enzyme, glutamine synthetase (GS), plays a central role in the assimilation of fixed nitrogen in plants. In combination with glutamate synthase, it catalyses the net biosynthesis of glutamate from 2-oxoglutarate and ammonia (reviewed by Miflin and Lea, 1980; 1982). Whereas GS isozymes are widely distributed in different organs and subcellular compartments in plants (Oaks and Hirel, 1985), the unique physiological environment of the nodule might well require the production of unique form of GS.

In *Phaseolus vulgaris*, a nodule-specific form of GS (GS_{n-1}) has been identified (Lara *et al.*, 1983; 1984). The corresponding gene has been cloned (Cullimore *et al.*, 1984) and shown to be expressed only in nodules (Gebhardt *et al.*, 1986). A nodule-specific GS gene has also been characterized in alfalfa (Dunn *et al.*, 1987). Thus, in the nodules of these legume species, the nodule-specific GS isoform may properly be considered a nodulin. In pea, however, it appears that though GS activity is elevated some 20-fold in nodules, compared to roots, the GS forms detectable in the nodule are not unique to that organ but are also present in small amounts in root and leaf tissue (Tingey *et al.*, 1987). Thus in pea, GS is a nodule-stimulated enzyme and strictly speaking, is not a nodulin.

In soybean, Sengupta-Gopalan and Pitas (1986) have shown that, as in *P. vulgaris* and alfalfa, enhanced GS activity in nodules is due to the synthesis of nodule-specific GS polypeptides. These workers used a cDNA clone for a *P. vulgaris* nodule-specific GS subunit to hybrid-select GS transcripts from soybean root and nodule RNA. Hybrid-selected translation products were immunoprecipitated with an anti-GS antibody and then subjected to 2-D gel electrophoresis. This procedure identified four immunoreactive polypeptides which were unique to the nodule. However, Hirel *et al.* (1987) recently found that the increase in soybean nodule GS activity appears to be due to NH_4^+-stimulated expression of GS isoforms common to root. No nodule-specific GS polypeptides were identified by the isoelectrofocusing of hybrid-selected GS mRNA translation products, nor were any nodule-specific clones isolated by extensive screening of a soybean nodule cDNA library with a *P. vulgaris* nodule-specific GS cDNA.

b. *Uricase II:* The second most abundant nodulin in soybean nodules is an approximately 125 kDa protein comprising four 35 kDa subunits identified as nodulin-35 (Legocki and Verma, 1979; Bergmann *et al.*, 1983). Using an immunofluorescent technique, nodulin-35 was shown to be present in the uninfected cells of soybean nodules (Bergmann *et al.*, 1983), and the enzyme was subsequently localized to the peroxisomes of these cells using immunogold labelling (Nguyen *et al.*, 1985; van den Bosch and Newcomb, 1986). Nodulin-35 has been shown to be a nodule-specific uricase, converting uric acid to allantoin in the ureide biosynthetic pathway (Bergmann *et al.*, 1983). A uricase enzyme which catabolizes uric acid in the purine degradation pathway is ubiquitous in living organisms. In soybean, this activity (detectable in uninfected root tissue and leaves) has been designated uricase I (Tayjima and Yomomato, 1975; Bergmann *et al.*, 1983). The antibodies against nodule-specific enzyme, uricase II, showed no immunological cross-reactivity with uricase I nor with any other proteins in uninfected roots and leaves. Confirming the nodule-specificity of uricase II, a nodulin-35 cDNA clone hybridized only to nodule RNA and not to RNA from roots or leaves (Nguyen *et al.*, 1985).

The nodulin-35 gene of soybean comprises almost 5 kb of DNA and contains 7 introns (Nguyen *et al.*, 1985). The processed mRNA encodes a 35,100 M polypeptide. The mRNA is translated on free polysomes, and the protein lacks a signal peptide, suggesting that it is not co-translationally processed into the endoplasmic reticulum but is instead post-translationally directed to the peroxisomes. The mechanism by which nodulin-35 is targeted to the peroxisomes has not been elucidated.

c. *Xanthine Dehydrogenase:* The ureides synthesized in the nodules of most tropical legumes are produced by the oxidative catabolism of purines and, accordingly, the enzymes of purine catabolism occur at high levels in nodules of ureide producers compared to the nodules of amide producers (Schubert, 1986). By analogy with nodulin-35 (uricase II), it might be expected that other enzymes involved in ureide biosynthesis are induced specifically in the nodule. One of these enzymes, xanthine dehydrogenase (XDH), catalyses the oxidation of hypoxanthine to xanthine and then to uric acid under microaerophilic conditions. XDH has recently been shown to be present in high concentrations in the infected cells of soybean nodules (Triplett, 1985). Significantly, soybean XDH showed antigenic cross-reactivity with nodule extracts from two ureide producers, cowpea and lima bean, but not from the amide producers alfalfa and lupin. However, XDH may not strictly speaking be nodule-specific since low levels of the enzyme were immunologically detectable in soybean leaves, roots, stems and pods (Triplett, 1985). The cloning of a legume XDH gene has not yet been reported.

d. *Purine Nucleosidase:* Another enzyme active in ureide biosynthesis, purine nucleosidase, is involved in the catabolism of the purine nucleosides, inosine and xanthosine, to hypoxanthine and xanthine respectively (Schubert, 1986). Purine nucleosidase activity is present in soybean roots and leaves but is elevated in nodules about 15-fold over root levels. Larsen and Jochimsen (1987) have purified the enzyme from soybean leaves, roots and nodules and their data indicates that leaves and roots contain one isozyme of purine nucleosidase (form I), whereas nodules contain, in addition to increased amounts of form I, approximately equal levels of a novel isoform of the enzyme (designated form II). Purine nucleosidase II is apparently a nodule-specific enzyme, though the application of more sensitive techniques is required to confirm that this isoform is indeed absent from other plant organs.

3. Enzymes Involved in Carbon Metabolism

The establishment of an effective symbiosis is an energy-intensive process and requires that abundant supplies of carbon substrates be delivered to the nodule. In addition to serving as the primary nutrients for both plant-derived nodule tissue and the bacteroids, carbohydrates are utilized as substrates in the assimilation of fixed nitrogen, and also provide carbon skeletons for the development of cellulosic structures in cell wall formation. In fact, the nodule constitutes one of the strongest sinks for carbon substrates, tapping up to 30% of the net photosynthate of the plant (Minchin *et al.*, 1981; Mahon, 1983). Thus, a number of enzymes involved in carbon metabolism (e. g., invertase, PEP carboxylase, sucrose synthase) show elevated activities in nodules (Robertson and Farnden, 1980; Verma and Nadler, 1984; Morell and Copeland, 1985), and it is conceivable that some of these enzymes occur in nodule-specific forms.

a. *Sucrose Synthase:* Sucrose is the major carbohydrate transported to nodules from the leaves (Antoniw and Sprent, 1978; Reibach and Streeter, 1983). In the nodules, sucrose is broken down to UDP-glucose and D-fructose in a reversible reaction catalyzed by sucrose synthase. This enzyme, a tetramer of 90,000 M_r subunits, is an abundant protein in the soybean nodule cytosol (Morell and Copeland, 1985). Recently, Thummler and Verma (1987) purified an abundant 90 kDa protein (nodulin-100) from soybean nodules, and assayed it for enzyme activities known to be increased in nodules. This approach identified the purified protein as sucrose synthase. Sequencing of a nodulin-100 cDNA clone isolated earlier (Fuller *et al.*, 1983) indicated 73% homology with the sequence of the maize sucrose synthase gene (Werr *et al.*, 1985), confirming that the nodulin-100 cDNA encodes sucrose synthase. This enzyme is thought to play a key role in controlling the flow of carbon substrates to the nodule, and the inactivation of this enzyme by heme, released from Lb during nodule senescense, has been postulated to shut off the supply of carbon to the bacteroids (see Fig. 1). Such a mechanism may prevent bacteria from becoming pathogenic on the host since the supply of carbon is maintained only for the duration of a fully effective symbiosis.

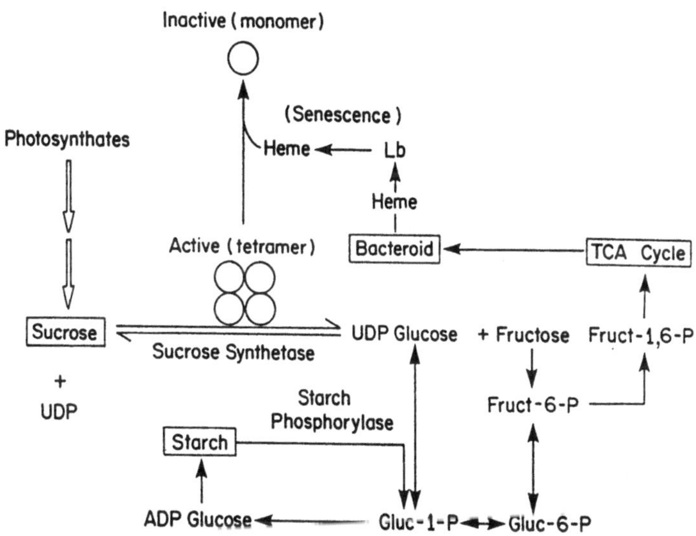

Fig. 1. Central role of sucrose synthase in the metabolism of sucrose in soybean nodules. Sucrose, the major carbon substrate delivered to the nodule, is broken down by sucrose synthase to UDP-glucose and D-fructose which are channeled into the tricarboxylic acid (TCA) cycle. The binding of heme, released from Lb during nodule senescence, is postulated to inactivate the enzyme by causing its dissociation into monomers. Starch accumulates in ineffective nodules when UDP-glucose is not rapidly catabolized via the TCA cycle to generate respiratory substrates for the bacteroids. (From Thummler and Verma, 1987)

The nodule form of sucrose synthase was not detectable in soybean leaves and stems, but an antigenically related protein was present at about 20-fold reduced levels in uninfected roots (Thummler and Verma, 1987). These findings are consistent with the levels of nodulin-100 mRNA detected in leaves and roots (Fuller and Verma, 1984), but it is not known whether the root and nodule enzymes are identical.

4. Nodulins of Unknown Function

Apart from the aforementioned nodulins whose functions have been defined, the functions of the majority of identified nodulins have not been ascertained. In a number of cases, however, structural features of the cloned nodulin genes have suggested possible functions of the corresponding proteins.

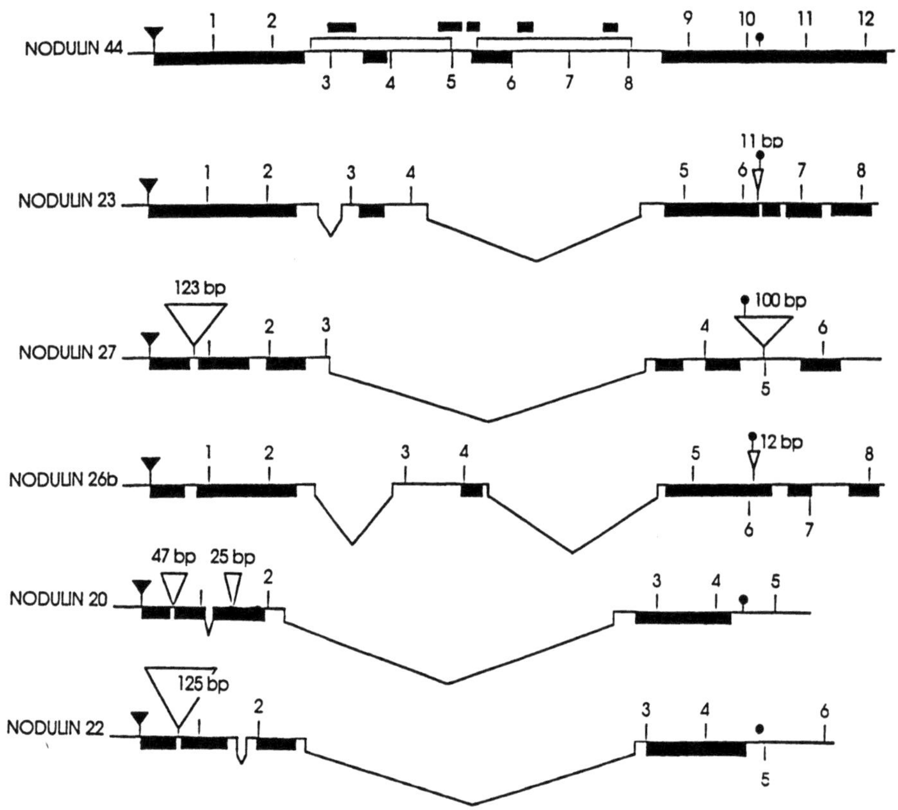

Fig. 2. Schematic comparison of six nodA cDNA sequences. Full-length cDNA sequences of nodulins 23 (Mauro *et al.*, 1985), 26b, 27 (Jacobs *et al.*, 1987), 20 and 22 (Sandal *et al.*, 1987) are aligned for maximum homology with the longest nodA sequence, nodulin-44 (designated "E27" in Sengupta-Gopalan *et al.*, 1986). Regions of >80% homology are indicated by solid bars below the lines, with the translational start and stop codons shown by closed triangles and circles respectively. Repeated elements are shown by open and closed bars above the nodulin-44 sequence, and insertions are indicated by open triangles

a. *The Nodulin-A Family:* One group of nodulin genes, designated the nodA class by Fuller *et al.* (1983), is abundantly expressed in soybean nodules, their transcripts representing about 6% of nodule polysomal mRNAs. This abundant family of nodulins has been studied by several different groups (see Fuller *et al.*, 1983; Sengupta-Gopalan *et al.*, 1986; Sandal *et al.*, 1987). Sequence analyses of four cDNAs encoding nodulin-23, -26b, -27 and -44 revealed the presence of two conserved domains sharing 70—95% homology, separated by a third domain unique to each cDNA (Jacobs *et al.*, 1987, see Fig. 2). Sandal *et al.* (1987) who compared the sequences for nodulin-23 and -44 with those for two other related nodulins, designated nodulin-20 and -22, found three domains conserved among the four nodulin genes, with highly divergent sequences separating the homologous regions. However, on closer examination it is clear that the 5′-terminal homologous domain identified by Jacobs *et al.* (1987) has been divided into two separate domains by insertions in the sequences of nodulin-20 and -22 (Sandal *et al.*, 1987; see Fig. 2).

By immunoprecipitating the hybrid-selected translation products of nodA cDNAs with antisera raised against pbm and soluble proteins, nodulin-23 and nodulin-27 were localized to the pbm and the nodule cytosolic compartments respectively (Jacobs *et al.*, 1987). The presence or absence of putatively functional signal sequences on the cDNA clones for nodulin-44 and nodulin-26b (see Fig. 3) suggested that nodulin-44 was membrane associated, whereas nodulin-26b was a soluble protein. Though the functions of the nodA proteins are unknown, Sandal *et al.* (1987) have noted that the structurally conserved domains all contain four cysteine residues which could be arranged in a configuration typical of metal-binding domains, implying that the binding of metal ions is important in the functioning of these proteins.

```
                                                  ⇩         ↓
Nodulin-23 (C51)        M E K M R V I V I T V F L F I G A A I A E D V G I G L L S
Nodulin-44 (E27)  M E E K I L M R V I V I T V F L F I G A A T A E D A A A E A
                                                                    ↓
Nodulin-27              M E K M R V V L I T L L L F I G A A V A E K A G N G K A A
Nodulin-26b                         M L I T L F L F I A A T V A E D A D N I G E A
Nodulin-20                    M R V V L I T L F L F I G A A V A E D A G I D A I T
Nodulin-22              M E K M R V V L I T L F L F I G A A V A E K A G N G K A A
```

Fig. 3. Alignment of the N-terminal signal sequences of the nodA proteins. The open arrow indicates the putative signal peptide cleavage site proposed by Sandal *et al.* (1987). This site is well conserved in the six nodulins suggesting that they are all similarly processed through the ER. However, the demonstration that some nodA proteins are pbm-associated whilst others are cytosolic led to the proposal that nodulins 23 and 44 have functional signal peptides, whereas nodulins 26b and 27 do not (see Jacobs *et al.*, 1987, for details). The putative cleavage sites in nodulins 23 and 44 suggested by Jacobs *et al.*, (1987) are indicated by solid arrows

b. *Peribacteroid Membrane (pbm) Nodulins:* The pbm which encloses the bacteroids inside the nodule plays a crucial role in regulating the flux of metabolites between the host plant and the endosymbiont. Thus, although derived from the plasma membrane, the pbm undergoes significant modifications to enable it to meet this unique structural and functional role. Not surprisingly, a number of the proteins associated with the pbm are unique to the nodule (Verma *et al.*, 1978; Fortin *et al.*, 1985; see for review Verma and Fortin, 1988).

Nodulin-24 was the first nodulin found to be localized in the pbm of soybean nodules (Katinakis and Verma, 1985; Fortin *et al.*, 1985). Sequence analysis of the nodulin-24 gene revealed a coding capacity of only 15,000 kDa, in contrast to the apparent M_r of 24,000 estimated by SDS-PAGE (Katinakis and Verma, 1985). The protein is extremely hydrophobic, which probably explains the discrepancy between its predicted and apparent M_r. Nodulin-24 contains three repeated hydrophobic domains comprising 18 amino acids each. Based on the hydrophobicity of these repeats, it was initially proposed that nodulin-24 may be a transmembrane protein (Katinakis and Verma, 1985). However, more detailed analysis of nodulin-24, including its possible secondary structure, revealed that the repeated domains could be folded into a markedly ampiphilic α-helix. Thus, nodulin-24 appears to be located on the inner surface of the pbm with the hydrophobic half of the α-helix embedded in the membrane and the hydrophilic half in the peribacteroid fluid (Fortin *et al.*, 1987).

Structural analysis of the nodulin-24 gene revealed the presence of five exons separated by four introns. The repeated amino acid domains encoded by the three central exons appear to have arisen by tandem duplication of a unit resembling an insertion sequence (Katinakis and Verma, 1985).

The characterization of another pbm nodulin, nodulin-26, was recently reported (Fortin *et al.*, 1987). This protein has an apparent M_r of 26,500 as estimated by SDS-PAGE; however the nodulin-26 cDNA clone encodes a polypeptide of only 22,500 M_r. Like nodulin-24, the discrepancy in real and apparent M_r may be due to the great hydrophobicity of nodulin-26. Based on secondary structure calculations, it was postulated that nodulin-26 is a transmembrane protein (Fortin *et al.*, 1987), which suggests that it may serve a transport function. Other characterized pbm-associated nodulins include nodulin-23 and possibly nodulin-44 (discussed above). However, the functions of these nodulins have not been deduced.

The targeting mechanism by which pbm nodulins are specifically sequestered into the pbm as opposed to the plasma membrane is not known. A pbm-specific signal sequence is apparently not involved since the pbm nodulins 23, 24 and 26, share no amino acid sequence homology, nor are there any significant similarities in their secondary structures (Fortin *et al.*, 1987). The possibility of targeting being mediated by protein glycosylation may be excluded since the signal sequence for N-glycosylation, N-X-T/S, is absent from nodulin-24. It is possible that pbm nodulins become membrane associated simply on the basis of their great hydrophobicity and the differential turnover of the pbm and the plasma membrane. It has been estimated that there is 20—40 times more pbm than plasma membrane in the infected cells of nodules (Verma *et al.*, 1978).

c. *Early Nodulins:* Several nodulin genes are expressed in the early stages of nodule morphogenesis, well before the onset of nitrogen fixation. A number of these co-called "early nodulins", identified in soybean nodules, have been categorized as class A nodulins by Gloudemans *et al.* (1987); in their classification scheme, other nodulins induced approximately concomitantly with the commencement of nitrogenase activity are designated class B nodulins.

One of the early nodulins from soybean has been studied in detail. Franssen *et al.* (1987) have isolated a cDNA clone for nodulin-75, one of two related early nodulins of apparent M_r 75,000. The cDNA clone hybridizes to a 1200 nucleotide mRNA species which has a maximum coding capacity for a 45,000 M_r protein. The discrepancy between the observed M_r and the predicted maximum M_r of nodulin-75 is very likely explained by the unusually high (45%) proline content of nodulin-75. This nodulin contains 20 repeated units of an 11-amino-acid sequence within which is a highly conserved heptapeptide sequence, P-P-X-E-K-P-P. Though nodulin-75 does not resemble any previously characterized plant hydroxyproline-rich glycoprotein, Franssen *et al.* (1987) postulate that nodulin-75 is a structural protein of the cell wall.

Early nodulins could conceivably be involved either in the early stages of nodule morphogenesis or in mediating the rhizobial infection process. Based on results indicating that nodulin-75 (and the pea homologue) is expressed in mutant-bacteria-elicited pseudonodules devoid of infection threads or internalized bacteria, Franssen *et al.* (1987) concluded that nodulin-75 is not involved in the infection process but rather in nodule morphogenesis.

d. *"Nodulin-25" of Alfalfa:* Very few nodulins have been characterized at the gene level in species other than soybean. Kiss *et al.* (1987) have recently reported preliminary data on an alfalfa nodulin gene originally designated nodulin-1. However, based on the M_r of the encoded polypeptide and the suggested nomenclature system for nodulins (van Kammen, 1984), nodulin-1 has since been redesignated nodulin-25 (G. Kiss, personal communication). An incomplete cDNA clone revealed two types of repeats, RA and RB. The RA repeats, present in two copies, coded for 12 amino acids separated by a region of 22 amino acids. The RB repeats comprised three tandem copies of a sequence coding for an 18 amino acid unit, reminiscent of the 18-amino-acid repeats observed in soybean nodulin-24 (Katinakis and Verma, 1985), although there is no homology between the nodulin-25 and nodulin-24 repeats. Sequencing of a nodulin-25 genomic clone revealed that the coding region of 738 bp was interrupted by 12 introns (G. Kiss, personal communication). The high hydrophilicity of the nodulin-25 sequence suggests a cytosolic location, but its function is unknown.

B. Regulation of Nodulin Gene Expression

Nodule morphogenesis is mediated by a highly coordinated series of changes in gene expression in both symbiotic partners, leading to the accumulation of various nodulins and bacteroidins at various phases of the differentiation process. The availability of a variety of *Rhizobium* mutants, defective at different stages of symbiosis (see Long, 1984, for a review), has provided an important tool for dissecting the hierarchy of developmental signals involved in nodulin gene expression. Figure 4 summarizes the classification of nodulin genes based on their possible mode of induction.

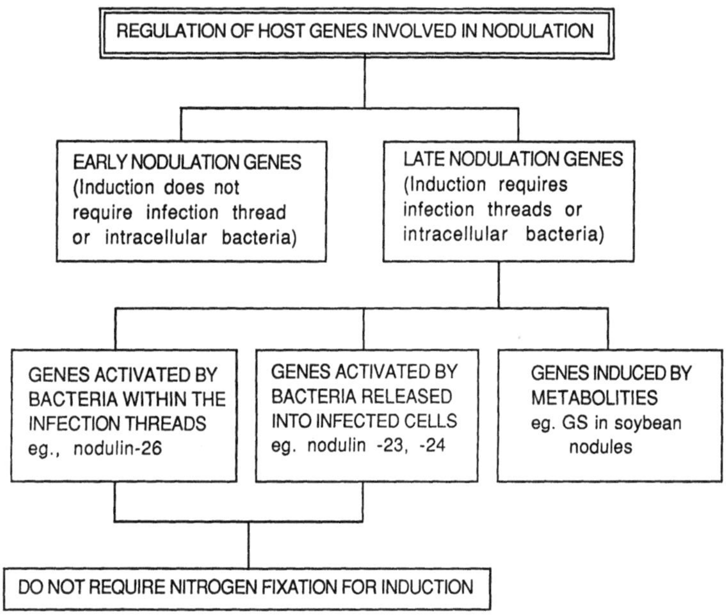

Fig. 4. Regulation of host genes induced during nodule development

1. Induction of Early Nodulin Genes Does Not Require Intracellular Bacteria or Infection Threads

Nodulation by compatible rhizobia is initiated by an exchange of regulatory signals between the two organisms. Legume-specific flavones released in root exudates induce the expression of host-specific and "common" nodulation (*nod*) genes of *Rhizobium* (Redmond *et al.*, 1986; Peters *et al.*, 1986; Firmin *et al.*, 1986). Conversely, it has recently been demonstrated that some flavones, identified in root-tip exudates, are inhibitory to *nod* gene induction which may be related to the root tip being refractory to infection (Djordjevic *et al.*, 1987). The expression of the *Rhizobium nod* genes triggers the induction of certain plant genes involved in the early phases of nodule morphogenesis. This "ping-pong" mechanism of gene regulation whereby signals from one symbiont (in this case, plant flavones) stimulate the expression of genes in its partner (the *nod* genes), which in turn induces gene expression in the other partner (early nodulin genes and the TSR (thick short root) phenotype, Van Brussel *et al.*, 1986) appears to be a recurrent theme in symbiotic gene control.

Induction of cortical root cell division and expression of early nodulin genes do not require the presence of internalized bacteria or infection threads. This is illustrated by the observation that an *Agrobacterium* strain harbouring the *R. leguminosarum sym* plasmid induced bacteria-free ("empty") pseudonodules on *Pisum sativum* in which the pea homologue(s) of the soybean early nodulin, nodulin-75, was expressed (Govers *et al.*, 1986a). The *Agrobacterium* strain lacking the *Rhizobium sym* plasmid did not itself elicit expression of the nodulin-75 homologue. Nodulin-75 was also expressed in empty soybean pseudonodules formed by *R. fredii* USDA 257 (Franssen *et al.*, 1987).

Similar results have been reported in alfalfa. Six early nodulins, normally first detectable when nodules are just visible, were expressed in pseudonodules induced by exopolysaccharide-deficient *R. meliloti* mutants (Lang-Unnasch *et al.*, 1985). Other nodulins identified in wild-type alfalfa nodules were not expressed in empty pseudonodules. Surprisingly, it was observed that these pseudonodules contained the nodule-specific form of GS which is normally expressed relatively late in nodule development, concomitantly with the commencement of nitrogenase activity. However, a subsequent report (Dunn *et al.*, 1987) indicated that no nodule-specific GS genes were expressed in empty pseudonodules.

2. *Expression of Late Nodulins Requires Infection Threads or Intracellular Bacteria*

Unlike the early nodulins whose expression is activated at a distance directly or indirectly by the *Rhizobium nod* gene products, the expression of late nodulins (i. e., those expressed just prior to or concomitantly with the onset of nitrogen fixation) appear to require regulatory signals from internalized bacteria. However, even among this group of nodulins it is apparent that different types of signals are involved.

Using two different *B. japonicum* mutants which had suffered deletions in the *nif* regions, Gloudemans *et al.* (1987) found that both types of fix^- mutant nodules contained the full complement of nodulins found in wild-type soybean nodules. Another fix^- *B. japonicum* mutant, T5-95, used by Morrison and Verma (1987) formed morphologically normal nodules (see Fig. 5), which contained near normal levels of six out of eight late nodulins assayed, but transcription of Lb and nodulin-35 genes was depressed by about half the level. All eight late nodulins were however absent from the empty pseudonodules produced by a *R. fredii* strain (Morrison and Verma, 1987).

An interesting class of nod^+ fix^- mutants form nodules which contain numerous infection threads but from which the bacteria are not released. In one such mutant, *B. japonicum* HS124, used by Gloudemans *et al.* (1987), the few nodule cells which do become infected appear to collapse in the center of the nodule. Analysis of nodulin gene expression in

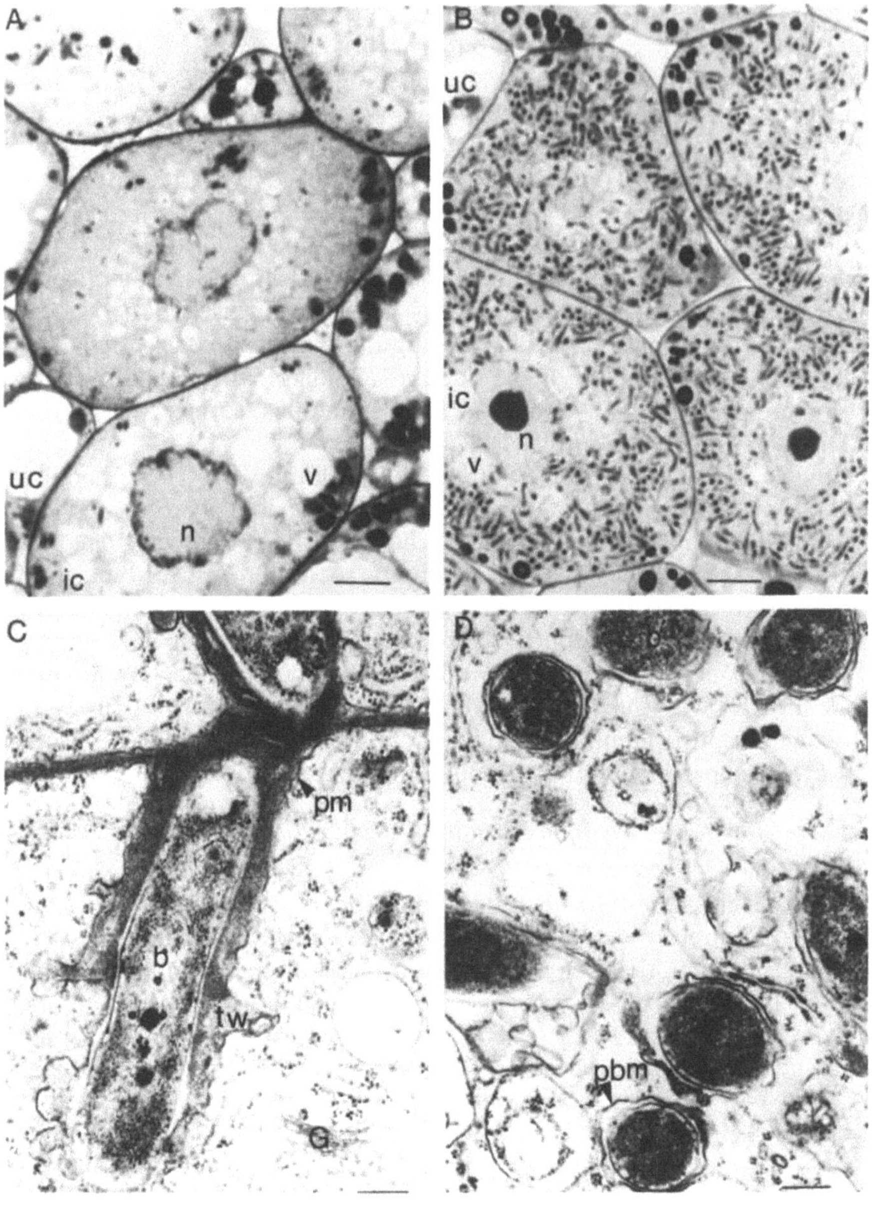

Fig. 5. Light (A and B) and electron (C and D) micrographs of soybean nodules elicited by the *Rhizobium* mutants T8-1 and T5-95. Nodules induced by T8-1 (A) contain normal-size, fully differential cell types but few, if any, bacteroids whereas T5-95 nodules (B) appear morphologically normal. Electron microscopy of T8-1 nodules (C) shows numerous profiles of vesicles fusing to the tip of the infection thread. T5-95 nodules (D) show abundant bacteroids enclosed in peribacteroid membranes with an essentially normal ultrastructure. Both types of nodules contain prominent amyloplasts which are characteristic of ineffective symbiosis. The bar represents 10 μm in A and B, and 1 μm in C and D. uc, uninfected cells; ic, infected cells; n, nucleus; v, vacuole, b, bacteroid; pm, plasma membrane; pbm, peribacteroid membrane; tw, thread wall

HS124-induced nodules revealed that whereas all the early nodulins were present, only half of the late nodulins were expressed. In a similar investigation by Sengupta-Gopalan *et al.* (1986), eight out of nine nodulin mRNAs assayed were present in HS124 nodules, though at generally reduced levels. One nodulin transcript was completely absent from the mutant nodules. These data suggest that, at the very minimum, two different signals originating from the invading bacteria are required for the activation of late nodulin genes. One signal is apparently transmissible from bacteria within the infection threads, whereas transmission of the other is dependent upon the release of bacteria into the infected cells. Further evidence in support of this hypothesis has recently been obtained by Morrison and Verma (1987) using the *B. japonicum* mutant, T8-1 (Rostas *et al.*, 1984). This mutant, like HS124 induces nodules with infection threads that fail to release the bacteriods (Fig. 5) and the process appears to be blocked at the level of endocytosis. It was found that transcription of several late nodulin genes, including genes for two pbm-associated nodulins, was strongly depressed. However, another pbm nodulin, nodulin-26, was expressed at essentially normal levels. These results indicate that the induction of pbm nodulin genes may require at least two signals.

3. Some Nodulins May Require Nitrogen Fixation for Induction

Most nodulin genes characterized to date are induced prior to and independent of the commencement of nitrogen fixation in nodules. The fact that these genes are also expressed in a variety of fix$^-$ nodules, suggests that they are induced by developmental signals independently of nitrogenase activity. However, the levels of nodulin gene transcripts in fix$^-$ nodules are frequently reduced to varying degrees (Sengupta-Gopalan *et al.*, 1986; Govers *et al.*, 1987; Morrison and Verma, 1987; Dunn *et al.*, 1987). This suggests that although nitrogen fixation is not a primary signal for activating most nodulin genes, the products of nitrogen fixation may have a regulatory role in maintaining the expression of these genes. It is also conceivable that the expression of certain nodulins is absolutely dependent upon an effective symbiosis.

Further evidence implicating nitrogen fixation in the triggering of specific gene expression in nodules comes from a recent study by Hirel *et al.* (1987) on the induction of GS genes in soybean nodules. This study suggested the elevated transcription of GS genes in nodules to be due to the high local concentration of ammonium. It was shown that the stimulation of GS expression could be fully reproduced in uninfected soybean roots by the application of external ammonium but not by nitrate. Conversely, nodule GS activity remained at basal (root) levels when nitrogen fixation was prevented by growing plants in an argon environment or by infecting them with a *B. japonicum* fix$^-$ mutant. However, Sengupta-Gopalan and Pitas (1986) have identified GS polypeptides induced specifically in the soybean nodule. They also found that the corresponding GS genes were induced at normal levels in nodules elicited by a *B. japonicum* fix$^-$ mutant.

Similarly, the expression of nodule-specific GS genes in alfalfa has been shown to be mostly developmentally regulated rather than being substrate-induced (Dunn *et al.*, 1987). Nodule-specific GS mRNA was found to occur at about 50% of wild-type levels in a variety of fix$^-$ *R. meliloti*-elicited nodules containing differentiated bacteroids. It is evident from these studies that the host plant may either produce nodule-specific forms of GS or elevate the level of constitutive GS in nodules in response to ammonia.

4. *Nodulin Gene Regulation at the Molecular Level: cis-Regulatory Sequences*

The use of symbiotically defective *Rhizobium* mutants has helped to delineate the stages of nodule development at which bacterial signals activate nodulin gene expression. Recently, studies have been directed at elucidating the molecular mechanisms of nodulin induction at the level of the gene.

In eucaryotes, coordinate expression of many genes is mediated by the interaction of a common *trans*-acting factor with conserved sequences in the 5′ promoter regions of the genes (see Davidson *et al.*, 1983). As increasing numbers of nodulin genes were cloned and characterized, their promoter regions were searched for conserved sequences which could conceivably serve as *cis*-regulatory motifs for nodulin gene induction. Analysis of the synchronously expressed genes for soybean Lbc_3, nodulin-23 and nodulin-24 revealed three blocks of conserved sequences in the 5′ flanking regions (Mauro *et al.*, 1985). The statistical probability that these conserved sequences occurred by chance were calculated to be vanishingly small, suggesting that they played a role in controlling the expression of these nodulins. In support of this hypothesis, the gene for nodulin-35 which is expressed in a different cell type (uninfected cells) at a later stage of nodule development did not contain the conserved motifs (Verma *et al.*, 1986). Sandal *et al.* (1987) have also identified conserved 5′ sequences common to the Lb genes of soybean and the non-legume, *Parasponia,* as well as four soybean nodulin genes coordinately induced with the Lb genes. However, these conserved sequences comprise only 5—6 nucleotides each, and the statistical significance of their occurrence is not clear.

To test the functionality of the conserved sequences, Mauro and Verma (1988) studied the *in vitro* transcription of various genes in isolated embryo nuclei with *E. coli* RNA polymerase. An actin and a root gene were found to be transcribed in this system, but several nodulin genes were not transcribed. However, when the nuclei were pretreated with nodule extract, transcription of the Lb, nodulin-23 and nodulin-24 genes was induced with similar kinetics, supporting the hypothesis of a common *trans*-activating factor. Induction was observed only with nodule-extract pretreatment and not with leaf or root extract. Significantly, the nodulin-35 gene was induced with different kinetics in the pretreated nuclei. Further evidence of a common *trans*-activation was provided by the demonstration that a protein factor present in nodule extract bound to the 5′ flanking region of the nodulin-23 gene. This binding was competitively inhibited by promoter sequences of nodulins 23 and 24 and Lbc_3 but not by sequences from a nodulin-35 gene, a soybean actin gene, nor a soybean root-specific gene (Mauro and Verma, unpublished observations).

While the *in vitro* activation experiments gave results consistent with the situation in nodules, these assays measure the accessibility of chromatin to an *E. coli* polymerase and cannot therefore be expected to accurately reflect the intricacies of the *in vivo* regulation of nodulin genes. More revealing insights into nodulin gene induction may be obtained by the transfer to appropriate legume hosts of putative regulatory gene sequences linked to a reporter gene whose activity can be easily monitored in the transgenic plants. Investigations along this line have been reported by Jensen *et al.* (1986b) who transferred a chimeric Lb-chloramphenicol acetyltransferase gene to *Lotus corniculatus* and found that the soybean Lb promoter was activated with the same tissue and temporal specificity as in soybean. The Lb 3′ terminator sequences were used in the chimeric construct transferred to *Lotus*, and it is possible that this 3′ flanking region contained sequences involved in gene regulation (cf. Thornburg *et al.*, 1987). However, as with most other plant genes studied in transgenic systems (e. g. Herrera-Estrella *et al.*, 1984; Jones *et al.*, 1985; Kaulen *et al.*, 1986; Fluhr *et al.*, 1986; Baumannn *et al.*, 1987), it is likely that the sequences responsible for nodule-specific expression are contained within the 5′ promoter region. Further studies using constructs with deleted promoter fragments are required to identify more precisely the *cis*-acting regulatory sequences in the Lb promoter and in the nodulin genes.

5. *Physiological Factors Involved in Induction of Nodulin Genes*

The specialized metabolism of the nodule creates an environment with an unique physiology (e. g., microaerophilic and alkaline conditions). Components of these unique physiological conditions may conceivably serve as signals for the regulation of nodulin genes. Govers *et al.* (1986b) found that none of the known pea nodulins were induced by microaerophyllic conditions. However, Larsen and Jochimsen (1987), measuring the levels in soybean callus tissue of various nodule enzymes in response to lowered-oxygen concentrations, found that whereas the expression of 5′-nucleotidase, XDH, allantoinase, GS and aspartate aminotransferase was not influenced by variations in oxygen concentration, the activity of two other enzymes increased with lowered oxygen levels. The activity of purine nucleosidase increased four-fold when the oxygen concentration was reduced to around 2 %, while uricase II activity increased two-fold, apparently due to increased *de novo* enzyme synthesis in tissue exposed to 4 – 5 % oxygen (Larsen and Jochimsen, 1986). These levels are still much lower than in mature fix^+ soybean nodules. Thus it is uncertain to what extent microaerophillic condition plays a role in activating these genes *in vivo*, especially considering that the microaerophillic environment is expected to be more pronounced in the infected cells of the nodule, whereas uricase II is expressed exclusively in the uninfected cells (Nguyen *et al.*, 1985).

It may also be noted that two different sucrose synthase genes in maize are induced under anaerobic conditions (McCarty *et al.*, 1986). This raises the possibility that the nodule form of sucrose synthase in soybean, nodulin-100, is similarly induced by the microaerobic environment of the nodule. However, this possibility has not yet been investigated.

The unique character of nodule tissue is also illustrated by the accumulation of certain metabolites in high concentrations. One such metabolite, ammonia, has been discussed above as an inducer of GS activity in soybean nodules. Another nodule metabolite potentially involved in nodulin gene regulation is heme. Jensen *et al.* (1986a) have reported that in *Saccharomyces cerevisiae*, heme regulates the expression of chimeric antibiotic resistance genes containing the 5′ untranslated region of the Lbc_3 mRNA. This effect is apparently exerted at the post-transcriptional level, probably by increasing the translational efficiency of the chimeric mRNAs. It seems plausible that heme plays a similar role in Lb mRNA translation *in vitro* in nodules, but no effect was found on *in vitro* translation of this mRNA in the presence of heme (Verma *et al.*, 1974).

Heme has also been implicated in the regulation of nodulin-100 (sucrose synthase) activity (Thummler and Verma, 1987; see Fig. 1). Active sucrose synthase is a tetrameric protein with an affinity for heme. Binding of heme rendered the enzyme non-functional by causing the dissociation of the tetramer into inactive monomers. That this does not normally occur in an effective nodule is due to the much higher affinity for heme exhibited by apo-Lb. However, during nodule senescence, heme released by the degradation of Lb may bind to and inactive sucrose synthase, thus cutting off the flow of carbon substrates to the bacteroids. It has been shown that sucrose synthase activity is indeed reduced in extracts from senescing nodules (Thummler and Verma, 1987).

IV. Rapid Evolution of Legume-*Rhizobium* Symbiosis

Two contrasting scenarios may be envisaged to account for the origin of the legume-*Rhizobium* symbiosis (see also Verma and Nadler, 1984). On the one hand, it is possible that an ancestral species of present-day *Rhizobium*, lacking the capability for nitrogen fixation, first established a pathogenic relationship with leguminous plants. The fact that the closely related *Agrobacterium* species are plant pathogens lends credibility to this scenario. Furthermore, certain *Rhizobium* strains induce nodules which fix nitrogen inefficiently or not at all. Such strains may be regarded as essentially pathogenic since they utilize the plant's resources while providing little reciprocal benefit to the host. Pathogenic rhizobia may have learned to breach the plant's frontline defenses while the plant, for its part, may have contained the infection by encapsulating the bacteria within membranous compartments and subsequently concentrating a counter-attack on the entrapped bacteria. Interestingly, it has been observed that in certain ineffective soybean nodules, the release of bacteroids directly into the host cytoplasm caused by premature disruption of the pbm provokes a stringent defensive response from the plant (Werner *et al.*, 1984; 1985). At a later stage in evolution, the *Rhizobium* progenitor may have acquired plasmid-borne genes confering the ability to fix nitrogen under microaerobic conditions. The potential of the "pathogen" to contribute reduced nitrogen to the plant would then have stimulated the modification of the pathogenic association towards a progressively more complex symbiotic relationship from which both partners benefited.

The alternative scenario supposes that the *Rhizobium* ancestor was one of a number of bacterial diazotrophs which existed before the evolution of higher plants. Such an organism might have used hydrogen as a reductant for carbon dioxide and nitrogen fixation, a capacity still retained by certain rhizobia (Lepo *et al.*, 1980). The high energy investment required for nitrogen fixation may have prompted the rhizobia to tap the plant's resources for respiratory substrates and reductant, "paying" for these commodities with fixed nitrogen (see also Verma and Nadler, 1984).

Once the fundamental foundations for a symbiotic relationship had been established, co-evolution of both the host and endosymbiont ensued (Verma and Stanley, 1987), leading to the elaboration of the structural and metabolic machinery necessary for a mutually beneficial association. There are suggestions from studies of the structure and expression of nodulin genes that the symbiotic apparatus is still subject to a process of dynamic evolution. Analysis of the nodA gene family revealed the existence of several closely related genes which were postulated to have recently evolved from a common ancestor (Jacobs *et al.*, 1987). Some members of the family are soluble nodulins, whilst others are apparently integrated into the pbm. Based on their observations of high sequence diversity and different subcellular locations of the nodA proteins, Jacobs *et al.* (1987) proposed that these proteins represented possible candidates for nodule-specific functions, none of which had yet been optimally configured, and that a dynamic process of evolutionary selection was still in progress. The regulation of GS expression in nodules of different legumes may also be a reflection of different evolutionary stages reached in these legumes. In certain legumes (e. g., alfalfa and *P. vulgaris*), a nodule-specific form of GS has been identified. In alfalfa, induction of the nodule-specific GS gene is mediated by a developmental signal and not by the presence of fixed nitrogen (Dunn *et al.*, 1987). In pea and possibly in soybean, on the other hand, the GS isozymes expressed in nodules are common to root tissue, and in soybean, GS induction in the nodule may be in direct response to the accumulation of fixed nitrogen (Hirel *et al.*, 1987). In the early phases of symbiotic evolution, native plant genes were no doubt recruited to serve nodule-specific roles. The expression of pre-existing GS genes in the plant may have been stimulated by the ammonia produced by the bacteroids.

Subsequently, certain GS isozymes may have undergone modifications to make them better suited to the special physiological conditions of the evolving nodule. Once a gene for a uniquely nodule form of the enzyme had evolved, expression of that gene may then have been brought within the regulatory mechanisms of nodule differentiation. The uricase II gene in soybean is an excellent example of this, since the normal uricase activity present in roots is reduced and a different uricase activity is induced in nodules. This enzyme has no immunological cross-reactivity with the root enzyme.

In mature nodules, the bacteroids exist and function essentially as cellular organelles, but are capable of an independent existence. An intriguing question (addressed by Verma and Long, 1983; Verma and Nadler, 1984) is whether *Rhizobium* in evolving towards an obligatory organellar existence or whether the physiological constraints on nitrogen fixation will restrict the development of an association significantly more intimate than the present symbiotic conditions.

A major physiological constraint on symbiotic nitrogen fixation stems from the very nature of the nitrogenase enzyme. Nitrogenase needs micro-aerobic conditions to function, a requirement essentially incompatible with that for oxidative phosphorylation. This dilemma has been resolved by compartmentalizing the bacteria within the infected cells of the nodule and utilizing Lb to deliver oxygen to the bacteroids at high flux but low partial pressures. Other reactions involved in the assimilation of fixed nitrogen have also had to be compartmentalized depending on their need for, or lability to, free oxygen (see Schubert, 1986). It seems unlikely that the requirements for this degree of compartmentalization could be satisfied in other plant organs. Other symbiotic diazotrophs such as the actinomycete *Frankia* and the cyanobacterium *Nostoc,* totally unrelated to *Rhizobium,* are also confined to separate organs in their associations with plant hosts (see Elmerich, 1984).

The containment of *Rhizobium* within specialized organs removed from the reproductive organs precludes the transmission of the bacteroids through the gametes, therefore placing a strong evolutionary pressure on *Rhizobium* to retain its capacity for independent growth. From the plant's perspective, the high energy cost of symbiotic nitrogen fixation does not favor an obligatory dependence on *Rhizobium* for reduced nitrogen. However, there is presumably much scope for further refinements of the present symbiotic association, in terms of more efficient nitrogen fixation by *Rhizobium,* improved selection of efficient strains, and more efficient assimilation of fixed nitrogen by the host. A detailed understanding of the molecular mechanisms underlying these processes may enable these refinements to be brought about by man rather than by natural evolution.

References

Antoniw, L. D., Sprent, J. I., 1978: Primary metabolites of *Phaseolus vulgaris* nodules. Phytochem. **17,** 675—678.

Appleby, C. A., 1984: Leghemoglobin and *Rhizobium* respiration. Ann. Rev. Plant Physiol. **35,** 443—478.

Appleby, C. A., Tjepkema, J. D., Trinick, M. J., 1984: Hemoglobin in a non-leguminous plant, *Parasponia.* Possible genetic origin and function in nitrogen fixation. Science **220,** 951—953,

Auger, S., Verma, D. P. S., 1981: Induction and expression of "nodule-specific" host genes in effective and ineffective root nodules of soybean. Biochemistry **20,** 1300—1306.

Bauer, W. D., 1981: Infection of legumes by rhizobia. Ann. Rev. Plant Physiol. **32,** 407—449.

Baumann, G., Raschke, E., Bevan, M., Schoffl, F., 1987: Functional analysis of sequences required for transcriptional activation of a soybean heat shock gene in transgenic tobacco plants. EMBO J. **6,** 1161—1166.

Bergersen, F. J., Appleby, C. A., 1981: Leghemoglobin within bacteroid-enclosing membrane envelopes from soybean root nodules. Planta **152,** 534—543.

Bergmann, H., Preddie, E., Verma, D. P. S., 1983: Nodulin-35: A subunit of specific uricase (uricase II) localized in uninfected cells of nodules. EMBO J. **2,** 2333—2339.

Bisseling, T., Been, C., Klugkist, J., Van Kammen, A., Nadler, K., 1983: Nodule-specific host proteins in effective and ineffective root nodules of *Pisum sativum.* EMBO J. **2,** 961—996.

Bisseling, T., Van den Bos, R. C., Van Kammen, A., 1986: Host-specific gene expression in legume root nodules. In: Broughton, W. J., Puhler, S. (eds.), Nitrogen Fixation: Vol. 4, Molecular Biology, pp. 280—312. Oxford: Clarendon Press.

Bojsen, K., Abiladsten, D., Jensen, E. O., Paluden, K., Marcker, K. A., 1983: The chromosomal arrangement of six soybean leghemoglobin genes. The EMBO J. **2,** 1165—1168.

Brisson, N., Verma, D. P. S., 1982: Soybean leghemoglobin gene family: Normal, pseudo and truncated genes. Proc. Natl. Acad. Sci. U.S.A. **79,** 4055—4059.

Brown, G. G., Lee, J. S., Brisson, N., Verma, D. P. S., 1984: The evolution of a plant globin gene family. J. Mol. Evol. **21,** 19—32.

Campos, F., Vazquez, M., Padilla, J., Enriguez, C., Sanchez, F., 1987: Nodule-specific genes in *Phaseolus vulgaris.* In: Verma, D. P. S., Brisson, N. (eds.), Molecular Genetics of Plant-Microbe Interactions, pp. 115—117. Dordrecht: Martinus Nijhoff.

Cullimore, J. V., Gebhardt, C., Saarelainen, R., Miflin, B. J., Idler, K. B., Barker, R. F., 1984: Glutamine synthetase of *Phaseolus vulgaris* L.: Organ specific expresion of a multigene family. J. Mol. Appl. Genet. **2,** 589—599.

Currier, W. W., Strobel, G. A., 1976: Chemotaxis of *Rhizobium* spp. to plant root exudates. Plant Physiol. **57,** 820—823.

Davidson, E. H., Jacobs, H. T., Britten, R. J., 1983: Very short repeats and coordinate induction of genes. Nature **301,** 468—470.

Dazzo, F. B., Gardiol, A. E., 1984: Host-specificity in *Rhizobium*-legume interactions. In: Verma, D. P. S., Hohn, T. H. (eds.), Genes Involved in Microbe-Plant Interactions, pp. 3—31. Wien - New York: Springer-Verlag.

De Lajudie, P., Huguet, T., 1987: Plant gene expression during effective and ineffective nodule development of the tropical stem-nodulated legume *Sesbania rostrata*. In: Verma, D. P. S., Brisson, N. (eds.), Molecular Genetics of Plant-Microbe Interactions, pp. 130—132. Dordrecht: Martinus Nijhoff.

Djordjevic, M. A., Redmond, J. W., Batley, M., Rolfe, B. G., 1987: Clovers secrete specific phenolic compounds which either stimulate or repress nod gene expression in *Rhizobium trifolii*. EMBO J. **6,** 1173—1179.

Dobereiner, J., Boddey, R. M., 1981: Nitrogen fixation in association with Gramineae. In: Gibson, A. H., Newton, W. E. (eds.), Current Perspectives in Nitrogen Fixation. Australian Academy of Science, Canberra.

Dunn, K., Peterman, T. K., Burnett, B. K., Goodman, H. M., Ausubel, F. M., 1987: Expression of nodule specific glutamine synthetase in alfalfa root nodules does not require bacteroid production of ammonia. Molec. Gen. Genet. (in press).

Elmerich, C., 1984: Molecular biology and ecology of diazotrophs associated with non-leguminous plants. Biotechnology **2,** 867—978.

Firmin, J. L., Rossen, L., Johnston, A. W. B., 1986: Flavonoid activation of nodulation genes in *Rhizobium* reversed by other compounds present in plants. Nature **324,** 90—92.

Fluhr, R., Kuhlemeier, C., Nagy, F., Chua, N.-H., 1986: Organ-specific and light-induced expression of plant genes. Science **232,** 1106—1112.

Fortin, M. G., Morrison, N. A., Verma, D. P. S., 1987: Nodulin-26, a peribacteroid membrane nodulin is expressed independently of the development of the peribacteroid compartment. Nucl. Acids Res. **15,** 813—824.

Fortin, M. G., Zelechowska, M., Verma, D. P. S., 1985: Specific targeting of membrane nodulins to the bacteroid-enclosing compartment in soybean nodules. EMBO J. **4,** 3041—3046.

Franssen, H. J., Nap, J.-P., Gloudemans, T., Stiekema, W., Van Dam, H., Govers, F., Louwerse, J., Van Kammen, A., Bisseling, T., 1987: Characterization of cDNA for nodulin-75 of soybean: a gene product involved in early stages of root nodule development. Proc. Natl. Acad. Sci. U. S. A. **84,** 4495—4499.

Fuchsman, W. H., Appleby, C. A., 1979: Separation and determination of the relative concentrations of the homogeneous components of soybean leghemoglobin by isoelectric focusing. Biochim. Biophys. Acta **579,** 314—324.

Fuller, F., Kunstner, P., Nguyen, T., Verma, D. P. S., 1983: Soybean nodulin genes: Analysis of cDNA clones reveals several major tissue-specific sequences in nitrogen-fixing root nodules. Proc. Natl. Acad. Sci. U.S.A. **80,** 2594—2598.

Fuller, F., Verma, D. P. S., 1984: Appearance and accumulation of nodulin mRNAs and their relationship to the effectiveness of root nodules. Plant Mol. Biol. **3,** 21—28.

Gebhardt, C., Oliver, J. E., Forde, B. G., Saarelainen, R., Miflin, B. J., 1986: Primary structure and differential expression of glutamine synthetase genes in nodules, roots and leaves of *Phaseolus vulgaris*. EMBO J. **5,** 1429—1435.

Gloudemans, T., de Vries, S., Bussink, H.-J., Malik, N. S. A., Franssen, H. J., Louwerse, J., Bisseling, T., 1987: Nodulin gene expression during soybean (*Glycine max*) nodule development. Plant Mol. Biol. **8,** 395—403.

Govers, F., Moerman, M., Downie, J. A., Hooykaas, P., Franssen, H. J., Louiwerse, J., Van Kammen, A., Bisseling, T., 1986a: *Rhizobium* nod genes are involved in inducing an early nodulin gene. Nature **323,** 564—566.

Govers, F., Moerman, M., Hooymans, J., Van Kammen, A., Bisseling, T., 1986b: Microaerobiosis is not involved in the induction of pea nodulin genes. Plant Mol. Biol. **8,** 425—435.

Govers, F., Nap, J.-P., Moerman, M., Franssen, H. J., Van Kammen, A., Bisseling,

T., 1987: cDNA cloning and developmental expression of pea nodulin genes. Plant Mol. Biol. **8,** 425—435.

Guerinot, M. L., Chelm, B., 1985: Bacterial cAMP and heme in the *Rhizobium*-legume symbiosis. In: Evans, H. J., Bottomley, P. J., Newton, W. E. (eds.), Nitrogen Fixation Research Progress, p. 220. Dordrecht: Martinus Nijhoff.

Hattori, J., Johnson, D. A., 1985: The detection of leghemoglobin-like sequences in legumes and non-legumes. Plant Mol. Biol. **4,** 285—292.

Herrera-Estrella, L., Van den Broeck, G., Maenhaut, R., Van Montagu, M., Schell, J., Timko, M., Cashmore, A., 1984: Light-inducible and chloroplast-associated expression of a chimaeric gene introduced into *Nicotiana tobacum* using a Ti plasmid vector. Nature **310,** 115—120.

Hirel, B., Bouet, C., King, G., Layzell, D., Jacobs, F., Verma, D. P. S., 1987: Glutamine synthetase genes are regulated by ammonia provided externally or by symbiotic nitrogen fixation. EMBO J. **6,** 1167—1171.

Hunt, L. T., Hurst-Calderone, S., Dayhoff, M. O., 1978: Globins. In: Dayhoff, M. D. (ed.), Atlas of Protein Sequence and Structure **5,** Suppl. 3, pp. 229—249. Washington: Natl. Biomed. Res. Found.

Hyldig-Nielsen, J. J., Jensen, E. O., Palndan, O., Wiborg, O., Carrett, R., Jorgensen, P., Marcker, K. A., 1982: The primary structures of two leghemoglobin genes from soybean. Nucl. Acids Res. **10,** 689—701.

Jacobs, F. A., Zhang, M., Fortin, M. G., Verma, D. P. S., 1987: Several nodulins of soybean share structural domains but differ in their subcellular localization. Nucl. Acids Res. **15,** 1271—1280.

Jensen, E. O., Marcker, K. A., Villadsen, I. S., 1986a: Heme regulates the expression in *Saccharomyces cerevisiae* of chimaeric genes containing 5′-flanking soybean leghemoglobin genes. EMBO J. **5,** 843—847.

Jensen, J. S., Marcker, K. A., Otten, L., Schell, J., 1986b: Nodule-specific expession of a chimeric soybean leghemoglobin gene in transgenic *Lotus corniculatus.* Nature **321,** 669—674.

Jones, J. D. G., Dunsmuir, P., Bedbrock, J., 1985: High level expression of introduced chimaeric genes in regenerated transformed plants. EMBO J. **4,** 2411—2418.

Katinakis, P., Verma, D. P. S., 1985: Nodulin-24 gene of soybean codes for a peptide of the peribacteroid membrane and was generated by tandem duplication of a sequence resembling an insertion element. Proc. Natl. Acad. Sci. U.S.A. **82,** 4157—4161.

Kaulen, H., Schell, J., Kreuzaler, F., 1986: Light-induced expression of the chimeric chalcone synthase-NPTII gene in tobacco cells. The EMBO Journal **5,** 1—8.

Kiss, G. B., Vincze, E., Vegh, Z., 1986: Isolation of nodule specific cDNA clones from *Medicago sativa.* In: Verma, D. P. S., Brisson, N. (eds.), Dordrecht: Martinus Nijhoff.

Kondorosi, E., Kondorosi, A., 1986: Nodule induction on plant roots by *Rhizobium.* Trends Biol. Sci. **11,** 296—299.

Lang-Unnasch, N., Ausubel, F. M., 1985: Nodule-specific polypeptides from effective alfalfa root nodules and from ineffective nodules lacking nitrogenase. Plant Physiol. **77,** 833—839.

Lara, M., Cullimore, J. V., Lea, P. J., Miflin, B. J., Johnston, A. W. B., Lamb, J. W., 1983: Appearance of a novel form of plant glutamine synthetase during nodule development in *Phaseolus vulgaris* L. Planta **157,** 254—258,

Lara, M., Porta, H., Padilla, J., Folch, J., Sanchez, F., 1984: Heterogeneity of

glutamine synthetase polypeptides in *Phaseolus vulgaris* L. Plant Physiol. **76,** 1019—1023.
Larsen, K., Jochimsen, B. U., 1986: Expression of nodule-specific uricase in soybean callus tissue is regulated by oxygen. EMBO J. **5,** 15—19.
Larsen, K., Jochimsen, B. U., 1987: Expression of two enzymes involved in ureide formation in soybean regulated by oxygen. Third International Symposium on The Molecular Genetics of Plant-Microbe Interactions. 162.
Lee, J. S., Brown, G. G., Verma, D. P. S., 1983: Chromosomal arrangement of leghemoglobin genes in soybean. Nucl. Acids Res. **11,** 5541—5553.
Legocki, R. P., Verma, D. P. S., 1979: A nodule-specific plant protein (Nodulin-35) from soybean. Science **205,** 190—193.
Legocki, R. P., Verma, D. P. S., 1980: Identification of "nodule-specific" host proteins (Nodulins) involved in the development of *Rhizobium*-legume symbiosis. Cell **20,** 153—163.
Leong, S. A., Ditta, G. S., Helinski, D. R., 1982: Heme biosynthesis in *Rhizobium.* J. Biol. Chem. **257,** 8724—8730.
Lepo, J. E., Hanus, F. J., Evans, H. J., 1980: Chemoautotrophic growth of hydrogen-uptake-positive strains of *Rhizobium japonicum.* J. Bacteriol. **141,** 664—670.
Lewin, R., 1981: Evolutionary history written in globin genes. Science **214,** 425—429.
Lim, G., Burton, J. C., 1982: Nodulation status of the leguminosae. In: Broughton, W. J. (ed.), Nitrogen Fixation. Vol. 2: *Rhizobium,* pp. 1—34. Oxford: Clarendon Press.
Long, S. R., 1984: Genetics of Rhizobium nodulation. In: Kosuge, T., Nester, E. W. (eds.), Plant-Microbe Interactions: Molecular and Genetic Perspectives, pp. 265—306. New York, London: Macmillan.
Mahon, J. D., 1983: Energy relationships. In: Broughton, W. J. (ed.), Nitrogen Fixation. Vol. 3: Legumes, pp. 299—325. Oxford: Clarendon Press.
Mauro, V. P., Nguyen, T., Katinakis, P., Verma, D. P. S., 1985: Primary structure of the soybean nodulin-23 gene and potential regulatory elements in the 5′ flanking regions of nodulin and leghemoglobin genes. Nucl. Acids Res. **13,** 239—249.
Mauro, V. P., Verma, D. P. S., 1988: Transcriptional-activation in nuclei from uninfected soybean of a set of genes involved in symbiosis with *Rhizobium.* Molecular Plant-Microbe Interactions **1,** 46—51.
McCarty, D. R., Shaw, J. R., Hannah, L. C., 1986: The cloning, genetic mapping, and expression of the constitutive sucrose synthase locus of maize. Proc. Natl. Acad. Sci. **83,** 9099—9103.
Miflin, B. J., Cullimore, J., 1984: Nitrogen assimilation in the legume-*Rhizobium* symbiosis: A joint endeavour. In: Verma, D. P. S., Hohn, T. (eds.), Genes Involved in Microbe-Plant Interactions, pp. 129—178. Wien - New York: Springer-Verlag.
Miflin, B. J., Lea, P. J., 1980: Ammonia assimilation. In: Miflin, B. J. (ed.), The Biochemistry of Plants, pp. 169—202. New York: Academic Press.
Miflin, B. J., Lea, P. J., 1982: Ammonia assimilation and amino acid metabolism. In: Boulter, D., Parthier, B. (eds.), Encyclopedia of Plant Physiology. Vol. 14A: Nucleic Acids and Protein in Plants 1, pp. 5—64. New York: Springer-Verlag.
Minchin, F. R., Summerfield, R. J., Hadley, P., Roberts, E. H., Rawsthorne, S., 1981: Carbon and nitrogen nutrition of nodulated roots of grain legumes. Plant Cell Environ. **4,** 5—26.

Mohapatra, S. S., Perlick, A., Pühler, A., 1987: Nodule-specific polypeptides of broadbean (*Vicia faba,* L.). Symbiosis (in press).

Morell, M., Copeland, L., 1985, Sucrose synthase of soybean nodules. Plant Physiol. **78,** 149—154.

Morrison, N., Verma, D. P. S., 1987: A block in the endocytosis of *Rhizobium* allows cellular differentiation in nodules but affects the expression of some peribacteriod membrane nodulins. Plant Mol. Biol. **9,** 185—196.

Nguyen, T., Zelechowska, M., Foster, V., Bergmann, H., Verma, D. P. S., 1985: Primary structure of the soybean nodulin-35 gene encoding uricase II localized in the peroxisomes of uninfected cells of nodules. Proc. Natl. Acad. Sci. U.S.A. **82,** 5040—5044.

Oaks, A., Hirel, B., 1985: Nitrogen metabolism in roots. Ann. Rev. Plant Physiol. **36,** 345—365.

Peters, N. K., Frost, J. W., Long, S. R., 1986: A plant flavone, luteolin, induces expression of *Rhizobium meliloti* nodulation genes. Science **233,** 977—980.

Postgate, J. R., 1978: The nitrogen cycle. In: Arnold, E. (ed.): Nitrogen Fixation, pp. 1—3. Southampton: The Camelot Press Ltd.

Redmond, J. W., Batley, M., Djordjevic, M. A., Innes, R. W., Kuempel, P. L., Rolfe, B. G., 1986: Flavones induce expression of nodulation genes in *Rhizobium.* Nature **323,** 632—635.

Reibach, P. M., Streeter, J. G., 1983: Metabolism of ^{14}-C-labelled photosynthate and distribution of enzymes of glucose metabolism in soybean nodules. Plant Physiol. **72,** 634—640.

Roberts, M. P., Jafar, S., Mullin, B. C., 1985: Leghemoglobin-like sequences in the DNA of non-leguminous plants. Plant Mol. Biol. **5,** 333—337.

Robertson, J. G., Farnden, K. J. F., 1980: Ultrastructure and metabolism of the developing legume root nodule. In: Miflin, B. J. (ed.), The Biochemistry of Plants, pp. 65—113. New York: Academic Press.

Robertson, J. G., Lyttleton, P., Bullivant, S., Grayston, G. F., 1978: Membranes in lupin root nodules. I. The role of Golgi bodies in the biogenesis of infection threads and peribacteroid membranes. J. Cell Sci. **30,** 129—149.

Robertson, J. G., Lyttleton, P., Topper, B. A., 1985: The role of peribacteroid membrane in legume root nodules. In: C. Verger, W. E. Newton (eds.), Advances in Nitrogen Fixation Research, Proc. 5th Int. Symp. Nitrogen Fixation, pp. 475—481. The Hague: Martinus Nijhoff.

Robertson, J. G., Wells, B., Bisseling, T., Farnden, K. J. F., Johnston, A. W. B., 1984b: Immunogold localization of leghemoglobin in cytoplasm in nitrogen-fixing root nodules of pea (*Pisum sativum*). Nature **311,** 254—256.

Rolfe, B. G., Shine, J., 1984: *Rhizobium*-leguminosae symbiosis: The bacterial point of view. In: Verma, D. P. S., Hohn, T. (eds.), Genes Involved in Microbe-Plant Interactions, pp. 95—128. Wien - New York: Springer-Verlag.

Rostas, K., Sista, P. R., Stanley, J., Verma, D. P. S., 1984: Transposon mutagenesis of *Rhizobium japonicum.* Mol. Gen. Genet. **197,** 230—235.

Sandal, N. N., Bojsen, K., Marcker, K. A., 1987: A small family of nodule-specific genes from soybean. Nucl. Acids Res. **15,** 1507—1519.

Schubert, K. R., 1986: Products of biological nitrogen fixation in higher plants: synthesis transport, and metabolism. Ann. Rev. Plant Physiol. **37,** 539—574.

Sengupta-Gopalan, C., Pitas, J. W., Thompson, D. V., Hoffman, L. M., 1986: Expression of host genes during root nodule development in soybeans. Mol. Gen. Genet. **203,** 410—420.

Sengupta-Gopalan, C., Pitas, J. W., 1986: Expression of nodule-specific glutamine synthetase genes during nodule development in soybeans. Plant Mol. Biol. **7**, 189—199.

Sutton, W. D., 1983: Nodule development and senescence. In: Broughton, W. J. (ed.), Nitrogen Fixation. Vol. 3: Legumes, pp. 144—212. Oxford: Clarendon Press.

Tayjima, S., Yomomato, Y., 1975: Enzymes of purine catabolism in soybean plants. Plant and Cell Physiol. **16**, 271—282.

Thornburg, R. W., An, G., Cleveland, T. T., Johnson, R., Ryan, C. A., 1987: Wound-inducible expression of a potato inhibitor II-chloramphenicol acetyl transferase gene fusion in transgenic tobacco plants. Proc. Nat. Acad. Sci. **84**, 744—748.

Thummler, F., Verma, D. P. S., 1987: Nodulin-100 of soybean is the subunit of sucrose synthetase regulated by the availability of free heme in nodules. J. Biol. Chem. **262**, 14730—14736.

Tingey, S. V., Walker, E. L., Coruzzi, G. M., 1987: Glutamine synthetase genes of pea encode distinct polypeptides which are differentially expressed in leaves, roots, and nodules. EMBO J. **6**, 1—9.

Tjepkema, J. C., 1983: Hemoglobins in the nitrogen-fixing root nodules of actinorrhizal plants. Can. J. Bot. **63**, 2924—2929.

Triplett, E. W., 1985: Intracellular nodule localization and nodule specificity of xanthine dehydrogenase in soybean. Plant Physiol. **77**, 1004—1009.

van Brussel, A. A. N., Zaat, S. A. J., Canter-Cremers, H. C. J., Wijffelman, C. A., Dees, E., Tak, T., Lugtenberg, B. J. J., 1986: Role of plant root exudate and *sym* plasmid. Localized nodulation genes the synthesis by *R. leguminosarum* of tsr factor which causes thick short roots on common vetch. J. Bact. **165**, 517—522.

Vance, C. P., Boylan, K. L. M., Stade, S., Somers, D. A., 1985: Nodule specific protein in alfalfa (*Medicago Sativa* L.). Symbiosis **1**, 69—84.

van Kammen, A., 1984: Suggested nomenclature for plant genes involved in nodulation and symbiosis. Plant Mol. Biol. Reporter **2**, 43—45.

van den Bosch, K. A., Newcomb, E. H., 1986: Immunogold localization of nodule-specific uricase in developing soybean nodules. Planta **167**, 425—436.

Verma, D. P. S., Nash, D. T., Schulman, H. M., 1974: Isolation and *In vitro* translation of soybean leghaemoglobin mRNA. Nature **251**, 74—77.

Verma, D. P. S., Bal, A. K., 1976: Intracellular site of synthesis and localization of leghemoglobin in root nodules. Proc. Natl. Acad. Sci. U.S.A. **73**, 3843—3847.

Verma, D. P. S., Ball, S., Guerin, C., Wanamaker, L., 1979: Leghemoglobin biosynthesis in soybean root nodules. Characterization of nascent and released peptides and the relative rate of synthesis of major leghemoglobins. Biochemistry **18**, 476—483.

Verma, D. P. S., Fortin, M. G., Stanley, J., Mauro, V. P., Purohit, S., Morrison, N., 1986: Nodulins and nodulin genes of *Glycine max*. Plant Mol. Biol. **7**, 51—61.

Verma, D. P. S., Stanley, J., 1987: Molecular interactions in endosymbiosis between legume plants and nitrogen-fixing microbes. New York Academy of Sciences **503**, 284—294.

Verma, D. P. S., Fortin, M. G., 1988: Peribacteroid compartment of root nodules. In: Vasil, I. K. (ed.); Molecular Biology of Plant Nuclear Genes. Academic Press. In press.

Verma, D. P. S., Kazazian, V., Zogbi, V., Bal, A. K., 1978: Isolation and characterization of the membrane envelope enclosing the bacteroids in soybean root nodules. J. Cell. Biol. **78**, 919—936.

Verma, D. P. S., Long, S., 1983: The molecular biology of *Rhizobium*-legume symbiosis. In: Jeon, K. (ed.), International Review of Cytology, Suppl. 14, pp. 211—245. New York: Academic Press.

Verma, D. P. S., Nadler, K., 1984: Legume-*Rhizobium* symbiosis: host's point of view. In: Verma, D. P. S., Hohn, T. (eds.), Genes Involved in Microbe-Plant Interactions, pp. 57—93. Wien - New York: Springer-Verlag.

Werner, D., Mellor, R. B., Hahn, M. G., Griesbach, H., 1985: Soybean root response to an ineffective type of soybean nodules with an early loss of the peribacteroid membrane. Z. Naturforsch. **40,** 179—181.

Werner, D., Morschel, E., Kort, R., Mellor, R. B., Bassarab, S., 1984: Lysis of bacteriods in the vicinity of the host cell nucleus in an ineffective (Fix^-) root nodule of soybean (*Glycine max*). Planta **162,** 8—16.

Werr, W., Frommer, W.-B., Maas, C., Starlinger, P., 1985: Structure of the sucrose synthase gene on chromosome 9 of *Zea mays* L. EMBO J. **4,** 1373—1380.

Wiborg, O., Hyldig-Nielsen, J. J., Hensen, E., Paludan, K., Marcker, K. A., 1982: The nucleotide sequences of two leghemoglobin genes from soybean. Nucl. Acids Res. **10,** 3487—3494.

Wiborg, O., Hyldig-Nielsen, J. J., Jensen, E. O., Paludan, K., Marcker, K. A., 1983: The structure of an unusual leghemoglobin gene from soybean. EMBO J. **2,** 449—452.

Chapter 11

Structure and Expression of Plant Genes Encoding Pathogenesis-Related Proteins

John F. Bol

Department of Biochemistry, Leiden University, Wassenaarseweg 64, NL-2333 AL Leiden, The Netherlands

With 4 Figures

Contents

I. Introduction

A. Occurrence of PR Proteins

In the early seventies Van Loon and Van Kammen (1970) and Gianinazzi *et al.* (1970) independently reported the *de novo* synthesis of four proteins in tobacco plants reacting hypersensitively to infection with tobacco mosaic virus (TMV). Later on, six additional components specific to TMV-infected tobacco were identified (Pierpoint *et al.*, 1981; Van Loon, 1982). Initially, these proteins were called "new components" or "b-proteins". The name "pathogenesis-related" proteins, or PRs, was proposed by Antoniw *et al.* (1980) to permit the grouping of similar proteins from different tobacco cultivars. The ten PR proteins induced by TMV-infection of the *Nicotiana tabacum* cultivars Samsun NN and Xanthi nc have been characterized in most detail. These are designated 1a (IV, b1), 1b (III, b2), 1c (II, b3), 2 (I, b4), N (b5), O (b6), P (b7), Q (b8), R (b9) and S in order of decreasing mobility in alkaline, non-denaturing gels (the notation given between parentheses is no longer commonly used). Proteins similar to one or more of these PRs have been detected in other *Nicotiana* species such as *N. debneyi, N. glutinosa, N. rustica, N. sylvestris* and *N. tomentosiformis* (Ahl and Gianinazzi, 1982; Ahl *et al.*, 1982).

PR proteins are not only induced by virus infection but also by other necrosis-inducing pathogens, such as bacteria and fungi, and by a number of specific chemicals. These treatments are now known to induce PR proteins in at least 20 plant species, including in addition to the *Nicotiana* spp., both dicotyledonous and monocotyledonous plants such as tomato, potato, bean, cowpea, cucumber, celery, lemon, *Gomphrena globosa, Gynura aurantiaca*, maize and barley (for references see Redolfi, 1983; Van Loon, 1985; White *et al.*, 1987). Most of this chapter is concerned with PR proteins from tobacco plants as these have been studied most intensively.

TMV-infection of tobacco results in a considerable increase in the activity of at least 25 enzymes many of which are involved in the biosynthesis of aromatic compounds (Van Loon, 1982). PR proteins can be distinguished from these enzymes by the following three criteria: (1) they are all secreted by the plant cells into the intercellular space of the leaf (Parent and Asselin, 1984), (2) they are protease resistant (Van Loon, 1982) and (3) they are soluble at low pH (Van Loon, 1976; Gianinazzi, 1977). Eight of the ten PRs from Samsun NN to tobacco have been purified to homogeneity by Jamet and Fritig (1986). Figure 1 shows the migration rate of proteins 1a, 1b, 1c, 2, N, O, P and Q in alkaline, non-denaturing gels and in SDS-containing gels. The molecular weights of these proteins, estimated by SDS-PAGE, are listed in Table 1. Three size classes can be recognized in Fig. 1B; proteins within each size class were found to be serologically related. One class is composed of the PR-1 proteins with mol. wts. of ~15 kDa, which were shown to be serologically related by several research groups (Matsuoka and Ohashi, 1984; Antoniw *et al.*, 1985; Hooft van Huijsduijnen *et al.*, 1985). Proteins 2, N and O constitute a second class of serologically related proteins (Fortin *et al.*, 1985) with an estimated mol. wt. of ~40 kDa. In gel permeation studies these proteins elute with an estimated mol. wt. of ~25 kDa (Jamet and Fritig, 1986). Recently, the PR proteins P and Q were shown to be serologically related (Hooft van Huijsduijnen *et al.*, 1987). Their mol. wt. estimated from SDS-PAGE is ~27 kDa; in gel permeation studies they co-elute with the PR-1 proteins with an estimated mol. wt. of ~14 kDa (Jamet and Fritig, 1986).

Proteins R and S from Samsun NN tobacco were recently purified by Van Loon *et al.* (submitted for publication). PR-R was found to be composed of two components of 13 and 15 kDa whereas PR-S migrated as a single component of ~23 kDa in SDS-gels. These authors reported that

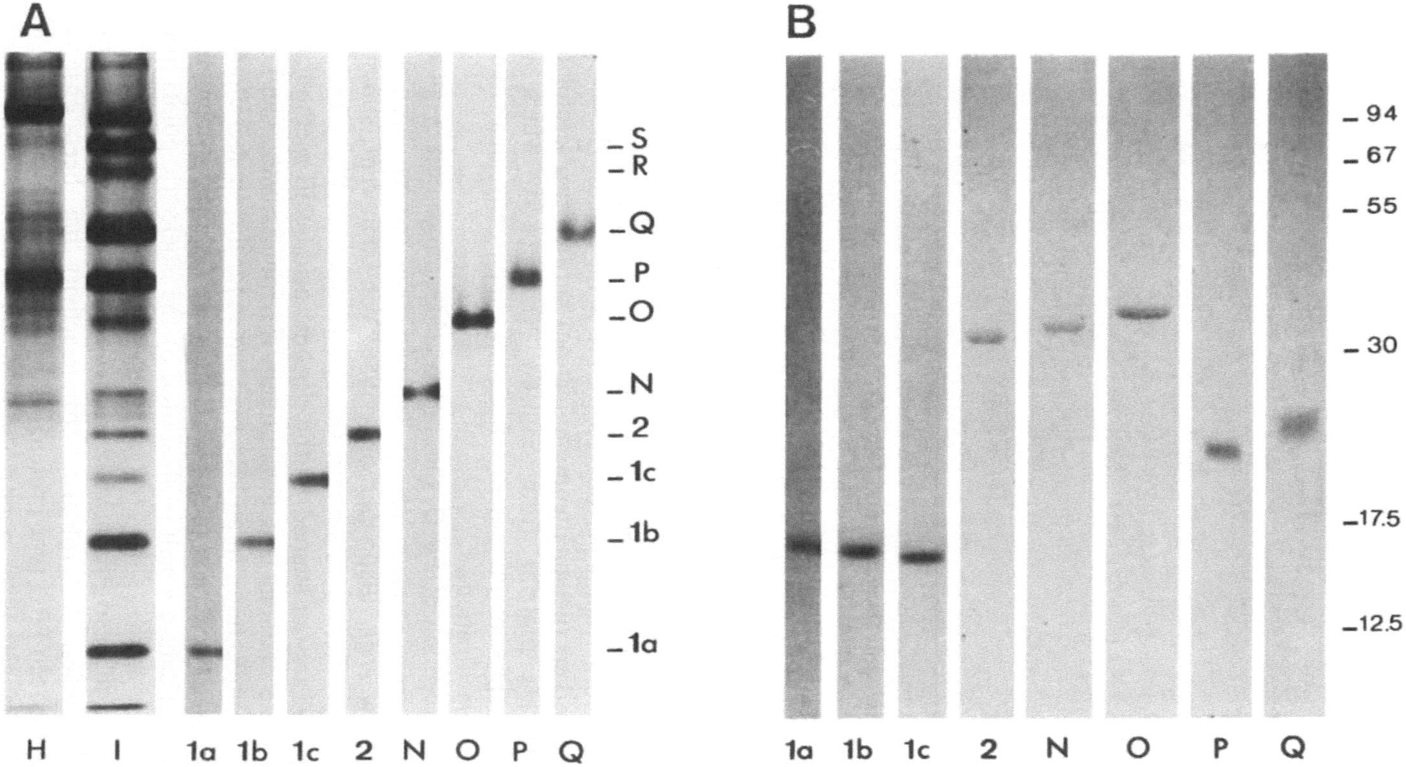

Fig. 1. PAGE under non-denaturing (A) and denaturing conditions (B) of 1 μg of each of the PR proteins 1a, 1b, 1c, 2, N, O, P and Q, purified from TMV-infected Samsun NN tobacco. Lanes H and I show the protein pattern obtained under non-denaturing conditions with 30 μg of proteins extracted at pH 2.8 from healthy and infected leaves, respectively. Figure is from Jamet and Fritig (1986); experimental details are found in this reference.

Table 1. Characteristics of PR proteins from Samsun NN tobacco

PR	Estimated molecular weight (kDa)	Isoelectric point
1 a	15.8	4.0
1 b	15.5	4.4
1 c	15.6	4.8
2	39.7	4.4
N	40.0	4.8
O	40.6	4.8
P	27.8	5.1
Q	29.5	5.6
R (R′)	13.0 + 15.0	6.7
S (R)	23.0	6.4

Nomenclature of PRs is according to Van Loon (1982); the names of proteins from Xanthi nc that are equivalent to R and S are given between parentheses (Pierpoint, 1986). Molecular weights were estimated by SDS-PAGE by Jamet and Fritig (1986) and Van Loon *et al.* (submitted for publication). The isoelectric points are from Van Loon *et al.* (submitted for publication).

PR-S is serologically related to the PR-1 group (1a, 1b, 1c) and that PR-R is serologically related to the PR-2 group (2, N, O). Proteins R and S from Samsun NN probably correspond to proteins R′ and R, respectively, that were purified from Xanthi nc tobacco by Pierpoint (1986). However, B. Fritig and coworkers (personal communication) purified two serologically related ~23 kDa proteins from Samsun NN tobacco which were considered to represent PR proteins R and S. Apparently, this situation needs further clarification.

With the exception of PR-O, the amino acid composition of all PRs, purified either from Samsun NN or Xanthi nc tobacco has been determined (Antoniw *et al.*, 1980; Pierpoint, 1986; Van Loon *et al.*, submitted for publication). These data support the classification of PRs mentioned above. The PRs are not stained by the periodic acid-Schiff method indicating that none of them is a glycoprotein (Pierpoint, 1986; Van Loon, submitted for publication). In all PRs analysed so far the number of acidic amino acids greatly exceeds the number of basic residues, in agreement with their acidic nature which is illustrated by the isoelectric points listed in Table 1.

Some of the tobacco PRs have been compared with PR proteins from other plant species. A serological cross-reactivity has been observed between the tobacco PR-1 type proteins and PRs from tomato and cowpea (Nassuth and Sanger, 1986), potato and *Solanum demissum* (White, 1983), *Gomphrena globosa, Chenopodium amaranticolor*, maize and barley (White *et al.*, 1987). At the amino acid sequence level a 60% homology was observed between PR-1 type proteins from tobacco and tomato (Lucas *et al.*, 1985; Cornelissen *et al.*, 1986a). Moreover, a 65% acid sequence homology was found between PR-S and thaumatin, the intensely sweet tasting protein from *Thaumatococcus daniellii* (Cornelissen *et al.*, 1986b; Pierpoint, 1987). This illustrates that the genes for PR proteins are highly conserved in the plant kingdom. They have served as genetic markers in phylogenetic studies in *Nicotiana* species (Gianinazzi and Ahl, 1983).

B. Induction of PR Proteins

PR proteins start to accumulate after the hypersensitive reaction induced by necrotizing pathogens such as viruses (e.g. TMV in Samsun NN tobacco; Van Loon and Van Kammen, 1970), fungi (e.g. *Thielaviopsis basicola* in Burley 49 tobacco; Gianinazzi *et al.*, 1980) and bacteria (e.g. *Pseudomonas syringae* in Samsun NN; Ahl *et al.*, 1981). The induction of PRs in tobacco is not dependent on the N-gene. Infection of Samsun nn, which lacks the N-gene, with tobacco necrosis virus (TNV) results in the accumulation of the same set of PRs that is induced in Samsun NN.

The level of PR-1 proteins increases 20,000-fold after infection of tobacco (Antoniw *et al.*, 1985) and it has been estimated that 7 days after inoculation, PR-1a may constitute about 1% of the soluble leaf protein (Antoniw and Pierpoint, 1978). A preliminary report (Carr *et al.*, 1982) suggested that the mRNAs for PR-1 proteins are present in equal amounts in healthy and diseased plants indicating that PR-1 synthesis is under translational control. However, this conclusion turned out to be wrong (Carr *et al.*, 1985). Using cDNA probes it was shown that the mRNAs for PR-1 proteins (Hooft van Huijsduijnen *et al.*, 1985) and several other classes of PRs (Hooft van Huijsduijnen *et al.*, 1986a) occur at a very low level in healthy plants and that this level is increased over 100-fold after infection. Recently, "run-on" assays with isolated nuclei confirmed that synthesis of PR-1 proteins is controlled at the level of transcription (J. Horowitz and R. B. Goldberg, personal communication). By Nothern blot analysis it was shown that in the primary infected leaves of tobacco, the mRNAs for PRs 1, P, Q and S become detectable 2 days after inoculation with TMV and reach a maximum after 4 to 5 days. From 8 days after inoculation the mRNAs became detectable in the virus-free upper leaves of the plant suggesting that induction of PR genes occurs by a "mobile compound" that is transported through the plant (Hooft van Huijsduijnen *et al.*, 1986a). An amphidiploid has been constructed of *N. glutinosa* and *N. debneyi* which produces constitutively the single PR encoded by the two parents. When the rootstock of this hybrid is used for grafting different other *Nicotiana* spp., the scions start to express PRs characteristic of the scion (Gianinazzi and Ahl, 1983). This demonstrates that the "mobile compound", which by definition can be called a plant hormone, moves through the plant but not the PR proteins (or their mRNAs) themselves. In TMV-inoculated tobacco leaves the largest amounts of PR-1a were found to be just outside the center of the local lesions, and the concentration then decreased with distance from the center (Antoniw and White, 1986). Apparently, the tissue around the necrotic lesion is most active in producing the "mobile compound".

Although PRs are present at very low levels in healthy plants, they accumulate in large amounts in older plants when these start to flower and have senescing lower leaves (Fraser, 1981). In some plants PRs are induced by plasmolysis (Wagih and Coutts, 1981). It has been suggested that expression of PR genes is a general response to stress. However, PRs are not produced under a number of stress conditions such as artificial aging due to leaf detachment, mechanical or chemical wounding, water, drought or salt stress (Van Loon, 1983) and heat-shock treatment (Hooft van Huijsduijnen, unpublished).

In addition to their induction by pathogens and certain stress conditions, PRs are also known to accumulate after application to the plant of a variety of chemicals (see Van Loon, 1983). Well known chemical inducers of PRs are polyacrylic acid (Gianinazzi and Kassanis, 1974), ethephon which metabolizes into the plant hormone ethylene (Van Loon, 1977), aromatic compounds like benzoic acid, salicylic acid and acetylsalicylic acid (White, 1979), amino acid derivates (Asselin *et al.*, 1985), the antiviral agents 2-thiouracil and dioxohexahydrotriazine, and barium and manganese salts (White *et al.*, 1986). The effects of these chemicals are considered to be specific. Application of a number of aromatic intermediates from the phenylpropanoid pathway, which more or less resemble salicylic acid in structure, did not induce PRs (Jamet, 1985). Also, out of a range of salts of ten metals only those of barium and manganese induced PRs although many other salts tested were phytotoxic and produced chlorotic and necrotic symptoms (White *et al.*, 1986), Finally, PRs are induced by a class of less defined agents, i. e. the elicitor preparations extracted from fungal or bacterial cultures (Leach *et al.*, 1983; Maiss and Poehling, 1983; Somssich *et al.*, 1986). Polysaccharides present in these preparations are held responsible for the effect. A peptide of 27 amino acids produced by virulent races of the fungus *Cladosporium fulvum* is known to induce PR-1 type proteins in susceptible tomato cultivars (De Wit and Buurlage, 1986).

Induction of PRs by salicylic acid and polyacrylic acid has been studied in some detail. Usually, these chemicals are injected in the leaf though in the case of salicylic acid, spraying of a 1—5 mM solution onto the plant is also very effective. As judged by the convential extraction procedures, the two chemicals are known to induce in tobacco a subset of the known 10 PRs, notably PRs 1a and 1b and low amounts of 1c, 2 and N (Antoniw and White, 1980; Hooft van Huijsduijnen *et al.*, 1986b). Recently, it was reported by Dumas *et al.* (1987) that an analysis of proteins in the intercellular fluid of tobacco demonstrated the induction of similar sets of proteins by TMV-infection and polyacrylic acid treatment. As will be shown below, spraying of tobacco plants with salicylic acid does induce the synthesis of PR-1 mRNAs but no increase in the level of mRNAs for PRs P, Q and S is observed. Salicylic acid induces the production of PRs in all tobacco cultivars tested, whereas the effect of polyacrylic acid is strongly cultivar dependent. A genetic analysis showed that the genetic determinant for the polyacrylic acid response behaves as a dominant character. The information coding for this response follows a monogenic segregation which is inherited independently of the N-gene (Dumas *et al.*, 1987).

C. PR Proteins and Acquired Resistance

Plants reacting hypersensitively to virus infection acquire a local resistance in the inoculated leaves and systemic resistance in the non-inoculated virus-free leaves to further infection (Ross, 1961, 1966). This resistance has a broad specificity. For example, the resistance acquired by tobacco after TMV-infection is effective against unrelated viruses (e. g. TNV), fungi (e. g. *Phytophtora parasitica*) and bacteria (e. g. *Pseudomonas tabaci*). Conversely, tobacco plants become resistant to TMV after their infection with necrosis-inducing bacteria, fungi or unrelated viruses (see Gianninazzi, 1983). Circumstantial evidence suggests a role of PR proteins in this acquired resistance (Kassanis *et al.*, 1974; Van Loon, 1975). All infections leading to acquired resistance are accompanied by the accumulation of PR proteins. The hypersensitive response to these pathogens triggers the production of ethylene which is thought to be responsible for the induction of PR synthesis and for an increase in the activity of at least 25 enzymes many of which are involved in the biosynthesis of aromatic compounds (Van Loon, 1982). Most of the chemicals that induce PR proteins also induce resistance, frequently associated with the production of ethylene (Van Loon, 1983). However, treatment of tobacco with salicylic acid does not result in ethylene production (Van Loon and Antoniw, 1982). Under these conditions, PR proteins are selectively induced and the plants become resistant to TMV-infection (White, 1979). Moreover, salicylic acid was found to inhibit the systemic replication of alfalfa mosaic virus in tobacco plants and the multiplication of this virus in cowpea protoplasts by up to 99 %. Salicylic acid treatment resulted in a block of viral plus- and minus-strand RNA-synthesis without interfering with the host metabolism at a detectable level (Hooft van Huijsduijnen *et al.*, 1986 b). These results indicate that the salicylic acid induced resistance is not effective at the level of cell-to-cell spread but is acting on primarily infected cells.

Another parallel between the induction of PR proteins and resistance was obtained in studies with polyacrylic acid. In tobacco cultivars in which polyacrylic acid induces PR proteins it also induces resistance, whereas in cultivars in which PR proteins are not induced no resistance is obtained (White *et al.*, 1983; Dumas *et al.*, 1987). The interspecific hybrid of *N. glutinosa* × *N. debneyi* which constitutively produces a PR-1 type protein, is highly resistant to infection with TMV and TNV (Ahl and Gianninazzi, 1982). When the rootstock of this hybrid is used for grafting other *Nicotiana* spp. the scions start to express both acquired resistance and PRs (Gianninazzi and Ahl, 1983).

Although the close parallel between the induction of PR proteins and acquired resistance suggests a causal relationship between the two phenomena, several observations indicate that PR proteins do not play a control role in the antiviral response. In a few cases no quantitative or temporal relationship between the onset of resistance and PR proteins production was found, and some chemicals are known to cause resistance without inducing PR proteins and *vice versa* (Fraser, 1982; Fraser and Clay, 1983; Sherwood, 1985). However, as will be discussed in more detail in section III, several PR proteins are probably involved in the induced resistance towards infection by fungi and other microorganisms. It was observed by Boller *et al.* (1983) that treatment of bean plants with ethylene resulted in the induction of a chitinase with an estimated mol. wt. of 30 kDa. While no endogenous substrate for purified chitinase was found in the plant, the enzyme readily attacked cell walls of a potential pathogenic fungus, *Fusarium solani*, and acted as a lysozyme on bacterial cell walls. This led the authors to the hypothesis that chitinase functions as a defence against chitinous pathogens. Because of its high isoelectric point the chitinase was considered to be different from the known acidic PRs. However, as we will see in section II C, a relationship between chitinase and PRs does in fact exist.

II. Characterization of PR mRNAs and Genes

A. cDNA Cloning of PR mRNAs

The determination of partial amino acid sequences of PR-1a from Samsun NN tobacco (Lucas *et al.*, 1985) permitted the synthesis of oligodeoxynucleotides complementary to the corresponding mRNA. These were used as primers to sequence the 5′-termini of PR-1 mRNAs (Hooft van Huijsduijnen, 1985) and as probes to isolate PR-1 clones from a cDNA library made to poly(A) RNA from TMV-infected Samsun NN tobacco (Cornelissen *et al.*, 1986a). Three classes of cDNA clones were obtained which were assigned to PR-1a, -1b and -1c on the basis of limited amino acid sequence information on these proteins. Although only the PR-1b clone was full-length, the data indicated that PR-1 proteins show an over 90% sequence homology and are made as precursors with 30 amino acids signal peptides that are probably removed from the mature proteins of 138 amino acids during their transportation from the plant cells into the intercellular space of the leaf. An extensive recent analysis of 32 cDNA clones confirmed that there are only three classes of PR-1 mRNAs and permitted the deduction of the complete amino acid sequences of the mature proteins 1a, 1b and 1c (Pfitzner and Goodman, 1987).

To isolate clones of other PR mRNAs, a cDNA library to poly(A) RNA from TMV-infected tobacco was screened by a differential hybridization procedure. Initially, cDNA clones corresponding to six classes of TMV-induced mRNAs were obtained, designated "clusters A to F" (Hooft van Huijsduijnen *et al.*, 1986a). As is illustrated in Figure 2, the mRNAs ranged in size from 650 to 1400 nucleotides. In addition to TMV-infection, mRNAs B and C were also strongly induced by spraying tobacco with salicylic acid whereas mRNAs A and F were weakly induced by this treatment. By using the clones in hybrid-selected translation experiments, it was shown that mRNAs B, D, E and F encode proteins that react with an antiserum to PR proteins. Group B cDNA clones were found to be homologous to the previously characterized PR-1 clones. A further analysis of these cDNAs yielded a clone encoding the complete PR-1a sequence. In addition, two clones were identified that cross-hybridized to the PR-1 cDNAs only when the hybridization was done at low stringency. These clones encoded a protein with a length of 173 amino acids excluding its signal peptide. The N-terminal 137 amino acids of this protein showed a 67 % homology to the PR-1 proteins (Cornelissen *et al.*, 1987). For reasons given below, the mRNA encoding this PR-1 like protein was considered to represent a seventh class of TMV-inducible mRNAs, group G. An amino acid sequence comparison showed that many acidic and neutral residues in PR-1 were replaced by neutral and basic residues, respectively, in the G-protein. The excess of basic residues over acidic residues in the G-protein

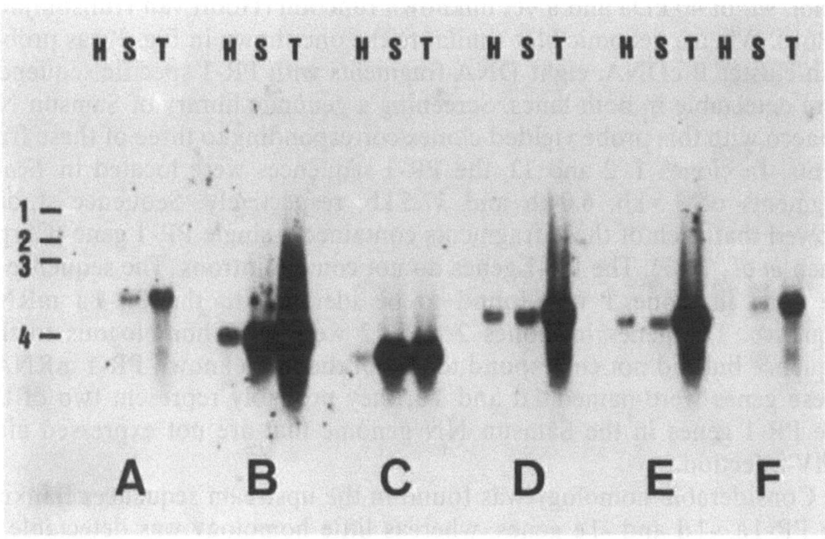

Fig. 2. Induction of mRNAs A to F in tobacco by salicylic acid spraying and TMV-infection. A Northern blot was loaded with 20 μg poly(A) RNA from healthy tobacco leaves **(lanes H)**, leaves sprayed with salicylic acid **(lanes S)** and TMV-infected leaves **(lanes T)**. The blot was cut into six stripes which were hybridized separately to 32P-labeled plasmids from clones belonging to clusters **A to F**. The stripes were realigned in their original orientation for autoradiography. The positions of AlMV RNAs 1 (3644 nt), 2 (2593 nt), 3 (2037 nt) and 4 (881 nt) are indicated on the left. Figure is from R. A. M. Hooft van Huijsduijnen *et al.* (1986a); experimental details are found in this reference

closely resembled the composition of the PR-1 type protein p14 from tomato, which has an isoelectric point of 10.7 (Camacho-Henriquez and Sanger, 1984). Examples of the occurrence of acidic and basic isoforms of other PR proteins are presented below. By a differential hybridization procedure, cDNA clones have been isolated to seven classes of mRNAs that are induced in peas following infection with the fungus *Fusarium solani* f. sp. *phaseoli* (Riggleman *et al.*, 1985). Some of these mRNAs may be homologous to the tobacco mRNAs described above as they are in the same size range (790—1980 nucleotides). However, the proteins corresponding to these mRNAs have not yet been characterized.

B. Genes Corresponding to Groups A, B, C and G

The cDNA clones of the TMV-inducible mRNAs A to G habe been used to analyse the corresponding genes in the Samsun NN genome by Southern blot hybridization (Bol *et al.*, 1987). Groups A, B, C and G were found each to correspond to a family of approximately eight genes, whereas the gene families corresponding to groups D, E and F showed a lower complexity. As an example, Figure 3 shows the pattern obtained by hybridizing labeled group A cDNA to a blot loaded with tobacco DNA that was digested with *Hind* III or *Eco*RI. The genes of the group A family encode a protein with a mol. wt. of 40 kDa and a yet unknown function (Hooft van Huijsduijnen, 1986 a). When a genomic blot similar to the one shown in Fig. 3 was probed with cluster B cDNA, eight DNA-fragments with PR-1 specific sequences were detectable in both lanes. Screening a genomic library of Samsun NN tobacco with this probe yielded clones corresponding to three of these fragments. In clones 1, 2 and 12, the PR-1 sequences were located in *Eco*RI fragments of 3.5 kb, 6.0 kb and 17.5 kb, respectively. Sequence studies showed that each of these fragments contained a single PR-1 gene (Cornelissen *et al.*, 1987). The PR-1 genes do not contain introns. The sequence of the gene in clone 1 was found to be identical to the PR-1a mRNA sequence. The genes in clones 2 and 12 were 90% homologous to this sequence but did not correspond to any of the three known PR-1 mRNAs. These genes were named 1d and 1e; they probably represent two of the five PR-1 genes in the Samsun NN genome that are not expressed after TMV-infection.

Considerable homology was found in the upstream sequences flanking the PR-1a, -1d and -1e genes, whereas little homology was detectable in the downstream sequences. Figure 4 shows a schematic representation of these three PR-1 genes. Compared to clone 1, clone 12 contains an insertion of 185 bp in the putative promoter region. This insertion, which is flanked by 18 bp repeats, may be responsible for the inactivation of the PR-1d gene. Moreover, the TAA termination codon in the PR-1a gene is changed into GAA in the PR-1d gene giving an extension of the reading frame with 16 triplets. Compared to clone 12, clone 2 contains a deletion of 522 bp in the promoter region, removing the putative CAAT-box. In addition, there is a deletion of 21 bp around the position corresponding to

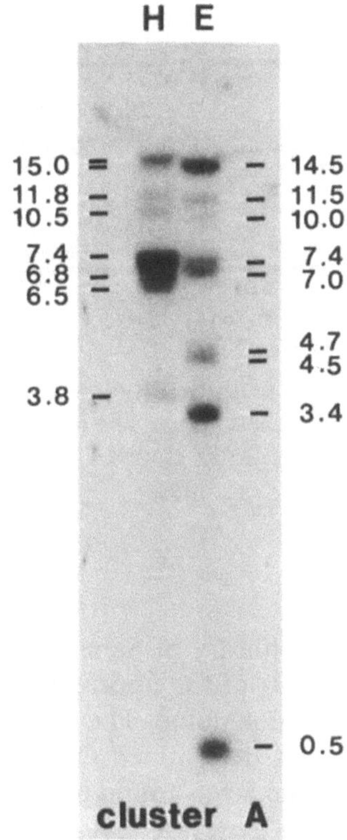

Fig. 3. Southern blot analysis of cluster A genes in genomic DNA of Samsun NN tobacco. A filter containing genomic DNA digested with *Hind III* **(lane H)** or *Eco*RI **(lane E)** was hybridized to the nick-translated insert (405 bp) that was cut out with *Hind III* and *Eco*RI from a pUC9 vector, containing cluster A cDNA (PROB 40). Experimental details will be described elsewhere (R. A. M. Hooft van Huijsduijnen *et al.*, submitted for publication). The estimated size of the fragments (kb) is indicated

the TAA termination codon in the PR-1a gene. These deletions corroborate the notion that PR-1e is a silent gene.

Probing a genomic blot with group G cDNA revealed a pattern that was completely different from the one obtained with the group B probe (Cornelissen *et al.*, 1987). The data indicate that in addition to eight genes encoding acidic PR-1 proteins, the Samsun NN genome contains also approximately eight genes encoding basic PR-1 like proteins. Samsun NN is amphidiploid, containing 2n chromosomes from *N. sylvestris* and 2n chromosomes from *N. tomentosiformis.* Probably, four genes of the cluster B and G families are derived from one parent and four from the other parent. It has been shown that the genes for PR-1a and -1c in Samsun NN tobacco originate from *N. sylvestris* whereas the PR-1b gene has been provided by *N. tomentosiformis* (Gianninazzi and Ahl, 1983).

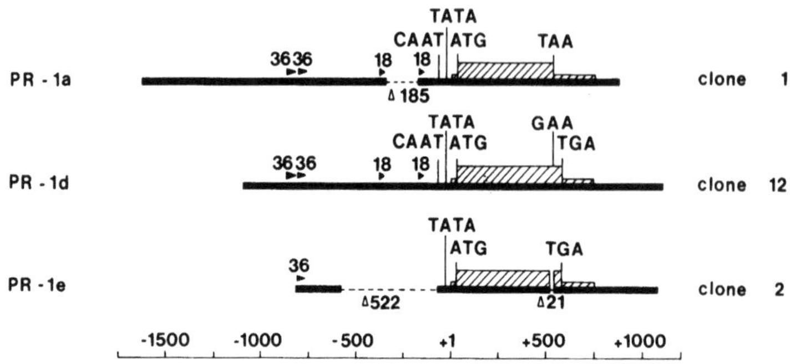

Fig. 4. Schematic representation of three PR-1 genes from Samsun NN tobacco. The genes are represented by hatched boxes. The location of direct repeats of 18 bp and 36 bp is indicated. Deletions (Δ) in clones 1 and 2 are represented by dotted lines

Screening of the genomic library of Samsun NN with group C cDNA as probe yielded clones of four of the approximately eight group C genes that are present in the tobacco genome. The sequence of the genes in two clones, numbers 4 and 8, was determined (Van Kan *et al.*, manuscript in preparation). The gene in clone 8 was found to contain an intron of 555 bp. Although the sequence of this gene was not identical to the two available cDNA sequences, S1-nuclease mapping studies strongly indicated that this gene is expressed after TMV-infection. Similar S1-studies suggested that clone 4 correspond to a group C gene that is not expressed after TMV infection. The intron in this gene was found to be 1954 bp in length. The reading frame of the gene in clone 8 encodes a protein of 109 amino acids. The N-terminal sequence of 26 amino acids has the characteristics of a signal peptide, indicating that, like the known PR proteins, the cluster C protein is translocated through a membrane. The putative mature protein of 83 amino acids contains 25 % glycine and 35 % of charged residues (Asp, Glu, Arg, Lys). A high proportion of glycin-residues is characteristic for the structural proteins from the seed coats and cell walls of a number of plant species (see Varner and Cassab, 1986). This may suggest that the cluster C protein is a structural component of tobacco cell walls.

C. Genes Corresponding to Groups D, E and F

By hybrid-selected translation it was shown that proteins encoded by mRNAs corresponding to clusters D, E and F had mol. wts. of 25 kDa, 25 kDa and 34 kDa, respectively (Hooft van Huijsduijnen *et al.*, 1986a). These mol. wts. are in the size range of those of PRs 2, N, O, P, Q and S. To further analyse their nature, the translation products of mRNAs selected with cDNA clones D, E and F were subjected to immunoprecipitation with antisera to purified PRs 2, P and Q (Hooft van Huijsduijnen *et al.*, 1987).

The protein corresponding to group E did not react with any of the antisera. However, the group D selected protein strongly reacted with the antisera to P and Q whereas the cluster F selected protein weakly reacted with these antisera. The available cDNA clones of clusters D and F were incomplete, encoding only the C-terminal half of the respective proteins. A comparison of these termini showed a 65 % amino acid sequence homology between the cluster D encoded 25 kDa protein and the cluster F encoded 34 kd protein. Moreover, it was found (Hooft van Huijsduijnen *et al.*, 1987) that the cluster F sequence was 76 % homologous to a basic chitinase from bean plants (Broglie *et al.*, 1986) and 100 % homologous to a basic chitinase that was identified in tobacco (Shinshi *et al.*, 1987). Recently, Legrand *et al.* (1987) analyzed chitinase activities induced by TMV-infection in Samsun NN tobacco. In addition to two basic chitinases corresponding cluster F, two acidic chitinases were identified that were shown to be identical to PRs P and Q. The serological relationship between the acidic and basic chitinases was confirmed in this study.

From genomic blots it was concluded that clusters D, E and F each correspond to two to four genes in the Samsun NN genome (Hooft van Huijsduijnen *et al.*, 1987). One of the group E cDNA clones was nearly full-length, encoding a protein of 226 amino acids. This protein was found to be 65 % homologous with the sweet-tasting protein thaumatin (Cornelissen *et al.*, 1986b). An alignment of its sequence with that of thaumatin indicated that the N-terminal sequence of 25 amino acids represents a signal peptide involved in the secretion of the cluster E protein into the intercellular space of the tobacco-leaf. Partial determination of the amino acid sequence of the ~23 kDa. PR protein that is called R by the group of Pierpoint (Harpenden, U. K.) and S by the group of Van Loon (Wageningen, NL) showed it to be homologous to the thaumatin-like protein of tobacco (Pierpoint, 1987). This sets the mol. wt. of this protein at 21,596 (Cornelissen *et al.*, 1986b).

III. General Conclusion

A. Role of PR Protein

For more than a decade, there has been much speculation about the possible role of PR proteins in healthy and diseased plants. Already in 1972 it was shown that PR proteins are not associated with 25 enzyme activities, a number of which are known to occur in the intercellular space of the leaf (see Antoniw and White, 1983). The assumption that PRs are involved in defence mechanisms was based merely on the close association between the induction of these proteins and the acquired systemic resistance. The recent demonstration that some of the PRs are in fact chitinases (see section II C) is the first solid evidence for a function of these proteins in the response of the plant to pathogens. Chitin, the substrate for chitinase, is a polymer of N-acetylglucosamine that is absent in plants but occurs in the cell walls of fungi, the larval style of nematodes and the insect exoskeleton. The purified basic chitinase from bean plants was shown to be a potent inhibitor of the growth of the fungus *Trichoderma viride* (Schlumbaum *et al.*, 1986). As this enzyme also affects bacterial cell walls (Boller *et al.*, 1983) it may have a role in the anti bacterial response too. Legrand *et al.* (1987) compared the specific activity of two basic chitinases that are induced by TMV-infection of tobacco with the specific activity of the acidic chitinases PR-P and -Q. The basic enzymes (estimated mol. wts. of 32 kDa and 34 kDa) were found to be six times more active in the *in vitro* assay than their acidic isoforms. However, due to the relative abundance of PR-R and -Q, it was estimated that the acidic proteins account for one-third of the chitinase activity that is induced by TMV-infection of Samsun NN tobacco, whereas two-third of the activity is contributed by the basic enzymes. The total increase in chitinase activity after TMV-infection was found do be 40-fold.

The relative fraction of chitin in the cell walls of fungi ranges from 0.3 % in *Phytophthora* to about 50 % in *Fusarium.* The other major component in the cell wall of these microorganisms is β-1,3-glucan. Already in the early seventies it was shown that TMV-infection induces an increase in the level of β-1,3-glucanase in *Nicotiana glutinosa* (Moore and Stone, 1972). The sequence of a basic β-1,3-glucanase of 322 amino acids was determined by H. Shinshi (quoted by Fincher *et al.*, 1986). Recently, it was found by B. Fritig and coworkers (personal communication) that TMV-infection of Samsun NN tobacco induces the accumulation of a basic β-1,3-glucanase with an estimated mol. wt. of 32 kDa and three acidic β-1,3-glucanases which turned out to be identical to PR-2, -N and -O. Thus, it now appears that five of the ten known acidic PR proteins are hydrolytic enzymes with a putative role in the induced resistance to fungal infection. Because PR proteins were mostly analyzed in alkaline gel systems, the basic isoforms of a number of these proteins had been overlooked so far. Table 2 summarizes our present knowledge of the function and properties of PRs and related proteins.

It seems unlikely that a single resistance mechanism could protect plants against such diverse pathogens as viruses, fungi and bacteria. It is possible that the hypersensitive reaction induced by a localized pathogen activates several, separate resistance mechanisms for the different types of pathogens, and different PR proteins may be involved in each of these. In view of the inhibition of virus multiplication by salicylic acid, PR proteins induced by this chemical may be involved in the antiviral response. As shown in Figure 2, the mRNAs for the PR-1 proteins (cluster B) and the mRNAs corresponding to cluster C are most strongly induced by salicylic acid. An analysis of PR-1 proteins in TMV-infected tobacco tissue by ELISA and immunofluorescence microscopy localized these proteins to the extra cellular spaces predominantly in regions adjacent to viral lesions (Antoniw and White, 1986; Carr *et al.*, 1987). Although this is consistent with a potential role of PR-1 proteins in limiting TMV-infection, it could

Table 2. Characteristics of proteins and mRNAs induced by TMV infection of Samsun NN tobacco

	PR	mol. wt. (kDa)	cDNA cluster	size of mRNA (nt)	induction of mRNA by salicylic acid	function
Acidic PR proteins	1a/1b/1c	15	B	850	++	?
	2/N/O	40	—	?	?	β-1,3-glucanase
	P/Q	28	D	1050	—	chitinase
	S (R)	23	E	1000	—	α-amylase inhibitor
	R (R′)	13/15	—	?	?	?
Basic PR-like proteins	—	19	G	?	?	? (homologous to PR-1)
	—	32	—	1700	?	β-1,3-glucanase
	—	32/34	F	1300	+	chitinase
Unclassified proteins	—	40	A	1400	+	?
	—	12	C	650	+++	?

Nomenclature of PR proteins is according to Van Loon (1982); (R) and (R′) refer to the nomenclature used by Pierpoint (1986).

merely reflect the site of production of the "mobile compound" that induces the synthesis of PRs. It was shown that PR-1 proteins produced in protoplasts are also exported into the medium (Carr *et al.*, 1987). This is difficult to reconcile with a role of PR-1 type proteins in the salicylic acid induced inhibition of viral RNA synthesis in protoplasts (Hooft van Huijsduijnen *et al.*, 1986b). The detection of basic and acidic PR-1 type proteins is reminiscent of the occurrence of basic and acidic chitinases and glucanases. This may suggest that PR-1 proteins have a similar function, being hydrolytic enzymes with a yet unknown substrate specificity.

It is tempting to speculate that the group C encoded protein has a role in the antiviral response. It is strongly induced by TMV-infection and salicylic acid treatment, it does not belong to the known PRs and it has not been detected in the intercellular fluid of salicylic acid treated tobacco so far (Hooft van Huijsduijnen *et al.*, 1986a, 1986b). If it turns out to be a cell wall component, as suggested by its high content of glycine, it may affect the physiological conditions in the cell by altering the cell wall structure, thereby inhibiting virus multiplication. The finding that half of the known PR proteins are probably involved in the antifungal response, supports the view that the other PR proteins or virus-induced non-PR proteins are involved in the defence against other pathogens. Recently, the thaumatin-like PR protein S was found to be homologous to a maize protein that is a bifunctional inhibitor of bovine trypsin and the α-amylase from *Tribolium castanetum* beetles (Richardson *et al.*, 1987). Possibly, this PR protein is involved in a defence against insects.

B. Prospects for Future Research

PR genes have a number of interesting features, including the function of their coding region, the structure of their promoter sequence and the way their signal peptides act in targeting these proteins. Insight into the function of their coding sequences can be obtained by transforming plants with PR genes that are fused to an efficient promoter. It will be interesting to see whether plants producing chitinases and/or glucanases constitutively, show an increased resistance to infection with fungi or other microorganisms. The physiological meaning of the occurrence of acidic and basic isoforms of these enzymes is not yet known. The construction of transgenic plants producing different combinations of these proteins will give an answer to the question whether the coordinate expression of acidic and basic isoforms is required for maximal protection against a wide range of microorganisms. Likewise, transgenic plants producing PR proteins with a yet unknown function can be assayed for resistance to other classes of pathogens such as viruses. Possibly, the phenotype of these plants could mimic that of the virus-resistant hybrid *N. glutinosa* × *N. debneyi.* The observation that several PR proteins correspond to rather complex gene families and the possibility that the expression of a specific combination of these genes is required in a given defence response may complicate this type of experiment.

At present only a few inducible plant promoters are available for studies on the molecular biology of plants. The characterization of promoters that can be induced at will by a simple chemical treatment of a plant or cell culture would be potentially useful and of fundamental interest. In addition to TMV infection, mRNAs B and C in Figure 2 are strongly induced by spraying tobacco plants with a low concentration of salicylic acid. Preliminary studies indicate that this induction also occurs in tobacco protoplasts (Hooft van Huijsduijnen, unpublished result). A comparison of the upstream sequences of the group B and C genes showed only limited homologies, mainly located between the cap-site and TATA-box (Van Kan *et al.*, manuscript in preparation). The putative promoter sequences can be fused to a reporter gene and the induction of this gene by salicylic acid can be studied after introduction of the construct in protoplasts e. g. by electroporation or in the plant genome by a T-DNA based transformation procedure. In this way it will be possible to identify the minimal sequences required for the response to salicylic acid. Except for the treatment with salicylic acid, all events that are known to induce PR proteins are accompanied by the production of ethylene. Conversely, treatment of the plants with ethylene or the ethylene releasing compound ethephon induces PR proteins (Van Loon, 1982, 1983). A comparison of the promoter sequences of the PR genes may reveal common regulatory elements involved in this response. On the other hand, it is possible that ethylene triggers the production of different effectors each being responsible for the activation of transcription of different (classes of) PR genes. Once the regulatory sequences involved in the induction of PR genes by pathogens, ethylene or salicylic acid have been identified by the transfection and transformation procedures mentioned above, the binding of nuclear proteins to these sequences can be studied. This will give further insight into the diversity of the pathways that lead to the induction of PR genes and may provide a link to the identification of the "mobile compounds" that act as long distance signals in this induction.

The data obtained so far support the notion that all PRs are derived from precursors by removal of an N-terminal signal peptide during their export into the intercellular space of the leaf. A more detailed study of this process will be of fundamental importance. Moreover, the combination of an inducible PR promoter and the coding sequence of the matching signal peptide may be useful to produce proteins of interest in plant systems as these proteins can be easily purified from the intercellular fluid of transgenic plants or the medium of plant cell cultures.

Acknowledgements

I am indebted to Drs. J. F. Antoniw, B. Fritig, S. Gianninazzi, M. Legrand and L. C. van Loon for sending me manuscripts prior to publication. This work was supported in part by the Netherlands Foundation for Chemical Research (SON) with financial aid from the Netherlands Organization for the Advancement of Pure Research (ZWO).

IV. References

Ahl, P., Benjama, A., Samson, R., Gianninazzi, S., 1981: Induction chez le tabac par *Pseudomonas syringae* de nouvelles protéines (protéines b) associées au développement d'une résistance non spécifique à une deuxième infection. Phytopath. Z. **102,** 201—212.

Ahl, P., Gianninazzi, S., 1982: *b*-Protein as a constitutive component in highly (TMV) resistant interspecific hybrids of *Nicotiana glutinosa* × *Nicotiana debneyi.* Plant Sci. Lett. **26,** 173—181.

Ahl, P., Cornu, A., Gianninazzi, S., 1982: Soluble proteins as genetic markers in studies of resistance and phylogeny in *Nicotiana.* Phytopathology **72,** 80—85.

Antoniw, J. F., Pierpoint, W. S., 1978: The purification and properties of one of the "b" proteins from virus-infected tobacco plants. J. Gen. Virol. **39,** 343—350.

Antoniw, J. F., White, R. F., 1980: The effects of aspirin and polyacrylic acid on soluble leaf proteins and resistance to virus infection in five cultivars of tobacco. Phytopathol. Z. **98,** 331—341.

Antoniw, J. F., Ritter, C. E., Pierpoint, W. S., Van Loon, L. C., 1980: Comparison of three pathogenesis-related proteins from plants of two cultivars of tobacco infected with TMV. J. Gen. Virol. **47,** 79—87.

Antoniw, J. F., White, R. F., 1983: Biochemical properties of the pathogenesis-related proteins from tobacco. Neth. J. Plant Pathol. **89,** 255—264.

Antoniw, J. F., White, R. F., Barbara, D. J., Jones, P., Longley, A., 1985: The detection of PR (b) protein and TMV by ELISA in systemic and localized virus infections of tobacco. Plant Mol. Biol. **4,** 55—60.

Antoniw, J. F., White, R. F., 1986: Changes with time in the distribution of virus and PR protein around single local lesions of TMV infected tobacco. Plant Mol. Biol. **6,** 145—149.

Asselin, A., Grenier, J., Cote, F., 1985: Light-influenced extra cellular accumulation of *b* (pathogenesis-related) proteins in *Nicotiana* green tissue induced by various chemicals or prolonged floating on water. Can. J. Bot. **63,** 1276—1283.

Bol, J. F., Hooft van Huijsduijnen, R. A. M., Cornelissen, B. J. C., Van Kan, J. A. L., 1987: Characterization of pathogenesis-related (PR) proteins and genes. In: Plant resistance to viruses. Evered, D., Harnett, S., Eds. J. Wiley and Sons Ltd., Chichester, p. 72—80.

Boller, T., Gehri, A., Mauch, F., Vogeli, U., 1983: Chitinase in bean leaves: induction by ethylene, purification, properties, and possible function. Planta **157,** 22—31.

Broglie, K. E., Gaynor, J. J., Broglie, R. M., 1986: Ethylene-regulated gene expression: molecular cloning of the genes encoding an endochitinase from *Phaseolus vulgaris.* Proc. Natl. Acad. Sci. U.S.A. **83,** 6820—6824.

Camacho Henriquez, A., Sanger, H. L., 1984: Purification and partial characterization of the major "pathogenesis-related" tomato leaf protein p14 from potato spindle tuber viroid (PSTV)-infected tomato leaves. Arch. Virol. **81,** 263—284.

Carr, J. P., Antoniw, J. F., White, R. F., Wilson, T. M. A., 1982: Latent messenger RNA in tobacco (*Nicotiana tabacum*). Biochem. Soc. Trans. **10,** 353—354.

Carr, J. P., Dixon, D. C., Klessig, D. F., 1985: Synthesis of pathogenesis-related proteins in tobacco is regulated at the level of mRNA accumulation and occurs on membrane-bound polysomes. Proc. Natl. Acad. Sci. U.S.A. **82,** 7999—8003.

Carr, J. P., Dixon, D. C., Nikolau, B. J., Voelkerding, K. V., Klessig, D. F., 1987: Synthesis and localization of pathogenesis-related proteins in tobacco. Mol. Cell. Biol. **7,** 1580—1583.

Cornelissen, B. J. C., Hooft van Huijsduijnen, R. A. M., Van Loon, L. C., Bol, J. F., 1986a: Molecular characterization of the messenger RNAs for "pathogenesis-related" proteins 1a, 1b and 1c, induced by TMV infection of tobacco. EMBO J. **5,** 37—40.

Cornelissen, B. J. C., Hooft van Huijsduijnen, R. A. M., Bol, J. F., 1986b: A tobacco mosaic virus-induced tobacco protein is homologous to the sweet-tasting protein thaumatin. Nature **321,** 531—532.

Cornelissen, B. J. C., Horowitz, J., Van Kan, J. A. L., Goldberg, R. B., Bol, J. F., 1987: Structure of tobacco genes encoding pathogenesis-related proteins from the PR-1 group. Nucleic Acids Res. **15,** 6799—6811.

De Wit, P. J. G. M., Buurlage, M. B., 1986: The occurrence of host-, pathogen- and interaction-specific proteins in the apoplast of *Cladosporium fulvum* (syn. *fulvia Fulva*) infected tomato leaves. Physiol. Mol. Plant Pathol. **29,** 159—172.

Dumas, E., Gianninazzi, S., Cornu, A., 1987: Genetic aspects of polyacrylic acid induced resistance to tobacco mosaic virus and tobacco necrosis virus in *Nicotiana* plants. Plant Pathol., in press.

Fincher, G. B., Lock, P. A., Morgan, M. M., Lingelbach, K., Wettenhall, R. E. H.,

Mercer, J. F. B., Brandt, A., Thomsen, K. K., 1986: Primary structure of the (1,3-1,4)-β-D-glucan 4-glucohydrolase from barley aleurone. Proc. Natl. Acad. Sci. U.S.A. **83,** 2081—2085.

Fortin, M. G., Parent, J. G., Asselin, A., 1985: Comparative study of two groups of b proteins (pathogenesis-related) from the intercellular fluid of *Nicotiana* leaf tissue infected by tobacco mosaic virus. Can. J. Bot. **63,** 932—937.

Fraser, R. S. S., 1981: Evidence for the occurrence of the "pathogenesis-related" proteins in leaves of healthy tobacco plants during flowering. Physiol. Plant Pathol. **19,** 69—76.

Fraser, R. S. S., 1982: Are "pathogenesis-related" proteins involved in acquired systemic resistance of tobacco mosaic virus? J. Gen. Virol. **58,** 305—313.

Fraser, R. S. S., Clay, C. M., 1983: Pathogenesis-related proteins and acquired systemic resistance: causal relationship or separate effects? Neth. J. Plant. Pathol. **89,** 283—292.

Gianinazzi, S., Martin, C., Vallee, J. C., 1970: Hypersensibilite aux virus, temperatures et proteines solubles chez le *Nicotiana* Xanthi-nc. Apparition de nouvelles macromolecules lors de la repression de la synthese virale. C. R. Acad. Sci. Paris D. **270,** 2383—2386.

Gianinazzi, S., Kassanis, B., 1974: Virus resistance induced in plants by polyacrylic acid. J. Gen. Virol. **23,** 1—9.

Gianinazzi, S., Prat, H. M., Shewry, P. R., Miflin, B. J., 1977: Partial purification and preliminary characterization of soluble proteins specific to virus infected tobacco plants. J. Gen. Virol. **34,** 345—351.

Gianinazzi, S., Ahl., P., Cornu, A., Scalla, R., Cassini, R., 1980: First report of host b-protein appearance in response to a fungal infection in tobacco. Physiol. Plant Pathol. **16,** 337—342.

Gianinazzi, S., 1983: Genetic and molecular aspects of resistance induced by infections or chemicals. In: Plant-Microbe Interactions Molecular and Genetic Perspectives. Nester, E. W., Kosuge, T. Eds., Macmillan Publ. Co., New York, vol. **1,** 321—342.

Gianinazzi, S., Ahl., P., 1983: The genetic and molecular basis of b-proteins in the genus *Nicotiana.* Neth. J. Plant Pathol. **89,** 275—281.

Hooft van Huijsduijnen, R. A. M., Cornelissen, B. J. C., Van Loon, L. C., Van Boom, J. H., Tromp, M., Bol, J. F., 1985: Virus-induced synthesis of pathogenesis-related proteins in tobacco. EMBO J. **4,** 2167—2171.

Hooft van Huijsduijnen, R. A. M., Van Loon, L. C., Bol, J. F., 1986b: cDNA cloning of six mRNAs induced by TMV infection of tobacco and a characterization of their translation products. EMBO J. **5,** 2057—2061.

Hooft van Huijsduijnen, R. A. M., Alblas, S. W., De Rijk, R. H., Bol, J. F., 1986b: Induction by salicylic acid of pathogenesis-related proteins and resistance to alfalfa mosaic virus infection in various plant species. J. Gen. Virol. **67,** 2135—2143.

Hooft van Huijsduijnen, R. A. M., Kauffmann, S., Brederode, F. Th., Cornelissen, B. J. C., Legrand, M., Fritig, B., Bol, J. F., 1987: Homology between chitinases that are induced by TMV infection of tobacco. Plant Mol. Biol. **9,** 411—420.

Jamet, E., 1985: Modification de la synthèse protéique chez des plantes hypersensibles à une infection virale: étude plus particulière des protéines PR ("pathogenesis-related") du tabac. Ph. D. Thesis, University Louis Pasteur, Strasbourg.

Jamet, E., Fritig, B., 1986: Purification and characterization of 8 of the pathogenesis-related proteins in tobacco leaves reacting hypersensitively to tobacco mosaic virus. Plant Mol. Biol. **6,** 69—80.

Kassanis, B., Gianinazzi, S., White, R. F., 1974: A possible explanation for the resistance of virus-infected tobacco plants to second infection. J. Gen. Virol. **23,** 11—16.

Leach, J. E., Sherwood, J., Fulton, R. W., Sequira, L., 1983: Comparison of soluble proteins associated with disease resistance induced by bacterial lipopolysaccharide and by viral necrosis. Physiol. Plant Pathol. **23,** 377—385.

Legrand, M., Kauffmann, S., Geoffroy, P., Fritig, B., 1987: Biological function of "pathogenesis-related" proteins: four tobacco PR-proteins are chitinases. Proc. Natl. Acad. Sci. U.S.A. **84,** 6750—6754.

Lucas, J., Camacho Henriquez, A., Lottspeich, F., Henschen, A., Sanger, H. L., 1985: Amino acid sequence of the "pathogenesis-related" leaf protein p14 from viroid-infected tomato reveals a new type of structurally unfamiliar proteins. EMBO J. **4,** 2745—2749.

Maiss, E., Poehling, H. M., 1983: Resistance against plant viruses induced by culture filtrates of the fungus *Stachybotrys chartarum.* Neth. J. Plant Pathol. **89,** p. 323.

Matsuoka, M., Ohashi, Y., 1984: Biochemical and serological studies of pathogenesis-related proteins of *Nicotiana* species. J. Gen. Virol. **65,** 2209—2215.

Moore, A. E., Stone, B. A., 1972: Effect of infection with TMV and other viruses on the level of a β-1,3-glucan hydrolase in leaves of *Nicotiana glutinosa.* Virology **50,** 791—798.

Nassuth, A., Sanger, H. L., 1986: Immunological relationship between "pathogenesis-related" leaf proteins from tomato, tobacco and cowpea. Virus Research **4,** 229—242.

Parent, J. G., Asselin, A., 1984: Detection of pathogenesis-related proteins (PR or b) and of other proteins in the intercellular fluid of hypersensitive plants infected with tobacco mosaic virus. Can. J. Bot. **62,** 564—569.

Pfitzner, U. M., Goodman, H. M., 1987: Isolation and characterization of cDNA clones encoding pathogenesis-related proteins from tobacco mosaic virus infected tobacco plants. Nucleic Acids Res. **15,** 4449—4465.

Pierpoint, W. S., Robinson, N. P., Leason, M. B., 1981: The pathogenesis-related proteins of tobacco: their induction by viruses in intact plants and their induction by chemicals in detached leaves. Physiol. Plant Pathol. **19,** 85—97.

Pierpoint, W. S., 1986: The pathogenesis-related proteins of tobacco leaves. Phytochemistry **25,** 1595—1601.

Pierpoint, W. S., Tatham, A. S., Pappin, D. J. C., 1987: Identification of the virus-induced protein of tobacco leaves that resembles the sweet-protein thaumatin. Physiol. Mol. Plant Pathol., in press.

Redolfi, P., 1983: Occurrence of pathogenesis-related (b) and similar proteins in different plant species. Neth. J. Plant Pathol. **89,** 245—254.

Richardson, M., Valdes-Rodriguez, S., Blanci-Labra, A., 1987: A possible function for thaumatin and a TMV-induced protein suggested by homology to a maize inhibitor. Nature **327,** 432—434.

Riggleman, R. C., Fristensky, B., Hadwiger, L. A., 1985: The disease resistance response in pea is associated with increased levels of specific mRNAs. Plant Mol. Biol. **4,** 81—86.

Ross, A. F., 1961: Systemic acquired resistance induced by localized virus infections in plants. Virology **14,** 340—358.

Ross, A. F., 1966: Systemic effects of local lesion formation. In: *Viruses of Plants,* Beemster, A. B. R., Dijkstra, J. Eds., North Holland Publ. Co., Amsterdam, pp. 127—150.

Schlumbaum, A., Mauch, F., Vogli, U., Boller, T., 1986: Plant chitinases are potent inhibitors of fungal growth. Nature **324,** 365—367.

Sherwood, J. L., 1985: The association of (pathogenesis-related) proteins with viral-induced necrosis in *Nicotiana sylvestris.* Phytopathol. Z. **112,** 48—55.

Shinshi, H., Mohnen, D., Meins, F., 1987: Regulation of a plant pathogenesis-related enzyme: inhibition of chitinase and chitinase mRNA accumulation in cultured tobacco tissues by auxin and cytokinin. Proc. Natl. Acad. Sci. U.S.A. **84,** 89—93.

Somsich, I. E., Schmelzer, E., Bollman, J., Hahlbrock, K., 1986: Rapid activation by fungal elicitor of genes encoding "pathogenesis-related" proteins in cultured parsley cells. Proc. Natl. Acad. Sci. U.S.A. **83,** 2427—2430.

Van Loon, L. C., Van Kammen, A., 1970: Polyacrylamide disc electrophoresis of the soluble leaf proteins from *Nicotiana tabacum* var "Samsun" and "Samsun NN". II. Changes in protein constitution after infection with tobacco mosaic virus. Virology **40,** 199—211.

Van Loon, L. C., 1975: Similarity of qualitative changes of specific proteins after infection with different viruses and their relationship to acquired resistance. Virology **67,** 566—575.

Van Loon, L. C., 1976: Specific soluble leaf proteins in virus-infected tobacco plants are not normal constituents. J. Gen. Virol. **30,** 375—379.

Van Loon, L. C., 1977: Induction by 2-chloroethylphosphonic acid of viral-like lesions, associated proteins, and systemic resistance in tobacco. Virology **80,** 417—420.

Van Loon, L. C., 1982: Regulation of changes in proteins and enzymes associated with active defence against virus infection. In: Wood, R. K. S., Ed., Active defense mechanisms in plants. Plenum Press, New York, pp. 247—273.

Van Loon, L. C., Antoniw, J. F., 1982: Comparison of the effects of salicylic acid and ethephon with virus-induced hypersensitivity and acquired resistance in tobacco. Neth. J. Plant Pathol. **88,** 237—256.

Van Loon, L. C., 1983: The induction of pathogenesis-related proteins by pathogens and specific chemicals. Neth. J. Plant Pathol. **89,** 265—273.

Van Loon, L. C., 1985: Pathogenesis-related proteins. Plant Mol. Biol. **4,** 111—116.

Varner, J. E., Cassab, G. I., 1986: A new protein in petunia. Nature **323,** p. 110.

Wagih, E. E., Coutts, R. H. A., 1982: Comparison of virus-elicited and other stresses on the soluble protein fraction of cucumber cotyledons. Phytopathol. Z. **104,** 364—374.

White, R. F., 1979: Acetylsalicylic acid (aspirin) induces resistance to tobacco mosaic virus in tobacco. Virology **99,** 410—412.

White, R. F., 1983: Serological detection of pathogenesis-related proteins. Neth. J. Plant Pathol. **89,** p. 311.

White, R. F., Antoniw, J. F., Carr, J. P., Woods, R. D., 1983: The effects of aspirin and polyacrylic acid on the multiplication and spread of TMV in different cultivars of tobacco with and without the N-gene. Phytopath. Z. **107,** 224—232.

White, R. F., Dumas, E., Shaw, P., Antoniw, J. F., 1986: The chemical induction of PR (b) proteins and resistance to TMV infection in tobacco. Antiviral Research **6,** 177—185.

White, R. F., Rybicki, E. P., Von Wechmar, M. B., Dekker, J. L., Antoniw, J F., 1987: Detection of PR1 type proteins in *Amaranthaceae, Chenopodiaceae, Gramineae* and *Solanaceae* by immunoelectroblotting. J. Gen. Virol. **68,** 2043—2048.

Chapter 12

Proteinase Inhibitor Gene Families: Tissue Specificity and Regulation

Clarence A. Ryan

Institute of Biological Chemistry, Washington State University, Pullman, WA 99164, U.S.A.

With 3 Figures

Contents

I. Introduction

Proteinase inhibtors are a multifamily group of proteins that are ubiquitous in nature (Laskowski, Jr. and Kato, 1984). Inhibitor proteins have been isolated that specifically inhibit each of the four known mechanistic classes of proteolytic enzymes, i. e. serine, thiol, aspartyl and metalloproteinases. Overall, the serine proteinase inhibitors comprise over ten unrelated protein families (Laskowski, Jr., 1986) that are found within the animal and plant kingdoms (Table 1). The functional role of these inhibitor proteins appears to be either to protect tissues or fluids from proteolysis by foreign proteases or to regulate the levels of proteases that are metabolically active in the tissues or fluids that they are associated with. The majority of proteinase inhibitor proteins that have been purified from plants have been inhibitors of serine endopeptidases such as the animal digestive enzymes trypsin, chymotrypsin and elastase or the bacterial proteinase subtilisin. In plants, the inhibitor proteins usually account for from 1—15 % or more of the proteins of various storage organs such as seeds and tubers (Ryan, 1974) and in some plant species, in leaves, in fruit, or in both (see below).

Table 1.
Families of Protein Inhibitors of Serine Proteinases

Animals
1. Bovine pancreatic trypsin inhibitor (Kunitz) family
2. Pancreatic secretory trypsin inhibitor (Kazal's) family
3. *Ascaris* inhibitor family
4. Serpin family (mechanistically distinct)
5. Hirudin family
Plants
6. Soybean trypsin inhibitor (Kunitz) family
7. Soybean proteinase inhibitor (Bowman-Birk) family
8. Potato I family
9. Potato II family
10. Barley trypsin inhibitor family
11. Squash inhibitor family
Microbial
13. *Streptomyces* subtilisin inhibitor (SSI) family
14. Other families

In the plant kingdom there are at least six families of serine proteinase inhibitors that have apparently originated independently from one another (cf. Table 1). Only one family, the Inhibitor I family (Balls and Ryan, 1963), unambiguously contains members that are present in both the animal and plant kingdoms. This family is present in tissues of plants from the Solanaceae (Balls and Ryan, 1963), Leguminosae (Fabaceae) (Svendsen *et al.*, 1984) and Graminae (Svendson *et al.*, 1980) families where its synthesis is regulated in a variety of different ways (see below) and is also present in the leech (Seemuller *et al.*, 1980). The leech inhibitor exhibits over 30 % sequence identity with its plant cousins. Thus, all members of the Inhibitor I family members share a common ancestral gene that must have existed nearly one billion years ago (Fig. 1) probably in a unicellular microorganism. Recently, a sequence of a protein from barley endosperm was deduced whose partial sequence was homologous with the alpha-1-antitrypsin inhibitor family (Hejgaard *et al.*, 1985). The primary structure of this protein suggests that it may be a proteinase inhibitor, with a Met-Ser at the active site, although its activity has not been established in enzymatic assays using known proteases.

From a practical standpoint, the plant proteinase inhibitors can comprise several percentages of the proteins of many seeds and tubers that are human foods, such as grains, beans and potatoes. The inhibitors can severely reduce the nutritional quality of foods if not completely denatured

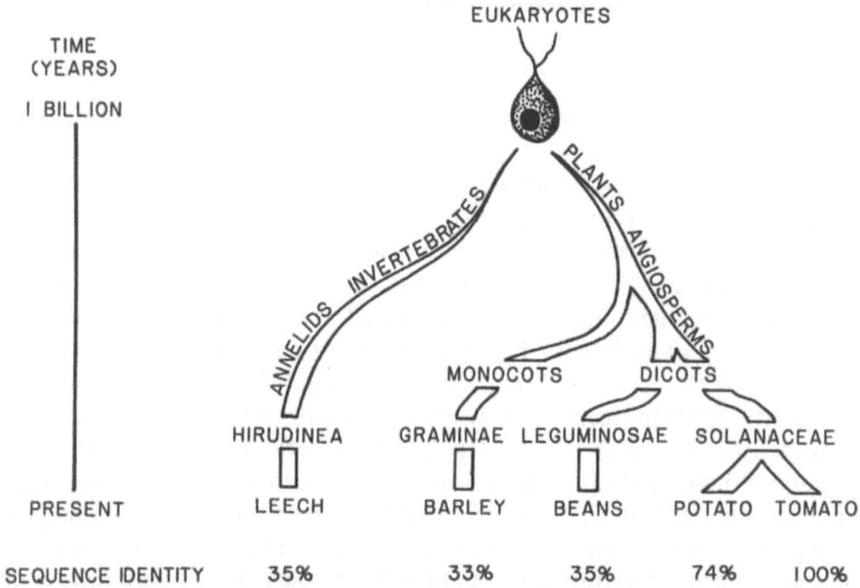

Fig. 1. Proposed evolutionary relationships between various members of the Potato Inhibitor I family. The percentages shown are based on amino acid sequence analyses and are compared to the sequence of the tomato leaf inhibitor I

by cooking (Rackis, 1981). More fundamentally, the plant proteinase inhibitors have been used as biochemical tools to study the mechanism of action and function of proteinases (Laskowski, Jr. and Sealock, 1971). Studies of their synthesis, function and regulation in plants have provided useful information concerning aspects of seed development (Waling *et al.*, 1986; Wilson *et al.*, 1986) and induced natural plant defensive mechanisms (Ryan *et al.*, 1978). These studies have provided basic information with which to further explore these systems at the molecular biological level and have provided attractive models to study the diversity of regulatory controls that have been imposed on closely related proteins. The possible roles of the proteinase inhibitors in plant protection and their roles in nutrition provide a backdrop of functionality that is useful in interpreting the diversity that is encountered in (i) their inhibitory activities; (ii) in their species and tissue specificities; and (iii) in their different modes of regulation of synthesis, i. e. developmental or environmental. This chapter will review recent research concerned with the regulation of expression of plant proteinase inhibitor genes with an emphasis on the possible relationships of gene structures with their environmental or developmental regulation in various plant tissues.

II. Developmentally Regulated Proteinase Inhibitor Genes in Seeds, Tubers and Fruit

The Kunitz trypsin inhibitor (KSTI) and the Bowman-Birk inhibitor (BBI) (Laskowski, Jr. and Kato, 1980) are among the less prevalent storage proteins in soybean seeds that are regulated during embryogenesis in developing legume seeds (Goldberg, 1986; Walling *et al.*, 1986). While not major storage proteins of legume seeds, they usually account for several percent of the seed proteins, depending upon the variety. Expression of the Kunitz trypsin inhibitor gene has been investigated in seeds and in mature plant organ systems, including 14 day postgerminating cotyledons, leaves, stems and roots (Walling *et al.*, 1986). The embryo, at 70 days after flowering, contained 18,000 molecules of Kunitz inhibitor mRNA per cell, which accounted for 3 % of the total mRNA. The mature organ tissues contained only about 5 molecules of inhibitor mRNA per cell. All of the seed protein gene families, including the Kunitz inhibitor, were shown to be regulated during early embryogenesis, at least in part, at the transcriptional level. As the seeds approached dormancy, their mRNAs became undetectable. From rates of transcription in the nucleus, and from mRNA levels in the cytoplasm, it was deduced that posttranscriptional events such as cytoplasmic entry rates for the mRNA and/or mRNA stabilities may also be important in regulating the expression of the seed protein genes, including the Kunitz inhibitor gene.

The transcriptionally regulated, tissue-specific expression of the Kunitz trypsin inhibitor gene and of the lectin gene can be explained by a model suggested by Goldberg (1986) (Fig. 2). In the model, the genes are regulated within different tissues (roots and embryos) with tissue-specific trans-acting factors that trigger transcription in response to developmental signals. Each seed protein gene would contain a hierarchy of cis-control

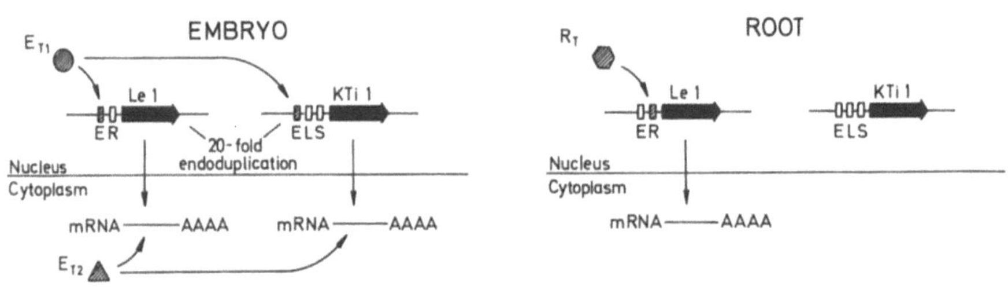

Fig. 2. Proposed model for the regulation of tissue specific expression of lectin and Kunitz trypsin inhibitor genes (Goldberg, 1986). E, R, L and S refer to embryo, root, leaf and stem cis-control elements, respectively. ET1 and RT represent embryo and root trans-acting factors that interact with E and R cis-control elements. ET2 represents a cytoplasmic factor that interacts with seed protein mRNAs to increase their stabilities

sequences that allow it to respond to developmental circumstances. In this model the trans-acting elements would include some that could serve to enhance the stability of mRNAs and/or to facilitate selective export of transcripts out of the nucleus.

In a recent study to identify *trans*-acting proteins and *cis*-acting sequences required for controlling the developmental expression of these seed protein genes (Jofuku *et al.*, in press), a DNA binding protein was identified that reacted with specific sequences of a developmentally regulated lectin gene. A 60 kDa protein from the nuclei of soybean embryos was identified that interacted with a fragment (Lel 5′) from the intronless lectin gene. This protein also interacted with the 5′ region of a Kunitz trypsin inhibitor gene, suggesting that the protein recognized specific sequences shared by both genes. A 7 nucleotide core motif, 5′ATT(A/T)AAT3′ was identified in the lectin gene by DNAase I digestion that was present in the 5′ region of the Kunitz trypsin inhibitor gene. Embryo nuclear proteins protect this motif in the Kunitz inhibitor gene from DNAase I digestion. Both the lectin and Kunitz trypsin inhibitor 5′ regions compete for the same nuclear DNA binding proteins. The developmentally-modulated binding of the 60 kDa nuclear protein with both genes correlated with seed protein gene transcriptional activity. While a function for the 60 kDa protein has not been established, it is an excellent candidate for a trans-acting factor that may interact with the 7 nucleotide *cis*-acting element to regulate the developmental expression of the Kunitz trypsin inhibitor.

The potato Inhibitor I and II families of proteinase inhibitors (Laskowski, Jr., 1986) exhibit considerable diversity in their modes of regulation of synthesis in the plant kingdom. They are both developmentally and environmentally (wound-induced) regulated in a variety of tissues from species of the Solanaceae family (Ryan, 1984) and, as mentioned earlier, Inhibitor I has also been found in storage organs of both legumes and grasses. In tissues of the Solanaceae family both Inhibitors I and II are frequently found together, regulated in an apparently coordinated manner (Plunkett *et al.*, 1982; Ryan *et al.*, in press; Pearce *et al.*, in press). The inhibitors are wound-inducible in tomato and potato leaves (Ryan, 1978), and developmentally regulated in potato tubers (Roschol *et al.*, 1986; Ryan *et al.*, in preparation) as well as in fruit of wild tomato species (Pearce *et al.*, in press). On the other hand, although Inhibitor I has been found in seeds of barley and bean, the presence of Inhibitor II in these tissues has not been reported.

In potato plants, proteinase Inhibitors I and II accumulate in tubers during their growth and development where they can eventually comprise up to 15 % of the soluble proteins of the cortical tissues (Ryan *et al.*, 1976). Two cDNA clones coding for Inhibitor II were prepared from mRNA obtained from tubers of a potato line HH80 1201/7 (Roschol *et al.*, 1986). A cDNA clone (pct800), derived from an Inhibitor II mRNA, was employed as a probe to assay levels of Inhibitor I mRNA in tubers, as well as in leaves, stems and roots. Inhibitor II mRNA was present in developing potato tubers at levels 200—500 fold higher than in the other tissues.

The synthesis of Inhibitor II mRNA in potato tissues (Roschol *et al.*, 1986) appears to be under both transcriptional and post-transcriptional regulation, a situation similar to that found for the Kunitz trypsin inhibitor in soybean embryos (Walling *et al.*, 1986). This conclusion was based on the observation that while the Inhibitor II mRNA is much higher in tuber cytoplasm than in leaf cytoplasm, the amount of mRNA in the nuclei of tubers is about equal to that found in leaves. The mechanism of this regulation is not understood, but it is anticipated that the genes coding for the tuber specific Inhibitor II mRNAs will provide further detailed information concerning the *cis-* and *trans-*acting elements that are involved in the regulation of expression of this small gene family in potato tubers and of sequences that may contribute to the regulation of the post-translational processing of the mRNA.

Inhibitor I and Inhibitor II are also present in tomato fruit. Until recently the two inhibitors had not been found in the fruit of the modern tomato, *Lycopersicon esculentum.* Within the past year the two inhibitors were found in the fruit of a wild tomato line of *L. peruvianum,* where they comprise over 40 % of the fruit proteins (Pearce *et al.*, in press). Fruit from lines of other wild tomato species were also found to contain the two inhibitors, but at much lower levels than the *L. peruvianum* line. The inhibitors are considered to be protective agents to discourage insect pests or small mammals from consuming them while the seeds develop in the fruit. However, in most wild species the inhibitors disappear as the fruit ripen, which may facilitate their consumption and therefore their seed dispersal (Pearce *et al.*, in press). When the mRNA isolated from developing fruit from *L. peruvianum* was translated *in vitro,* the newly translated Inhibitor I and II preproteins predominated the protein profiles. A library of *L. peruvianum* genes has been prepared in bacteriophage lambda and a cDNA library has been prepared from the *L. peruvianum* fruit mRNA. Fruit Inhibitor I and II genes are presently being identified using tomato leaf Inhibitor I and II cDNAs as probes (Wingate, V., Broadway, R. M., Ryan, C. A., in preparation). Inhibitors I and II have been purified from the fruit and partial amino acid sequences (about 25 residues of their N-termini) have been determined for several of the isoinhibitors (Pearce *et al.*, in preparation). Thus, the identities of fruit specific Inhibitors I and II genes can be established. The isolation and characterization of these inhibitor genes will provide the beginning of studies to identify nucleotide sequences that are involved in the regulation of their fruit specific expression during fruit development.

A member of the Inhibitor I gene family has recently been identified (Lincoln *et al.*, 1987) in the modern tomato (*L. esculentum*) fruit. This novel Inhibitor I is one of the ethylene induced proteins in tomato fruit whose amino acid sequence was deduced from a cDNA isolated from a library of cDNAs prepared from ethylene-induced mRNAs. The ethylene-induced Inhibitor I, while clearly a member of the Inhibitor I family, exhibits only 67 % identity at the nucleotide level, and 50 % identity at the protein level, with the wound-induced tomato leaf Inhibitor I. The inhibitor in fruit was probably not detected earlier by immunological techniques because of its lack of immunological cross reactivity when assayed with rabbit anti-tomato leaf Inhibitor I in Ouchterloney assays. The ethylene-induced Inhibitor I is synthesized at a stage in fruit development coinciding with ripening. It is not yet known if the ethylene-induced proteinase Inhibitor I is present in fruit of wild tomato species.

III. Wound-Inducible Proteinase Inhibitor Genes in Leaves

The wound-induction of proteinase inhibitors in leaves has been documented in species of several plant families. The most extensively studied wound-inducible inhibitors have been the Inhibitor I and II families in tomato and potato from the Solanaceae (Ryan, 1978; Plunkett *et al.*, 1982) and, more recently the Bowman-Birk inhibitor in leaves of alfalfa (Brown *et al.*, 1985; Brown and Ryan, 1986), from the Fabaceae family.

Wound-inducible genes coding for members of the Inhibitor I and II families have been isolated from both tomato and potato gene libraries (Lee *et al.*, 1986; Cleveland *et al.*, 1987; Thornburg *et al.*, 1987; Sanchez-Serrano *et al.*, 1987). The general features of members of these two gene families are shown in Figure 3. Inhibitor I and II genes are typical of many other eucaryotic genes, having CAT and TATA boxes, polyadenylation signals and introns. Each gene contains an intron in the signal, or transit sequence whereas the Inhibitor I genes contain an additional intron within the coding region of the native protein. The Inhibitor I primary amino acid

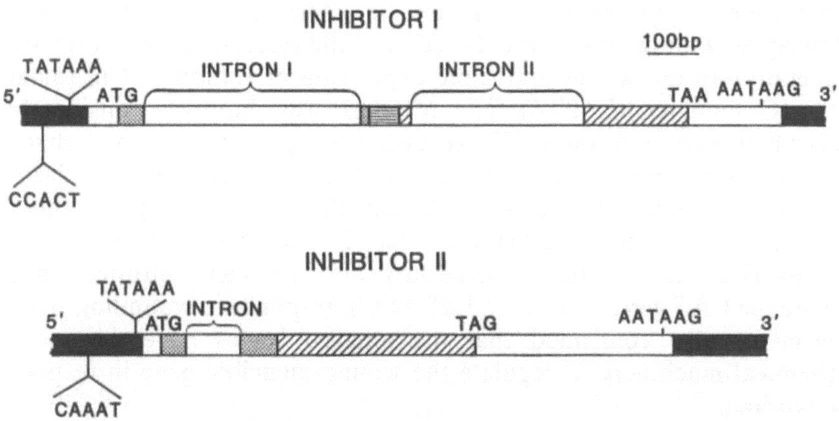

Fig. 3. Linear maps showing the general features of the Inhibitor I and II families of serine proteinase inhibitor genes from potato and tomato. Black areas are untranscribed 5′ and 3′ regions of the genes; dotted areas are the regions of the signal or transit sequences; areas with horizontal lines represent a 'pro'-sequence found only in the Inhibitor I genes; areas with slanted lines represent the open reading frames of the native proteins. Various other sequences shown are putative 'CAT' boxes, TATA boxes, polyadenylation signals, translation initiation and termination codons, and introns

sequence is apparently further processed beyond the cleavage of its transit sequence, as evidenced by a highly charged nineteen amino acid segment that is present in the translated protein between the transit sequence and the native protein sequence. The cellular location where the processing takes place is not known.

The Inhibitor II gene from potato has been studied in more detail. This inhibitor gene was isolated from the Russet Burbank variety (Thornburg *et al.*, 1987) and from the variety HH80 1201/7 (Sanchez-Serrano *et al.*, 1987). In one study (Thornburg *et al.*, 1987), a chimeric gene was constructed containing 1000 bp of the 5′ and 3′ regions of a wound-inducible gene (Inhibitor IIK) fused with the open reading frame of the chloramphenicol acetyltransferase gene, or CAT, to produce a sensitive, reporter enzyme to monitor wound-inducible gene expression in tobacco or other plants transformed with the fused gene. A second chimeric gene was constructed using 1000 bp of the 5′ region of the Inhibitor IIK gene fused with the CAT open reading frame, but terminated with the 3′ region of a constitutive gene, the 6b gene of the Ti plasmid. This 3′ region was used previously to terminate a fusion between the *nos* promoter and the CAT open reading frame in which CAT was constitutively expressed in transformed tobacco plants (An, 1986; 1987). When transferred into tobacco cells via a binary Ti vector system (An, 1986), and plants regenerated from the transformed cells, the CAT gene was expressed in leaves of the plants in response to wounding only in plants transformed with the fused gene containing both the 5′ and 3′ from Inhibitor II. Plants transformed with the construct containing the 6b terminator expressed the CAT gene weakly, either constitutively or when wounded. The wound-inducible CAT expression was similar, both in the time course of the response (maximizing at about 12—16 hr following wounding) and in the direction of the systemic effects (primarily acropetally), to the wound-inducible expression of Inhibitor II in tomato plants (Graham *et al.*, 1986). The response was, however considerably weaker than expected; the CAT expression being only about 10 % that of CAT expression under the influence of the *nos* promoter and the 6b terminator. Overall, the data confirmed that the 1000 bp of the 3′ noncoding region plus 1000 bp of the 5′ noncoding region of the Inhibitor IIK gene contained necessary and sufficient information to regulate the CAT gene in tobacco leaf cells in response to wounding. These experiments also confirmed that the tobacco leaves have the proper biochemical machinery to regulate the wound-incucible gene in response to wounding.

An Inhibitor II gene has been isolated (Sanchez-Serrano *et al.*, 1987) from the genome of potato variety HH80 1201/7 that is nearly identical to the Inhibitor II gene from the Russet Burbank potato genome (i. e. Fig. 3, bottom). The gene, containing 3 kb of the 5′ region and 1.45 kb of the 3′ region, was transferred into tobacco plants intact. Little or no expression of the gene could be detected in unwounded tobacco plants whereas wounding, or treatment of detached leaves with the chitosan and oligogalacturonic acid, led to a systemic induction of Inhibitor II mRNA.

Wounding the transformed plants led to the systemic accumulation of Inhibitor II mRNA in nonwounded leaves, stems and roots. The Inhibitor II gene was not found in the tobacco genome in untransformed plants, even though the plants have the capacity to regulate the transferred potato gene in response to wounding.

The Bowman-Birk proteinase inhibitor is strongly induced to accumulate in alfalfa leaves by wounding (Brown and Ryan, 1984). The response is similar in virtually all respects as the wound-induction of Inhibitors I and II in tomato and potato leaves (Brown *et al.,* 1984; 1985). Detached leaves supplied with either pectic fragments or chitosan respond by synthesizing the inhibitor for several hours. A cDNA library has recently been prepared from alfalfa leaf mRNA that had been enriched in Bowman-Birk inhibitor mRNA by wounding. A full length cDNA coding for the inhibitor has been isolated and characterized (Mukherjee, S., McGurl, B., Ryan, C. A., in preparation). A complete library of alfalfa genes has been prepared in bacterophage lambda (Mukherjee S., Ryan, C. A., unpublished data) and is being screened for the wound-inducible Bowman-Birk gene. Since the complete sequence of the wound inducible alfalfa leaf inhibitor is already known (Brown *et al.,* 1985) the identity of the gene can be verified. Thus, three unrelated proteinase inhibitor family genes will be available to compare the nucleotide sequences of wound-inducible promoters and to transform a variety of plants with heterologous proteinase inhibitor genes to test the functionality of different wound-inducible systems and to test the natural defensive capabilities of the different plant species having different proteinase inhibitors that have different inhibitory specificities.

IV. Summary

The diversity of serine proteinase inhibitor proteins in plants, in their evolutionary origins, specificities toward serine endopeptidases, tissue specificities and modes of regulation, as well as their importance in natural plant protection and their nutritional significance in agriculture and human health, have made their genes attractive for studies of the *cis*-acting and *trans*-acting elements that regulate their tissue specific developmental and environmental expression. Several proteinase inhibitor genes have been isolated and characterized, and in one gene, the developmentally regulated soybean Kunitz trypsin inhibitor, putative *cis*- and *trans*-acting elements have been identified. Research on other genes, both developmentally and environmentally regulated, is rapidly progressing along similar lines. Such studies are bringing these systems toward an understanding of the fundamental mechanisms of proteinase inhibitor gene regulation. Knowledge of these new research areas should significantly advance the potentials for technological applications in both agriculature and human health.

Acknowledgements

The author wishes to thank his colleagues, coworkers and students for reading the manuscript and for their helpful comments and suggestions. The support of NSF, the USDA and Washington State University of the research performed in my laboratory is gratefully acknowledged.

V. References

An, G., 1986: Development of plant promoter expression vectors and their use for analysis of differential activity of nopaline synthase promoter activity in transformed tobacco cells. Plant Physiol. **81,** 86—91.

An, G., Ebert, P. R., Yi, B.-Y., Choi, C.-H., 1986: Both TATA box and upstream regions are required for the nopaline synthase promoter activity in transformed tobacco cells. Mol. Gen. Genet. **203,** 245—250.

Balls, A. K., Ryan, C. A., 1963: Concerning a chymotryptic inhibitor from potatoes and its binding capacity for the enzyme. J. Biol. Chem. **238,** 2976—2982.

Brown, W. E., Koji, T., Titani, K., Ryan, C. A., 1985: Wound-induced trypsin inhibitor in alfalfa leaves: Identity as a member of the Bowman-Birk inhibitor family. Biochemistry **24,** 2105—2108.

Cleveland, T. E., Thornburg, R. W., Ryan, C. A., 1987: Molecular characterization of a wound-inducible inhibitor gene from potato and the processing of its mRNA and protein. Plant Mol. Biol. **8,** 199—207.

Goldberg, R. B., 1986: Regulation of plant gene expression. Phil. Trans. R. Soc. Lond. B **314,** 343—353.

Graham, J. G., Hall, G., Pearce, G., Ryan, C. A., 1986: Regulation of synthesis of proteinase inhibitors I and II mRNAs in leaves of wounded tomato plants. Planta **196,** 399—405.

Hejgaard, J., Rasmussen, S. K., Brandt, A., Svendsen, I., 1985: Sequence homology between barley endosperm protein Z and protease inhibitors of the alpha-1-antitrypsin family. FEBS Lett. **180,** 89—94.

Jofuku, D. K., Okamuro, J. K., Goldberg, R. B., in press. An embryo DNA binding protein that interacts with a soybean lectin gene upstream region.

Laskowski, M., Jr., 1986: Protein inhibitors of serine proteinases-mechanism and classification. In: Nutritional and Toxicological Significance of Enzyme Inhibitors in Foods (ed. M. Freidman), pp. 1—17. Plenum, New York.

Laskowski, M., Jr., Kato, I., 1980: Protein inhibitors of proteinases. Ann. Rev. Biochem. **49,** 593—629.

Laskowski, M., Jr., Sealock, R. W., 1971: Protein proteinase inhibitors-molecular aspects. The Enzymes, Vol. III (P. Boyer, ed.), pp. 375—483. Academic Press, New York.

Lee, J. S., Brown, W. E., Graham, J. S., Pearce, G., Fox, E., Dreher, T. W., Ahern, K. G., Pearson, G. D., Ryan, C. A., 1986: Molecular characterization and phylogenetic studies of a wound-inducible proteinase inhibitor gene in *Lycopersicon* species. Proc. Natl. Acad. Sci. U.S.A. **83,** 7277—7281.

Lincoln, J. E., Cordes, S., Read, E., Fischer, R. L., 1987: Regulation of gene expression by ethylene during *Lycopersicon esculentum* (tomato) fruit development. Proc. Natl. Acad. Sci. U.S.A. **84,** 2793—2797.

Pearce, G., Liljegren, D., Ryan, C. A., in press. Isolation and characterization of proteinase Inhibitors I and II from fruit of a wild tomato species (*Lycopersicum peruvianum*).

Plunkett, G., Senear, D. F., Zuroske, G., Ryan, C. A., 1982: Proteinase inhibitor I and II from leaves of wounded tomato plants: Purification and properties. Arch. Biochem. Biophys. **213,** 463—472.

Rackis, J. J., 1980: Protease inhibitors: Physiological properties and nutritional significance. In Antinutrients and Natural Toxicants in Foods (R. L. Ory, ed.), pp. 203—238. Food and Nutrition Press Inc. Westport, Conn.

Rosahl, S., Eckes, O., Schell, J., Willmitzer, L., 1986: Organ-specific gene expression in potato: Isolation and characterization of tuber-specific cDNA sequences. Mol. Gen. Genet. **202,** 368—373.

Ryan, C. A., 1973: Proteolytic enzymes and their inhibitors in plants. Ann. Rev. Plant Physiol. **24,** 173—196.

Ryan, C. A., 1978: Proteinase inhibitors in plant leaves: A biochemical model for pest-induced natural plant protection. TIBS **5,** 148—151.

Ryan, C. A., 1984: The defense mechanisms in plants. In: Plant Gene Research (E. S. Dennis, B. Hohn, Th. Hohn, P. King, J. Schell, D. P. S. Verma, eds.), pp. 375—386. Wien - New York: Springer-Verlag.

Ryan, C. A., Pearce, G., An, G., Thornburg, R., in press. The regulation of expression of proteinase inhibitor genes in food crops. Acta Horticulturae.

Sanchez-Sorrano, J. J., Keil, M., O'Connor, A., Schell, J., Willmitzer, L, 1987: Wound-induced expression of a potato proteinase inhibitor II gene in transgenic tobacco plants. EMBO J. **6,** 303—306.

Seemuller, U., Eulitz, M., Fritz, H., Strobl, A., 1980: Structure of the elastase-cathepsin G inhibitor of the leech *Hirudo medicinalis*. Hoppe-Seyler's S. Physiol. Chem. **361,** 1841—1846.

Svendsen, I., Hejgaard, J., Chavan, J. K., 1984: Subtilisin inhibitor from seeds of broad bean (*Vicia faba*); purification, amino acid sequence and specificity of inhibition. Carlsberg Res. Commun. **49,** 493—502.

Svendson, I., Jonassen, I., Hejgaard, J., Boisen, S., 1980: Amino acid sequence homology between a serine protease inhibitor from barley and potato inhibitor I. Carlsberg Res. Commun. **45,** 389—395.

Thornburg, R. W., An, G., Cleveland, T. E., Johnson, R., Ryan, C. A., 1987: Wound-inducible expression of a potato inhibitor II-chloramphenicol acetyltransferase gene fusion in transgenic tobacco plants. Proc. Natl. Acad. Sci. U.S.A. **84,** 744—748.

Walling, L., Drews, G. N., Goldberg, R. B., 1986: Transcriptional and posttranscriptional regulation of soybean seed protein mRNA levels. Proc. Natl. Acad. Sci. U.S.A. **83,** 2123—2127.

Wilson, K., 1986: Role of proteolytic enzymes in the mobilization of protein reserves in the germinating dicot seed. In: Plant Proteolytic Enzymes, Vol. II (M. J. Dalling, ed.), pp. 411—468. CRC Press, Inc., Boca Raton, Fla.

Chapter 13

Cell Wall Extensin Genes

James B. Cooper

Department of Biological Sciences, Stanford University, Stanford, California, U.S.A.

With 3 Figures

Contents

I. Cell Walls

Multicellular organisms are composed of coordinately regulated cells embedded in a secreted extracellular matrix. This secreted cell wall integrates the biophysical structure of groups of cells to form tissues, and regulates the growth and development of the entrapped cells by a variety of mechanisms. An excess of fixed carbon allowed plant cells to evolve walls with a high content of structural polysaccharides, the properties of which caused plants to evolve unique cellular strategies for regulating cell growth and development.

A general consensus regarding the structure of the cell wall is based on the studies of plant walls by chemists and biologists for the past 150 years. Crystalline cellulose microfibrils are assembled at the outer surface of the plasmamembrane and extruded into a wall matrix composed of polysaccharide "hemicelluloses" and "pectins", proteoglycans, and glycoproteins.

The cellulose microfibrils associate laterally into fiber bundles, and "hemicellulosic" xyloglucans hydrogen bond to the surface of the fibrils to form a hydrophilic coat. These oriented cellulose fibers provide the plant cell wall with a structural element of high tensile strength. This reinforces the wall matrix and allows plant cells to develop large turgor pressures which provide the driving force for cell growth. Neutral and acidic pectins accumulate on the outer surface of the wall to form an adhesive slime which immobilizes plant cells to each other. The middle lamella thus formed between neighboring cells prevents cell-cell movement during development.

Plant cell size and shape are both thought to be determined primarily by the spatial orientation of the cellulose microfibrils and the rheological properties of the wall matrix. Thus, wall structure plays an important role in plant morphogenesis. The role of the wall as a protective barrier for plant cells is equally important to the life of the plant. An effective strategy for disease resistance in plants is to rapidly modify the cell wall in ways which prevent further pathogen invasion and limit cell damage.

The cell wall also determines the economic value of many plant products. The use of plants for textiles and paper is based directly on the properties and purity of the cellulosic cell wall. Food texture is determined in large part by the properties of plant cell wall polymers; for example, the softness of many fruits is a consequence of the programmed degradation of the fruit cell walls. Pectins and gums are routinely added to processed foods to modify texture.

The occurrence of structural polypeptides outside the plasmamembrane of plants was implied by the 1960 discovery that the cell wall contains most of the hydroxyproline in plant cells. The implications of this discovery were profound: plant extracellular matrices might contain structural polypeptides, functionally analogous to animal extracellular matrix components, woven into the polysaccharide matrix. Unfortunately, this observation contradicted the established dogma that cell walls were composed entirely of carbohydrate, and general acceptance of cell wall polypeptides as important structural elements has been slow. Only with the recent advances in characterizing one class of cell wall hydroxyproline-rich polypeptides (the extensins) has the view that cell walls contain important structural proteoglycans become generally accepted. Before discussing extensin in detail, mention should of made of the other classes of cell wall polypeptides which have been described. 1. A large number of extracellular enzymes have been reported to be present in plant cell walls, including peroxidases, ascorbate oxidase, acid phosphatase, invertase, and a number of cell wall hydrolytic enzymes (see Lamport and Caat, 1981). The presence of enzymes and enzyme systems in the wall emphasizes the fact that the cell wall is a dynamic extracellular compartment of the plant cell. 2. Arabinogalactan-proteins, or AGPs, are large neutral and acidic arabinogalactans, 0-linked to hydroxyproline-rich polypeptides, which are thought to be ubiquitous in green plants but have no known function. AGPs are easily distinguished from extensins by amino acid and carbohydrate composition (Fincher *et al.*, 1983). 3. A novel class of glycine-rich structural cell wall proteins has recently been discovered about which virtually nothing is known (Condit *et al.*, 1986).

II. Extensin Networks

A. Insoluble Extensin

The discovery that cell walls contain most of the hydroxyproline in plants was rapidly followed by the general observation that the bulk of the hydroxyproline is contained in a wall polymer which remains firmly bound in the cell wall using nonproteolytic extraction methods. Obviously, the inability to extract the hydroxyproline-rich material without hydrolysis greatly hampered the characterization of this cell wall component. Despite this handicap, Lamport named the hydroxyproline-rich cell wall glycoprotein extensin in 1963, and began to characterize fragments of this insoluble wall polymer from suspension culture cell walls. Lamport's early biochemical analyses indicated clearly that extensin is an unusual glycopolypeptide with an extremely biased amino acid composition (Hyp, Ser, Tyr, Lys, and Val) and a simple pattern of glycosylation (Ser-O-Gal_{1-3} and Hyp-O-Ara_{3-4}). Using mild acid hydrolysis to deglycosylate the wall-bound extensin polypeptide, followed by trypsin digestion, Lamport solubilized and characterized five hydroxyproline-rich peptides from tomato cell walls which account for about 35 % of the cell wall hydroxyproline (Lamport, 1977). Remarkably, all five peptides contain copies of the pentapeptide Ser-Hyp-Hyp-Hyp-Hyp, indicating that extensin has a repeating polypeptide structure (Fig. 1 A). Two of the peptides also contain diphenyl-ether-linked isodityrosine (Fry, 1982), as an intra-molecular crosslink (Epstein and Lamport, 1984).

The prolonged failure to isolate and characterize a soluble precursor form of insoluble extensin led to an incomplete understanding of this cell wall component, and led many to question the importance of this cell wall protein in wall structure and function. Recently, however, independent approaches have converged to provide an exciting and sophisticated picture of the structure of this cell wall polymer system. The picture of extensin which emerges is that of a covalently crosslinked three-dimensional network formed by the intra- and inter-molecular crosslinking of a family of soluble extensin monomers. This view is based on 1. the isolation and characterization of several soluble hydroxyproline-rich glycoproteins from a variety of plant tissues which closely resemble the insoluble tomato glycopeptides, 2. the demonstration that these soluble glycoproteins are slowly insolubilized following secretion into the cell wall, and 3. the direct visualization of monomeric extensin and extensin polymers in the electron microscope.

Extensin Peptide Sequences

```
A. Tomato Cell Wall
                                  ┌-O-┐
      S O O O O S O S O O O O     Y Y Y K

                  S O O O O       S O K

                  S O O O O       K

                  S O O O O       V Y K
                                    ┌-O-┐
                  S O O O O       V Y K Y K

B. Tomato P1
                  S O O O O       T O V Y K

                  S O O O O       V K P Y H P T O V Y K

C. Tomato P2
                  S O O O O       V Y K (Y K)
```

Fig. 1. Repeating amino acid sequence motifs identified in extensin polypeptides. A. Insoluble cell wall extensin (Lamport, 1977). B., C. Soluble extensin monomers P1 and P2 (Smith *et al.*, 1986)

B. Soluble Extensins

The observation that carrot roots deposit large amounts of hydroxyproline in the cell wall following slicing and aeration (Chrispeels, 1969) played an important role in the elucidation of the structure and biosynthesis of extensin. A series of papers by Chrispeels and his colleagues describes the biosynthetic pathway of this hydroxyproline-rich cell wall component. In wounded carrot root cells, this cell wall polymer is synthesized by the sequential translation of extensin mRNA on rough endoplasmic reticulum, hydroxylation of peptidyl proline by a prolyl hydroxylase which resembles the animal prolyl hydroxylase, glycosylation of Hyp by oligo-arabinosides and of Ser by galactose in the golgi apparatus, and secretion into the cell wall (see Chrispeels *et al.*, 1974). Varner's laboratory subsequently purified and characterized the hydroxyproline-rich glycoprotein from wounded carrot root (Stuart and Varner, 1980 Van Holst and Varner, 1984). Most of the hydroxyproline accumulates in a single soluble 80 kilodalton glycoprotein containing about one-third protein and two-thirds carbohydrate. The composition of the carrot glycoprotein resembles the composition of Lamport's insoluble extensin tryptides: the 33 kilodalton hydroxyproline-rich polypeptide has a biased amino acid composition (Hyp, Ser, Tyr, Lys, His, and Val), and is heavily glycosylated with short tri- and tetra-arabinosides 0-linked to hydroxyproline, and small amounts of galactose. The isoelectric point of carrot extensin is extremely high (pI = 10—11) due the high content of Lys and His.

Studies on the secondary structure of this soluble carrot extensin using circular dichroism revealed that 100% of the peptide bonds are in the polyproline II helical conformation. The structure of soluble extensin was predicted to have an extended rod shape, and electron micrographs of the native glycoprotein confirmed this hypothesis: soluble extensin appears as an 80 nm long rod with kinks along its length (Van Holst and Varner, 1984; Stafstrom and Staehelin, 1986a). Only 50% of the amino acids have the polyproline II conformation in the deglycosylated molecule indicating that the carbohydrate side chains play a role in maintaining the polypeptide conformation (Van Holst and Varner, 1984). The single polypeptide polyproline II helix, is thought to be stabilized by hydrogen bonding between the oligo-arabinoside chains, attached to Hyp, and the peptide backbone. This model is supported by electron microscopic visualization of deglycosylated extensins which have lost the extended structure and appear as an amorphous globular mass (Stafstrom and Staehelin, 1986b).

Soluble extensins have now been isolated from a number of plant species and tissues. "Bacterial agglutinins", purified from potato tubers (Leach *et al.*, 1982) and tobacco callus (Mellon and Helgeson, 1982), are virtually identical in composition to the carrot glycoprotein. It is probable that these glycoproteins are soluble extensin monomers which act as nonspecific bacterial agglutinins because of their high isoelectric point. Two different types of soluble extensins (named P1 and P2) have been isolated and characterized from tomato cell suspension cultures (Smith *et al.*, 1984, 1986). Both types contain Ser-Hyp-Hyp-Hyp-Hyp pentapeptide repeats but differ in the repeating sequence motifs which flank this common pentapeptide (Fig. 1B, C).

C. Extensin Crosslinking

In both carrot and tomato, soluble extensins are slowly insolubilized following secretion into the cell wall, having soluble half lives of about 12 hours (Cooper and Varner, 1983; Smith *et al.*, 1984). Isodityrosine, synthesized in the cell wall during insolubilization (Cooper and Varner, 1983), is thought to form both intra- and inter-molecular crosslinks which serve to entangle the extensin monmers with cell wall polysaccharides and covalently crosslink the monmers to each other to form a three dimensional glycoprotein network. Direct support for a network model of extensin structure was obtained using electron microscopy to visualize size-fractionated carrot extensin (Stafstrom and Staehelin, 1986a). This work clearly demonstrates the existence of soluble extensin dimers, trimers, tetramers, and higher oligomers which form an open network structure by forming inter-molecular crosslinks primarily between the ends of glycoprotein rods (Stafstrom and Staehelin, 1986a). The observation that the two different tomato extensins are both incorporated into the insoluble extensin network with comparable kinetics demonstrates that plant cells can construct different types of extensin networks depending upon which types of extensin monomers are secreted into the wall.

The formation of this insoluble network is inhibited by antioxidants, especially ascorbate, and free radical scavengers (Cooper and Varner, 1983). This supports a phenolic crosslinking mechanism for the formation of insoluble extensin. *In vitro* experiments indicate that the crosslinking is inhibited by acid in the physiological pH range (Cooper and Varner, 1984). Plant cells may thus regulate the polymerization of extensin by secreting ascorbate or protons, or by secreting enzymes, such as ascorbate oxidase (Lamport, 1965), or extensin peroxidase (Fry, 1987).

Despite substantial progress made in the biochemical characterization of cell wall extensin, the repetitive blocks of amino acids and the intra- and inter-molecular crosslinking reactions have hindered all efforts to obtain the primary sequence of extensin by the peptide sequencing approach. An alternative strategy based on the power of recombinant DNA technology has recently been utilized to obtain the complete primary structure of carrot extensin. This approach is leading to a much broader understanding of the structure and function the cell wall extensin network.

II. Extensin Genes

A. Cloning Extensins

In addition to obtaining the complete primary structure of extensin's polypeptide backbone, the cloning of extensin has opened the doors for a variety of investigations on the structure, function and biosynthesis the extensin network, and on the genetic regulation of cell wall structure in general. Early attempts to identify extensin mRNAs in wounded carrot root tissues utilized strategies based on extensin's repeating Ser-Hyp-Hyp-Hyp-Hyp pentapeptide sequence. Smith (1981) unsuccessfully attempted to immunoprecipitate an extensin precursor from the *in vitro* translation products of wounded carrot root mRNA by raising polyclonal antibodies against the synthetic peptide Ser-Pro-Pro-Pro-Pro (precursor to Ser-Hyp-Hyp-Hyp-Hyp). Stuart *et al.* (1982) reasoned that since the codon for proline is CCX, the extensin mRNA should contain repeats of CCXCCXCCXCCX and thus might hybridize to an oligo-dG cellulose column. A hybrid-selected mRNA was obtained by this method which translated into a putative extensin precursor, a 55 kilodalton proline-rich, leucine-poor polypeptide, but the yield of putatitive extensin mRNA obtained by this approach was too low to permit cDNA cloning.

Chen and Varner (1985 a) succeeded in isolating an extensin cDNA clone from a cDNA library prepared from size-fractionated wounded carrot mRNA enriched in the mRNA encoding a putative extensin polypeptide precursor. The library was laboriously screened by using the clones to hybrid-select wounded carrot mRNA which was then used for parallel *in vitro* translations with [^{3}H]-proline and [^{3}H]-leucine. One clone, pDC5, was identified which hybrid selected an mRNA encoding a 33 kilodalton proline-rich, leucine-poor polypeptide. The DNA sequence of the 730 base pair cDNA insert of pDC5 revealed an open reading frame of 462 nucleotides encoding a polypeptide with two distinct domains. The C-terminal domain contains four repeats of the canonical extensin pentapeptide Ser-Pro-Pro-Pro-Pro, while the N-terminal portion of the polypeptide contains nine repeats of a different proline-rich pentapeptide, His-Lys-Pro-Pro-Val/Ile. In a subsequent cloning experiment, pDC5 was used to screen another carrot cDNA library. Several positive clones were characterized by sequence analysis. Two of the clones were homologous only the 5′ half of the clone and one clone was homologous only to 3′ half of the clone. Hybridization to pDC5 was used to isolate a carrot genomic clone, pDC5A1 (Chen and Varner, 1985b). The DNA sequence of pDC5A1 shows perfect homology with the 3′ half of pDC5, except for a putative intron in the 200 base pair 3′-nontranslated region, and diverges completely from the sequence of the 5′ half of pDC5. Chen and Varner concluded that the pDC5 insert is a cloning artefact resulting from the ligation of two distinct cDNAs, one of which encodes extensin.

The genomic extensin clone pDC5A1 has now been used by a number of laboratories to isolate homologous sequences from different plant species, including petunia, tomato, tobacco, potato, melon, bean, pea, and sunflower (Personal communications), although only a few of these reports have been published. Showalter *et al.* (1985) screened a tomato genomic library with pDC5 and isolated pTom5 which includes an 1100 base pair open reading frame containing 36 Ser-Pro-Pro-Pro-Pro repeats. Neither the carrot nor the tomato extensin genes contain introns in the polypeptide coding region. It will be interesting to see whether this is a general feature of all extensin genes. An extensin cDNA clone from wounded tomato stems (pTom17-1) was isolated by homology to pTom5. This clone contains an open reading frame encoding 150 amino acids, and contains Ser-Pro-Pro-Pro-Pro pentapeptides in large repeating domains which differ substantially from the repeats in pTom5 (Showalter and Varner, 1987).

Extensin clones hybridize to several bands on Southern blots of genomic DNA from a number of dicot plants species, and to several bands on Northern blots from different plant tissues. Since there is clear biochemical evidence for the existence of several closely related extensin proteins containing different repeating sequence motifs (Fig. 1), it is generally agreed that extensin polypeptides are encoded by a mutigene family. Using carrot and tomato probes, Corbin *et al.* (1987) have isolated gene-specific cDNA clones for three members of the extensin gene family of french bean. Under high stringency, each of the bean clones, pHyp3.6, pHyp2.13, and pHyp4.1, hybridizes a different restriction fragment of genomic DNA, and to distinct RNA transcripts (4.4 kb, 3.6 kb, and 2.5 kb respectively). Isolation and characterization of gene-specific probes is an important and necessary first step for investigating the structure, function and regulation of the extensin multigene family. The sequences of all three bean cDNA inserts contain multiple copies of the Ser-Pro-Pro-Pro-Pro pentapeptide, and are closely related to each other. Two of the clones, pHyp3.6 and pHyp2.13, contain 3′ polyA tails and distinct break points in the sequence which separate the repeating proline-rich domain from a globular domain (pHyp2.13) or from a region containing numerous translation stop codons (pHyp3.6). The third bean cDNA clone, pHyp4.1, lacks a polyA tail completely and appears to encode the N-terminal peptide of an extensin. Thus, all three bean cDNAs may be the result of artifactual cloning fusions between extensin and some cDNAs, as was the case for the original carrot extensin clone pDC5. Perhaps the repetitive nature of the extensin coding region leads to sequence rearrangements in *E. coli.*

B. Functional Domains

Glycosylation

The quintessential peptide of extensin is the repeated Ser-Hyp-Hyp-Hyp-Hyp pentapeptide found by Lamport in all of the insoluble extensin peptides. Ser-Pro-Pro-Pro-Pro serves as the biosynthetic precursor of Ser-Hyp-Hyp-Hyp-Hyp, and is thus the common repeating peptide in all of the cloned extensins (Fig. 2). There is no doubt that this repeated peptide functions as the major glycosylation domain of extensin. Following the post-translational hydroxylation of the peptidyl-proline residues by prolyl hydroxylase, the hydroxylated pentapeptide is recognized by golgi-localized glycosyltransferases which add the arabinoside and galactoside substituents. Generally, more than 90 % of the Hyp residues in extensin are glycosylated with tri- and tetra-arabinosides. As discussed above, the arabinosyl side chains are thought to stabilize the polyproline II conformation of the polypeptide. Thus these repeated peptides serve as short, rigid, structurally reinforced hydrophilic rods (about 1.5 nm in length per pentapeptide). The pentapeptide glycosylation domains are distributed throughout the length of the polypeptide backbone of the extensins which have been sequenced. All of the DNA-derived extensin sequences contain a higher order repeat composed of two pentapeptide glycosylation domains separated by specific repeated dipeptides (Fig. 2). Both bean and tomato extensins contain the repeat Ser-Pro-Pro-Pro-Pro-Ser-Pro-Ser-Pro-Pro-Pro-Pro (hydroxylated in the peptide sequences) which almost certainly forms a longer reinforced rod domain, about 3.5 nm long. The occurrence of the Lys-His dipeptide separating pentapeptide blocks may provide a "pectin-binding" domain by maintaining the spacing between the positive charges on the amino acid side chains at an appropriate distance to bind adjacent carboxyl groups on polygalacturonic acids.

The occurrence of tetraproline stretches in the extensin backbone is particularly interesting in light of the proposed role of proline as a rate limiting step in protein folding (Creighton, 1978). *Cis-trans* isomerization of peptide bonds involving proline is slower than the folding of other peptide bonds. Thus the isomerization of blocks of polyproline in extensin should be quite slow, allowing the possibility that post-translational modi-

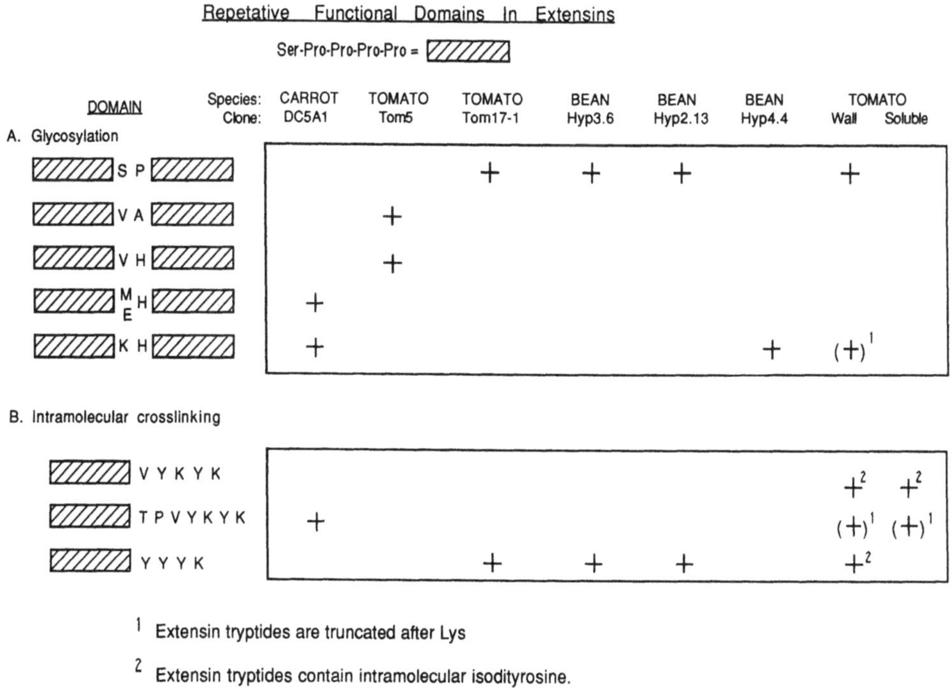

Fig. 2. Occurrence of higher order peptide repeats in extensins which involve the pentapeptide building block Ser-Pro-Pro-Pro-Pro, as deduced from DNA sequences. A. Glyosylation domains. B. Intramolecular crosslink domains

fication could further influence the form and function of extensin. Variations in the length of the polyproline blocks have been observed in most of the extensin sequences which might have some functional significance. The occurrence of Ser-Pro-Pro-Pro-Pro-Pro and Ser-Pro-Pro-Pro-Pro-Pro-Pro blocks may provide slow points in the folding of extensin which allow the final folding steps to occur long after the glycoprotein monomers have been secreted into the cell wall and perhaps even after crosslinking.

Phenolic Crosslinking

The insoluble extensin peptides characterized by Lamport contained a tyrosine derivative which was later identified as isodityrosine (Lamport, 1977; Fry, 1982). Isodityrosine occurs as an intra-molecular crosslink in at least two repeating peptides, Val-Tyr-Lys-Tyr-Lys and Tyr-Tyr-Tyr-Lys, which are each found in several of the deduced extensin sequences. This diphenylether crosslink is formed between tyrosine side chains in a single polypeptide which are separated by Lys or Tyr. In carrot extensin, the Val-Tyr-Lys-Tyr-Lys peptides are distributed throughout the length of the extensin rod, which correlates with the distribution of kinks in the glycoprotein rods observed in electron micrographs (Stafstrom and Staehelin, 1986). On this basis, it has been proposed that these intramolecular crosslinks cause kinking of the extensin rods.

The existence of inter-molecular isodityrosine crosslinks is an important feature of current models for extensin networks. It is therefore disconcerting that no inter-molecularly crosslinked extensin glycopeptides have yet been characterized. The existence of these (intermolecular) crosslinks is clearly visualized in the electron micrographs of extensin oligmers. Careful measurements of the localization of the intermolecular crosslinks along the length of extensin molecules in their micrographs, allowed Stafstrom and Staehelin (1986a) to show that most of the intermolecular crosslinks occur at the ends of extensin rods. Comparison of the deduced C-terminal sequences of different extensins reveals a conserved nonapeptide, Tyr-Ser-Tyr-Ala-Ser-Pro-Pro-Pro-Pro, followed by a tri- or penta-peptide containing one or two Tyr residues (Fig. 3). In three of the sequences, Tyr is the C-terminal amino acid (Fig. 3). A similar nonapeptide also occurs at the N-terminal end, preceding the first Ser-Pro-Pro-Pro-Pro pentapeptide, in the two extensin sequences which extend to the 5′ end of the coding region (Fig. 3). Comparison of the C-terminal polypeptide sequences deduced from the two tomato extensin clones is also enlightening. The polypeptide backbones of these two extensins share no homology beyond the Ser-Pro-Pro-Pro-Pro repeats (Fig. 2), yet the C-terminal 32 amino acids of these two extensins are quite homologous (Fig. 3). (In the case of the extensin polypeptide encoded by pTom5, the only two Tyr residues in the entire coding region occur in this C-terminal region of homology.) It is attractive to speculate that this C-terminal peptide

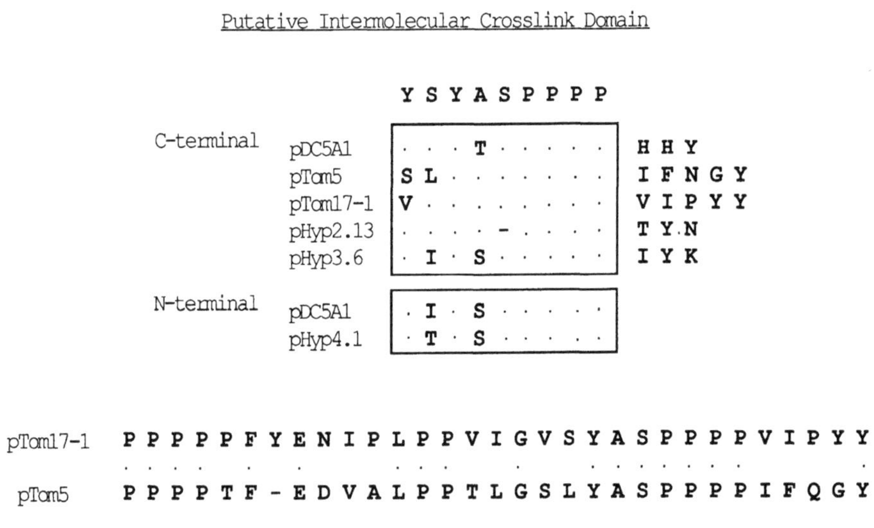

Fig. 3. Putative intermolecular crosslink domains in the C-terminal and N-terminal peptides of extensins

is functionally important, specifically as the site of intermolecular crosslinks formation.

One more peptide should be mentioned as a potential site for intermolecular crosslink formation. A predominant peptide of tomato extensin P1, (Ser-Hyp-Hyp-Hyp-Hyp-Val-Lys-Pro-Tyr-His-Pro-Thr-Hyp-Val-Tyr), is unusual among all of the extensin peptides which have been characterized because it contains two unhydroxylated Pro residues. Extensin P1 is known to be insolubilized in tomato cell walls (Smith *et al.*, 1984), and this peptide has not been recovered among Lamport's insoluble extensin peptides.

Globular Domains

All of the cloned extensins have a nonrepetitive peptide coding domain at the 5′ end of the sequence of the cloned insert. In the case of clones which include the 5′ end of the polypeptide coding region, (e. g. pDC5A1, pTom5, pHyp4.1), it is assumed that this represents the hydrophobic secretory signal peptide. In the case of the cDNA pHyp2.13, however, it is not clear whether this domain represents a unique functional cassette attached to a rigid extensin rod, or whether this represents a cloning artefact as occurred in pDC5. The hydroxyproline-rich potato lectin should be mentioned as one example of a protein which uses a rigid extensin domain attached to a globular cysteine-rich sugar binding domain (see Allen *et al.*, 1978).

C. Gene Regulation

Developmental Regulation

The deposition of cell wall extensin, as measured by the levels of wall hydroxyproline, has long been known to be developmentally regulated (Lamport, 1965; Cassab *et al.*, 1985). Although several observations have been made concerning the developmental expression of extensin genes, detailed studies on the molecular basis of this regulation have not yet been undertaken. Several transcripts which hybridize to extensin probes are present in rapidly growing shoot and root tissues (Chen and Varner, 1984, Showalter and Varner, 1987; Corbin *et al.*, 1987, Cooper, unpublished). The organ-specific expression of extensin mRNA, as measured by hydrization to Northern blots, generally appears to parallel the distribution of extensin glycoprotein. Thus, callus and roots contain high amounts of cell wall extensin and high levels of extensin mRNA, while leaves contain very low amounts of extensin protein and mRNA and shoots contain intermediate levels (Showalter and Varner, 1987; J. B. Cooper, unpublished).

The potential involvement of extensin in controlling cell elongation has not received any attention at the molecular level. A large number of studies have indicated that extensin may play some role in growth control. The current model for the structure of the extensin network predicts that cell wall extensibility should depend importantly on both the total amount of cell wall extensin, and on the amount of intermolecular crosslinking.

Studies with peas (Klis, 1976) and beans (Van Holst *et al.*, 1981) both demonstrate an inverse linear relationship between growth rate and extensin deposition. An isolated report claims that extensin deposition is also regulated by phytochrome (Pike *et al.*, 1979). Despite the importance of extensin in cell wall structure, and the potential role of extensin in regulating plant growth and development, molecular studies in this area have lagged far behind the molecular studies of the regulation of extensin gene expression by stresses.

Regulation by Wounding/Ethylene

The observation that wounded carrot root cells synthesize large amounts of hydroxyproline eventually led to the biochemical characterization and the molecular cloning of extensin. Because of this, the regulation of extensin synthesis in carrot roots has been studied in some detail (Chrispeels *et al.*, 1974; Chen and Varner, 1985a, 1985b; Ecker and Davis, 1987). Synthesis and secretion of the glycoprotein is induced within 4—8 hours after wounding and aeration. The carrot genomic clone pDC5A1 hybridizes to several transcripts from actively growing carrot roots: moderate amounts of a 1.5 kb extensin transcript and much lower amounts of 1.8 kb and 4.0 kb transcripts (Chen and Varner, 1985; Ecker and Davis, 1987). Following prolonged storage at 4C, all three extensin transcripts decrease in abundance, such that only weak hybridization to the 1.5 kb transcript is observed. Wounding and aeration of cold-stored roots leads to a dramatic 50—100X increase in the levels of all three transcripts in cold stored roots. The effects of wounding on actively growing carrot roots is qualitatively the same, but quantitatively only a 2—3X increase was observed. Extensin transcript levels increase 10X within 1.5 hours of wounding (Ecker and Davis, 1987).

The effect of wounding on extensin gene expression has also been investigated in tomato and bean tissues using Northern blots. Changes in the pattern of extensin gene transcripts are observed in wounded tomato stems, and depend upon which hybridization probe is used (Showalter and Varner, 1987). A 1.7 kb transcript with homology to pTom5 and pDC5A1 accumulates in tomato stem tissues following wounding, with a concomitant decrease in the level of a 1.2 kb transcript. A 4.9 kb transcript homologous to the tomato cDNA clone pTom17-1 also accumulates following wounding. No effect of wounding on extensin gene expression was observed in tomato leaves. In bean hypocotyls, increased levels of at least three extensin transcripts (corresponding to the three gene-specific probes) are induced by wounding (Corbin *et al.*, 1987). Within 1.5 hours, the levels of a 4.4 kb transcript homologous to pHyp3.6 increase dramatically. The abundance of this transcript remains constant for 12—18 hours. By 24 hours, the level of 4.4 kb transcript has nearly returned to level found in unwounded hypocotyls. Two transcripts, 3.3 kb and a 2.2 kb, which are homologous to pHyp2.13 and pHyp4.1, respectively, begin to accumulate after 4 hours, and reach a maximum within 12 hours. This maximum level is maintained through 24 hours. Nuclear runoff transcription assays demonstrate that the wound induction of extensin mRNA accumulation in bean hypocotyls is a result of transcriptional activation (Lawton and Lamb, 1987). Thus, these three extensin genes are differentially regulated by wounding. The length of the pHyp4.1 transcript (2.2 kb) is about 300 bases shorter than the homologous transcript induced by fungal elicitor (see below).

Ethylene is synthesized in response to wounding, and is known to cause the accumulation of cell wall extensin glycoprotein in pea and melon seedlings (Ridge and Osborne, 1970; Esquerre-Tugaye *et al.*, 1979). The effect of ethylene on extensin gene expression in carrot root tissues has recently been examined (Ecker and Davis, 1987). Treatment of growing roots with 10 ppm ethylene induces the accumulation of the 4.0 kb transcript. In contrast, treatment of cold stored carrot roots with ethylene and oxygen induces the accumulation of both the 4.0 kb and 1.8 kb transcripts, but not the 1.5 kb transcript which is induced by wounding. Interestingly, both the 1.5 kb and 1.8 kb transcripts are transcribed from different transcription start sites located within the promoter of a single extensin gene (Chen and Varner, 1985 b). This promoter contains a tandem set of transcription initiation signals separated from each other by a 300 bp region containing numerous long (18 bp) and short (9—12 bp) direct and inverted repeats. Since wounding induces the accumulation of both 1.5 kb and 1.8 kb transcripts from the same extensin gene, while only the longer 1.8 kb transcript is induced by ethylene, the signals which turn on extensin biosynthesis in wounded carrot roots must be more complicated than a simple "wound ethylene" switch. Clearly, this carrot extensin gene is regulated differentially by wounding and ethylene.

Regulation by Other Abiotic Stresses

Klis *et al.* (1983) reported that a variety of abiotic stresses which cause the formation of lesions on bean hypocotyls (e. g. heavy metals, cell wall hydrolytic enzymes, and bean cell wall fragments), also lead to the accumulation of cell wall extensin. Extensin deposition has also been reported to be a heat-shock response (Stermer and Hammerschmidt, 1985). The extent to which these responses to abiotic stresses are coordinated by ethylene remains to be seen.

Regulation by Pathogens/Elicitors

The structure of the cell wall changes dramatically in response to pathogen attack as the plant cell attempts to prevent or limit the invasion and spread of the pathogen. This plant defense response is induced either by the interaction of plant cells with the pathogen itself, or by treatment of the plant cells with soluble signal molecules called elicitors. The deposition of extensin is one of the changes in wall structure which is thought to play a role in disease resistance (Esquerre-Tugaye *et al.*, 1979 a, 1979 b; Hammerschmidt *et al.*, 1983), perhaps by acting as a highly charged agglutinin network, and/or by nucleating the polymerization of lignin precursors. The molecular basis of the induction of extensin biosynthesis by fungal infection and elicitors has been investigated in the *Phaseolus vulgaris-Colletrotrichum lindemuthianum* system (Showalter *et al.*, 1984; Lawton and Lamb, 1987; Corbin *et al.*, 1987).

Infection of *P. vulgaris* hypocotyls with an incompatible race of *C. lindemuthianum* evokes the resistant hypersensitive response at the inoculation site. After a lag of about 50 hours, the levels of two extensin mRNA transcripts (2.7 kb and 1.6 kb) increase rapidly by ten to twenty fold (over uninoculated controls) (Showalter *et al.*, 1985). A modest increase (about 5X) is also induced in cells which are distant from the inoculation site, indicating that a physical/chemical defense signal is released from the infection site. In contrast, in the compatible (susceptible) interaction, the levels of the same two extensin transcripts increase to about the same final level, but the rate of induction is significantly slower, such that the maximum is not reached until disease lesions have started to form. Systemic induction of extensin transcript accumulation is seen in the compatible interaction as well.

Using gene specific probes in the same system, Corbin *et al.* (1987) demonstrate that each of the three bean genes which they have cloned is regulated differentially during the host-pathogen interaction. All three genes are induced by both compatible and incompatible strains of the fungus. A 2.5 kb transcript homologous to pHyp4.1 was strongly induced above uninoculated controls in both the compatible and the incompatible interactions, and the transcript levels in the incompatible interaction decrease nearly to control levels after 134 hours. In contrast, the levels of this 2.5 kb transcript remain high in the compatible interaction for at least 139 hours. Hyp 3.6 is expressed in uninfected bean hypocotyls and was only weakly induced by the fungus. Hyp2.13 is actually induced more strongly in the compatible interaction, and the transcript levels remained high through the experiment (139 hours). As was observed using the heterologous tomato probe (Showalter *et al.*, 1985), the rate of induction of all three transcripts is faster in the incompatible (resistant) interaction. Hyp3.6 and Hyp4.1 were both induced systemically, while Hyp2.13 was only marginally induced in tissues distant from the physical site of host-pathogen interaction.

Bean cells treated with fungal elicitors also accumulate extensin mRNAs. Using the tomato probe Showalter *et al.* (1985) demonstrate that, following a 4 hour lag, the 2.7 kb transcript accumulates rapidly until 12 hours and levels out. The two other transcripts with homology to the tomato probe (1.6 kb and 5.6 kb) show the same pattern of transcript accumulation with a much lower total abundance. The three bean cDNAs were isolated from a cDNA library prepared from elicitor treated bean cells. Although all three transcripts are present in untreated cells, transcript levels increase following elicitor treatment, particularly for pHyp2.13. The increased abundance of extensin transcripts in bean cells following either fungal attack or elicitor treatment have been shown to be due to transcriptional activation using nuclear run-off experiments (Lawton and Lamb, 1987).

IV. Prospects

Despite the tremendous importance of the cell wall to both plants and to man, we know very little about how plant cells regulate the assembly of their cell walls, or about how the cell wall functions to regulate plant development. The adaptive strategy of using nuclear gene products to organize the extracellular matrix is well established in animal systems, but because the matrix embedding plant cells is so fundamentally different from that embedding animal cells, this concept has not been generalized to the plant kingdom. Extensins are one class of nuclear gene products which may function to organize other wall polymers. The cloned extensin genes should provide the starting materials for gene transfer experiments designed to 1. genetically dissect the functional domains of extensin, and 2. to genetically manipulate the structure of the cell wall. These approaches should lead to a better understanding of how the plant extracellular matrix regulates plant growth and development and the interactions between plants and microorganisms.

V. References

Allen, A. K., Desai, N. N., Neuberger, A., 1978: Progerties of potato lectin and the nature of its glycoprotein linkages. Biochem. J. **171,** 665—674.

Cassab, G. I., Nieto-Sotelo, J., Cooper, J. B., Van Holst, G.-J., Varner, J. E., 1985: A developmentally regulated hydroxyproline-rich glycoprotein form the cell walls of soybean seed coats. Plant Physiol. **77,** 532—535.

Chen, J., Varner, J. E., 1985: An extracellular matrix protein in plants: characterization of a genomic clone for carrot extensin. EMBO J. **4,** 2145—2151.

Chen, J., Varner, J. E., 1985: Isolation and characterization of cDNA clones for carrot extensin and a proline-rich 33-kDa protein. Proc. Natl. Acad. Sci. **82,** 4399—4403.

Chrispeels, M. J., 1969: Synthesis and secretion of hydroxyproline containing macromolecules in carrots. I. Kinetic analysis. Plant Physiol. **44,** 1187—1193.

Chrispeels, M. J., Sadava, D., Cho, Y. P., 1974: Enhancement of extensin biosynthesis in ageing carrot storage tissue. J. Exp. Bot. **25,** 1157—1166.

Condit, C. M., Meagher, R. B., 1986: A gene encoding a novel glycine-rich structural protein of petunia. Nature **323,** 178—181.

Cooper, J. B., Varner, J. E., 1983: Insolubilization of hydroxyproline-rich cell wall glycoprotein in aerated a carrot root slices. Biochem. Biophys. Res. Comm. **112,** 161—167.

Cooper, J. B., Varner, J. E., 1984: Crosslinking of soluble extensin in isolated cell walls. Plant Physiol. **76,** 414—417.

Corbin, D. R., Sauer, N., Lamb, C. J., 1987: A hydroxyproline-rich glycoprotein gene family in bean is differentially regulated by infection and wounding. In press.

Creighton, T. E., 1978: Possible implications of many proline residues for the kinetics of protein folding and unfolding. J. Mol. Biol. **125,** 401—406.

Ecker, J. R., Davis, R. W., 1987: Plant defense genes are regulated by ethylene. Proc. Nat. Acad. Sci. **84,** 5202—5206.

Epstein, L., Lamport, D. T. A., 1984: An intramolecular linkage involving isodityrosine in extensin. Phytochem. **23,** 1241—1246.

Esquerre-Tugaye, M.-T., Lamport, D. T. A., 1979: Cell surfaces in plant-microorganism interactions. 1. A structural investigation of cell wall hydroxyproline -rich glycoproteins which accumulate in fungus infected plants. Plant Physiol. **64,** 314—319.

Esquerre-Tugaye, M.-T., Lafitte, C., Mazau, D., Toppan, A., Touze, A., 1979: Cell surfaces in plant-microorganism interactions. II. Evidence for the accumulation of hydroxyproline-rich glycoproteins in the cell wall of diseased plants as a defense mechanism. Plant Physiol. **64,** 320—326.

Fincher, G. B., Stone, B. A., Clarke, A. E., 1983: Arabinogalactanproteins: structure, biosynthesis, and function. Ann. Rev. Plant Physiol. **34,** 47—70.

Fry, S. C., 1982: Isodityrosine, a new cross-linking amino acid from plant cell-wall glycoprotein. Biochem. J. **204,** 449—455.

Fry, S. C., 1987: Formation of isodityrosine crosslinks by peroxidase isozymes. J. Exp. Bot. **38,** 853—862.

Hammerschmidt, R., Lamport, D. T. A., Muldoon, E. P., 1984: Cell wall hydroxyproline enhancement and lignin desposition as an early event in the resistance of cucumber to *Cladosprium cucumerinum.* Physiological Plant Path. **24,** 43—47.

Klis, F. M., 1976: Glycosylated seryl residues in wall protein of elongating pea stems. Plant Physiol. **57,** 224—226.

Klis, F. M., Rootjes, M., Groen, S., Stegwee, D., 1983: Accelerated accumulation of wall-bound hydroxyproline in artificially induced lesions on bean hypocotyls. Zeitschrift Pflanzenphysiol. **110,** 301—307.

Lamport, D. T. A., 1965: The protein component of primary cell walls. Adv. Bot. Res. **2,** 151—218.

Lamport, D. T. A., 1977: Structure biosynthesis and significance of cell wall glycoproteins. Recent Adv. Phytochem. **11,** 79—111.

Lamport, D. T. A., Caat, J. W., 1981: Glycoproteins and enzymes of the cell wall. Encyclopedia Plant Physiol. **13B,** 133—165.

Lawton, M. A., Lamb, C. J., 1987: Transcriptional activation of plant defense genes by fungal elicitor, wounding, and infection. Mol. Cell. Biol. **7,** 335—341.

Leach, J. E., Cantrell, M. A., Sequeira, L., 1982: A hydroxyproline-rich bacterial agglutinin from potato: extraction, purification, and characterization. Plant Physiol. **70,** 1353—1358.

Mellon, J. E., Helgeson, J. P., 1982: Interaction of a hydroxyproline-rich glycoprotein from tobacco callus with potential pathogens. Plant Physiol. **70,** 401—405.

Pike, C. S., Un, H., Lystash, J. C., Showalter, A. M., 1979: Phytochrome control of cell wall-bound hydroxyproline content in etiolated pea epicotyls. Plant Physiol. **63,** 444—449.

Ridge, I., Osborne, D. J., 1970: Hydroxyproline and peroxidases in cell walls of *Pisum sativum*: regulation by ethylene. J. Exp. Bot. **21,** 843—856.

Showalter, A. M., Bell, J. N., Cramer, C. L., Bailey, J. A., Varner, J. E., Lamb, C. J., 1985: Accumulation of hydroxyproline-rich glycoprotein mRNAs in response to fungal elicitor and infection. Proc. Natl. Acad. Sci. **82,** 6551—6555.

Showalter, A. M., Varner, J. E., 1987: Molecular details of plant cell wall hydroxyproline-rich glycoprotein expression during wounding and infection. In: Molecular Strategies for Crop Protection. Arntzen, C., Ryan, C. (eds.). pp. 375—392, Alan R. Liss, Inc., New York.

Smith, J. J., Muldoon, E. P., Lamport, D. T. A., 1984: Isolation of extensin precursors by direct elution of tomato cell suspension cultures. Phytochem. **23,** 1233—1239.

Smith, J. J., Muldoon, E. P., Willard, J. J., Lamport, D. T. A., 1986: Tomato extensin precursors P1 and P2 are highly periodic structure. Phytochem. **25,** 1021—2030.

Smith, M., 1981: Characterization of carrot cell wall protein. II. Immunological study of cell wall protein. Plant Physiol.

Stafstrom, J. P., Staehelin, L. A., 1986: Crosslinking patterns in salt-extractable extensin from carrot cell walls. Plant Physiol. **81,** 234—241.

Stafstrom, J. P., Staehelin, L. A., 1986: The role of carbohydrate in maintaining extensin in an extended conformation. Plant Physiol. **81,** 242—246.

Stermer, B. A., Hammerschmidt, R., 1984: The induction of disease resistance by heat shock. In: Cellular and molecular biology of plant stress. Key, J. L., Kosuge, T. (eds.), pp. 291—302, Alan R. Liss, Inc., New York.

Stuart, D. A., Varner, J. E., 1980: Purification and characterization of a salt-extractable hydroxyproline-rich glycoprotein from aerated carrot discs. Plant Physiol. **66,** 787—792.

Stuart, D. A., Mozer, T. J., Varner, J. E., 1982: Cytosine-rich messenger RNA from carrot root disc. Biochem. Biophys. Res. Comm. **105,** 582—588.

Van Holst, G.-J., Klis, F. M., Bouman, F., Stegwee, D., 1980: Changing cell-wall compositions in hypocotyls of dark-grown bean seedlings. Planta **149,** 209—212.

Van Holst, G.-J., Varner, J. E., 1984: Reinforced polyproline II conformation in a hydroxyproline-rich cell wall glycoprotein from carrot root. Plant Physiol. **74,** 247—251.

Chapter 14

The Expression of Heat Shock Genes — A Model for Environmental Stress Response

Fritz Schöffl, Götz Baumann, and Eberhard Raschke

Universität Bielefeld, Fakultät für Biologie, Genetik, D-4800 Bielefeld 1, Federal Republic of Germany

With 4 Figures

Contents

I. Introduction

A. The Heat Shock Response

When cells, tissues, or whole organisms are subjected to grossly elevated temperatures or heat shock (hs), they respond transiently by synthesizing a number of new proteins, the prevalent heat shock proteins (hsps). Hsps are operationally defined as those proteins whose synthesis is immediately and dramatically induced at high temperatures. The induction of hsp synthesis is the major feature of the hs response; other aspects of the reprogramming of cellular activities are less well understood (for recent reviews see Burdon, 1986; Lindquist, 1986; Nagao *et al.*, 1986; Schlesinger, 1986; Schöffl *et al.*, 1986). Some of the general characteristics relevant to plants are briefly summarized here.

In plants, both an abrupt shift and a gradual increase of temperature seem to effectively elicit the hs response. The maximum response is usually achieved at temperatures 10—15°C above optimum growth temperature, correlating with the range of exposure under natural conditions. The induction of hsp synthesis is very rapid with a maximum activity within 1—2 hours and a reduced rate continuing for several hours. Sustained synthesis may only be observed when the severity of the hyperthermic stress is increased and cells begin to die. The threshold temperatures sublethal to plants may also vary depending on the normal range of environmental growth conditions, the duration of heat stress, and eventually on a conditioning pretreatment at high temperatures. These temperatures range from 37°C to about 45°C for most plants and several hours of treatment; however, desert succulents (Cacti and Agaves) may survive much higher temperatures for extended periods of time (Kee and Nobel, 1986).

Short pulses of extreme and otherwise lethal hyperthermia induce tolerance to these temperatures for longer periods, provided the shocks are intermitted by a sufficiently long recovery period (usually more than two hours) at lower temperatures (Lin *et al.*, 1984). The acquired thermotolerance in most cases protects plants in the natural environment from seasonal and diurnal hot spells. Plants in the field respond to the natural changes in ambient temperature in summer in a similar way as to hs simulation in the laboratory; however, only when the water supply is limited (Kimpel and Key, 1985; Burke *et al.*, 1985). Leaf temperatures of irrigated plants are usually below the threshold temperatures required for the elicitation of the hs response, probably due to the cooling effect of stomatal transpiration.

B. Groups of Related Heat Shock Proteins

The hsps in plants can be divided into several groups based on molecular masses, homologies of derived amino acid sequences, and subcellular compartmentation:

(i) High molecular mass (hmm) hsps ranging from 80 to 100 kDa. Two different hsps from this group usually occur in any given plant species. Genes with homologies to the *Drosophila* hsp83 have been recently isolated from soybean (Roberts and Key, 1985), *Brassica oleracea* (Kalish *et al.*, 1986), *Arabidopsis thaliana,* and *Zea mays* (Dietrich *et al.*, 1986). These proteins are probably also related to the other hsp83-homologous proteins from chicken (hsp90 and hsp108) and yeast (hsp90) (Blackman and Meselson, 1986; Kleinsek *et al.*, 1986; Sargan *et al.*, 1986) and with the hmm-hsps in human, mouse, and frog (Schlesinger *et al.*, 1982). The most interesting property of proteins in this group may be their capability to reversibly bind other proteins in the cytoplasm as shown by the interaction of hsp89 with pp50 and $pp60^{src}$ (Brugge *et al.*, 1981; Oppermann *et al.*, 1981), hsp90 with the 8S progesterone receptor (Catelli *et al.*, 1985) and hsp90 and hsp100 with polymerizing actin from rabbit skeletal muscle (Koyasu *et al.*, 1986), suggesting that these hsps may serve to carry important proteins around the cell in their inactive form.

(ii) The hsp70 family with molecular masses ranging from 68 to 70 kDa in different plants. Hsp70 is perhaps the most conserved gene product in nature as indicated by the 73% homology between human and *Drosophila* and 50% homology between the fly hsp70 and the *E. coli dna*K protein. Maize hsp70 genes share regional homology (75% in the N-terminal part of the protein coding region) with their counter part in *Drosophila* (Rochester *et al.*, 1986) which is also homologous to hsp70 genes in soybean (Roberts and Key, 1985), and *Arabidopsis thaliana* (Coldrion-Raichlen *et al.*, 1986).

Molecular models for the function of hsp70 are based on the capacity of the protein to bind and hydrolyze ATP, its reversible interaction with other proteins, and its localization in the nucleus/nucleolus during heat shock (Pelham, 1986). It is postulated that hsp70 has a general affinity for denatured or abnormal proteins, probably by hydrophobic interaction. This property may limit the formation of insoluble aggregates and protect proteins from proteolytic degradation.

(iii) Small or so-called low molecular mass (lmm) hsps, 15—20 kDa, are particularly prominent and heterogenous in plants (for review see Schöffl *et al.*, 1986). Sequence homology between lmm-hsps of different plants was shown by the cross hybridization of soybean cDNA probes with hs-induced mRNAs from pea, sunflower, millet, corn, parsley, and broad bean. These proteins tend to form higher order structures (cytoplasmic granules in tomato and soybean) or become associated with nuclei and other subcellular fractions containing organelles. Similar properties are also described for the *Drosophila* lmm-hsps (hsp22, -24, -26, -27). Features of the polypeptide structure common to plant and animal hsps are discussed later.

Other small hsps with molecular masses between 21 and approximately 30 kDa are less abundant in soybean, pea, maize, and other plants. Some of these proteins are transported into chloroplasts and mitochondria; two polypeptides, 21 and 24 kDa, become stably associated with mitochondria in soybean seedlings during heat shock and recovery (Lin *et al.*, 1984). Other precursor hsps with molecular masses between 27 kDa and 30 kDa are translocated into chloroplasts of pea, soybean, maize, and *Petunia* (Kloppstech *et al.*, 1985; Vierling *et al.*, 1986; Vierling *et al.*, 1987a); the molecular masses of the mature proteins within the chloroplasts are 22 kDa in pea, 20—22 kDa in soybean, and 20—23 kDa in *Petunia*. The biological function of these proteins is still obscure but they appear to be structurally related to the other small cytoplasmic hsps and they may be involved in similar processes inside and outside the chloroplast (Vierling *et al.*, 1987b). Efficient *in vitro* import of a cytosolic soybean hsp into pea chloroplasts occurred only when the protein (Gmhsp17.5-E) was genetically fused to the transit peptide of the small subunit of ribulose 1,5-bisphosphate carboxylase (Lubben and Keegstra, 1986). Although the imported hsp appeared to be targeted to the stroma, this may not be its final destination during heat shock. *In vitro* translocated native pea hsp22 is also targeted to the stroma in chloroplasts from normal plants, but it becomes associated with thylakoid membranes in chloroplasts from heat-stressed tissues (Glaczinski and Kloppstech, 1987). Although all the hsps described so far are encoded in the nucleus, there is evidence that *Brassica* and maize hs-induced proteins (62 kDa and 60 kDa) may be encoded in the mitochondrial genome (Sinibaldi and Turpen, 1985).

C. *Heat Shock and Other Environmental Stresses*

When plants are subjected to adverse environmental conditions other than heat shock, their cells respond to these stresses by changes in gene expression in a stress-specific manner (for a review see Sachs and Ho, 1986). However, certain stressors induce also some of the hsps in different plant species. This observation evokes two important questions:

(i) Is there a common molecular mechanism of hsp synthesis induction (gene activation) provoked by different environmental stresses?

(ii) Is there a functional interrelationship among the common hsps induced by the different stresses?

Several hypotheses have been put forward for the signal transduction induced by heat shock and a bewildering variety of stress agents in animal cells and bacteria (for a review see Nover *et al.*, 1984). One theory was based on the coinduction of hsps and unusual bisnucleotides, termed "alarmones", by many stressors; therefore it was proposed that alarmones trigger the hs response. This hypothesis has been discredited recently by srutinizing the effect of hyperthermia on hsp synthesis and alarmone concentration in *Xenopus* oocytes and cultured hepatoma cells (Guedon *et al.*, 1986). It was shown that hsps are already induced under conditions which do not significantly change the level of alarmones.

Mitochondria were also implicated in playing an important role in the hs response, since many other physico-chemical stresses, most of which interfere with oxidative phosphorylation of electron transport, induce hsp synthesis in *Drosophila.* Although anaerobic stress or recovery from anaerobiosis also induces hsps in many organisms, anaerobiosis in plants induces a different set of proteins (for a review see Sachs and Ho, 1986). Hsp synthesis, oxidative stress, and alarmones are not obviously linked in plants. Only arsenite and cadmium, two out of 20 tested physical and chemical stresses, induce mRNA and hsp synthesis in the soybean hypocotyl (Czarnecka *et al.*, 1984; Lin *et al.*, 1984). It seems conceivable that abnormal and denatured proteins which accumulate in the cells during stresses serve as stress signals and trigger the activation of hs genes. One model proposes that the interaction of ubiquitin with the proteolytic degradation system of the cell leads somehow to the activation of the hs transcription factor, HSTF (Munro and Pelham, 1985; Anathan *et al.*, 1986). HSTF subsequently recognizes and binds to hs promoter elements (HSE) upstream from hs genes (for a review see Pelham, 1985). According to this model, heat shock and arsenite trigger the hs response in plants in a very similar way. Heat-inducible transcription of soybean hs genes was demonstrated by run-off transcription in isolated nuclei (Schöffl *et al.*, 1987), and the DNA sequence analysis revealed multiple copies of HSE-like sequences in the upstream promoter regions (discussed later). Cadmium may possibly activate hs genes via a different transcription factor that also interacts with other promoter binding sites. Such conserved binding sites, termed metalresponsive elements or MRE, were identified in the promoter of metallothionine genes (Karin *et al.*, 1984), but homologous sequences are also present in the upstream regions of plant hs genes (Czarnecka *et al.*, 1985). The activation of certain human hsp70 genes by different stresses seems to require different promoter elements, e. g. for serum stimulation and cadmium induction (Wu and Morimoto, 1985). The developmental induction of the *Drosophila* hsp26, -28, and -83 genes seems to be regulated by another independent form of interaction between promoter element and protein (Hoffman *et al.*, 1987; Mestril *et al.*, 1986).

In plants, cell-specific differences and developmental induction of hsps have not yet been critically examined. Different tissues of a plant species usually synthesize identical sets of hsps with the exception of pollen tubes and grains (Xiao and Mascarenhas, 1985; Zhang *et al.*, 1984; Cooper *et al.*, 1984). The inability to induce a chimeric hs-*npt*II gene in pollen from transgenic tobacco (Spena and Schell, 1987) and the generally lower thermotolerance of pollen is in accordance with the view that the ability to synthesize hsps is crucial to the development of thermotolerance. Several lines of evidence support this hypothesis:

(i) There is a tight correlation between the kinetics of hsp synthesis and the development of thermotolerance in many organisms.

(ii) Conditioning heat treatment of the soybean hypocotyl can be replaced by incubation at normal temperature in the presence of arsenite, a chemical inducer of hsp synthesis (Lin *et al.*, 1984). The overlap in the stress protein patterns (analysed by 2-D PAGE) induced by arsenite and heat shock is incomplete; but, remarkably, it includes the abundant lmm-hsps. Ethanol, hypoxia, and other agents affect some animal cell lines in a very similar manner and conversely, heat shock induces tolerance to these other physical and chemical stresses (see Lindquist, 1986 and refs. therein).

(iii) Eukaryotic cell lines selected for increased survival at high temperatures constitutively produce hsps, whereas mutants of *E. coli* and *Dictyostelium* which are unable to acquire thermotolerance are defective in hsp synthesis (see Lindquist, 1986 and refs. therein). Varietal differences between high-temperature-tolerant and susceptible lines of *Sorghum* correlate with temporal differences in the capacity to synthesize hsps during early germination (Ougham and Stoddart, 1986).

Despite overwhelming correlative evidence, exceptions seem to contradict a protective role for hsps and their causal connection to thermotolerance. For example, germinating pollen of *Tradescantia* which fail to synthesize hsps nevertheless acquire a certain level of thermal tolerance by a conditioning heat treatment (Xiao and Mascarenhas, 1985). It seems conceivable that cells are able to cope with high temperature stress by different but concerting processes. Hsp synthesis may be crucial, but changes in lipid composition in membranes may also be important in plants (Bery and Björkman, 1980; Süss and Yordanov, 1986). Conversely hsps seem to protect the cells from harmful effects induced by other environmental stresses discussed above. The molecular mechanisms of such processes are not yet understood and it is unknown whether hsps act via the same mechanism during hyperthermia and chemical stresses.

Heat shock gene activation may be a good model for other environmental gene expression. The present data suggest common principles of induction for hs genes and in a broader sense also for the regulation of gene expression induced by other environmental stresses unrelated to heat stress and hs genes (for a review see Sachs and Ho, 1986). Investigations of these stress responses have received increased attention for scientific and agricultural reasons, but the preference of hs response as a model system is based on the accessibility (abundance of mRNAs and hsps) of the molecular processes of transcription, translation, etc., the rapidity of induction (within minutes following temperature rises), the coordinated expression of homologous genes (related gene families), and the advanced state of the molecular analysis of hs genes in both native and transgenic plants.

Soybean was the first plant used for these investigations and most molecular studies were performed on hs genes of this organism. Here we concentrate on this system with emphasis on the lmm-hsps, which are of paramount interest for plant-specific aspects of the hs response. The recent investigations of hsp70 genes in other plants are also discussed briefly.

II. Molecular Biology of Heat Shock Genes

A. Sequence Homology Among Small Heat Shock Proteins

Current knowledge of the amino acid sequences of hsps and their deduced secondary structures is based exclusively on DNA sequencing. Several cDNA and genomic clones for lmm-hsps (17—18 kDa) have been sequenced to date; six of the seven genes belong to the same gene family, class I, and one is a member of another family, class VI (see Table 1). The sequence homology within class I is about 90%, and the same degree of homology exists in class VI between the gene Gmhsp17.9-D (Raschke *et al.*, 1988) and a full-length cDNA clone pEC75 (Czarnecka *et al.*, 1984; Key *et al.*, 1985). The largest open reading frames encode polypeptides with calculated molecular masses of between 17.3 kDa and 18.5 kDa, assuming translation starts at the first ATG codon downstream from the 5′-terminus of the mRNA. The molecular masses of polypeptides translated *in vitro* from synthetic, SP6-polymerase-generated RNAs of Gmhsp17.3-B and Gmhsp17.6-L genes are consistent with the proposed translational start sites (F. Schöffl and S. Krusekopf, unpublished results). The 5′ untranslated leader sequences are remarkably divergent and unusually long (75—104 bp) in proportion to their respective coding sequences (459—483 bp). The leader sequences are probably involved in the selective translation during heat shock as has been reported for *Drosophila* hsp22 and hsp70 (Hultmark *et al.*, 1986; McGarry and Lindquist, 1985; Klemenz *et al.*, 1985), but except for being A-rich (40%) there are no striking homologies between 5′ sequences of plant and animal hs-mRNA.

Table 1. Small Soybean Heat Shock Genes Characterized by DNA Sequencing

hsp-gene	synonymous designation 1)	length of polypeptide (amino acid residues)	length of 5' nontranslated leadersequences (bp)	References
class I				
Gm hsp 17.3-B	hs 6871	153	104±2	Schöffl et al. 1984
Gm hsp 17.5-M	-	153	88,93	Nagao et al. 1985
Gm hsp 17.6-L	-	154	93,96	Nagao et al. 1985
Gm hsp 17.5-E	-	154	82	Czarnecka et al. 1985
Gm hsp 18.5-C	hs 53	161	76±2	Raschke et al. 1987
class VI				
Gm hsp 17.9-D	-	159	72±2	Rascke et al. 1987

The DNA homology within the coding regions of soybean class I hs genes varies between 83—94% and 90—95% of the amino acid sequence (see Fig. 1). The average amino acid homology between class I and class VI

```
                             10        20        30        40        50        60        70        80
CONSENSUS               MSLIPSFFGGRRSNVFDPFSLDVWDPFKDFHFPT/SLSA-------ENSAFVSTRVDWKETPEAHVFKADIPGLKKEEVK

Gm hsp 17.3-B (I)       .............S................P..S-.............................................
Gm hsp 17.5-M (I)       ......I...........................-.................N..............F............
Gm hsp 17.6-L (I)       ......I...P...........M........V..S.V...............N........Q....L.............
Gm hsp 17.5-E (I)       .....G................M........V..S.V...........................................
Gm hsp 18.5-C (I)       .....N......N.................P..NTLS..SFPEFSR..................................

Gm hsp 18.0-c75 (VI)MDFRV.G.E.PL.HTLQ-HMM.-M.E.AAGEN.TYSA..R.YVR.......DAK.MAA.PA.V..Y.NSY..EI.M....SGDI.
Gm hsp 17.9-D  (VI)MDFRV.G.ESPL.HTLQ-HMM.-M.E.GAGDN.THNA..W.YVR.......DAK.MAA.PA.V..Y.NSY..EI.M....SGDI.
                        * *    *       +* +* *     *   +*   +         +  *+  *  * **+    *+  * ****  ++*

                                 90        100       110       120       130       140       150       160
CONSENSUS               VQIEDDRVLQISGERNVEKEDKNDTWHRVERSSGKFMRRFRLPENAKVDQVKASMENGVLTVTVPK---EEVKKPDVKAIEISG

Gm hsp 17.3-B (I)       LE.Q.G............................LV..................................I........D...
Gm hsp 6834-A (I)           ???I..............................S....D..........................I.............
Gm hsp 17.5-M (I)       ................L.................N.............E...................................
Gm hsp 17.6-L (I)       .............................D..................E....C.........I..........S..P.....
Gm hsp 17.5-E (I)       ....................................T...........NE..................................
Gm hsp 18.5-C (I)       ......K.........................................E...................................

Gm hsp 18.0-c75 (VI)    ..V...N..L.....KRDE.KEGVKYL.M..RV.....K.V.....NT.AIS.VCQD...S...Q.LPPP.P...--RT..VKVF
Gm hsp 17.9-D  (VI)    ..V...NL.L.C...KDRE.KEGAKYL.M..RV..L..K.V.....NT.AIS.VCQD...S...Q.LPPP.P.L.--RT..VKVA
                        + + * ++* * *** ++ *       * +*  * + *+* **+**  + + *    ***+**+ *    *+*    + *++
```

Fig. 1. Comparison of the amino acid sequences of small soybean hsp. + : Conservative amino acid substitutions, *: invariable amino acid positions (for origin of DNA sequence data see Table I)

hsps is only 38% and is more pronounced in the C-terminal portions of the polypeptides. Secondary structure predictions and the distribution of hydrophobic and hydrophilic domains reveal a high degree of conservation in this region (Raschke *et al.*, 1988; Schöffl *et al.*, 1986). The conserved structural features are:

(i) A hydrophilic domain at the amino acid positions 80—90 with the potential for α-helix formation, followed by (ii) a hydrophobic region (positions 130—135) with the almost invariant sequence GlyValLeuThrVal and a potential for β-sheet formation, and (iii) another hydrophilic and probably α-helical region towards the C-terminus formed from charged and polar amino acid residues.

These three domains may be the important determinants for the structure and function of the small hsps in plants. Similar domains occur also in related hsps in *Drosophila* (Southgate *et al.*, 1983; Ingolia and Craig, 1982; Ayme and Tissières, 1985), *Caenorhabditis* (Johnes *et al.*, 1986), humans (Hickey *et al.*, 1986) and in the α-crystallins of the vertebrate eye lens (de Jong, 1982; van den Heuvel *et al.*, 1985). It seems that the selection pressure conserving the secondary structure of small hsps nevertheless allows a considerable variation in DNA and amino acid sequences in different organisms. The small hsps from one species are highly homologous in their C-terminal halves (>90% within soybean class I and class VI, up to 75% in *Drosophila*), but homology is markedly lower for different hsp families in one species or between hsps in different species (Fig. 2). The two hsp families I and VI in soybean diverge in amino acid sequence from each other to the same degree as from the *Drosophila* small hsps. The structural differences between these soybean hsps seem to be insignificant and do not imply different functional properties of these proteins. It seems possible that these two gene families originated from structurally and functionally related ancestors by gene duplication or gene conversion at a time late in evolution. The numbers of hs genes within each family exceed the potential numbers calculated on the basis of ploidy level of present day *Glycine max* plants which have tetraploid genomes (Singh and Hymowitz, 1985; Lee and Verma, 1984). Thus, allelic variation cannot be the chief cause of the existence of gene families in soybean. Additional data are required to investigate the evolution of hs genes in plants. It is noteworthy that hs-induced mRNAs of other dicot plants cross hybridize with both class I and class VI cDNA probes (Key *et al.*, 1983; F. Schöffl, unpublished results).

The small hsps from *Drosophila* and *Caenorhabditis* share a higher degree of homology with vertebrate α-crystallin proteins than with each other. The occurrence of introns is an other possible criterion for establishing the family connection of small hsps in different organisms. Introns were found in *Caenorhabditis* and human but not in soybean and fly hsp genes. The presence of introns may possibly impair the expression of these genes under severe heat stress as shown for the hsp83 gene in *Drosophila*

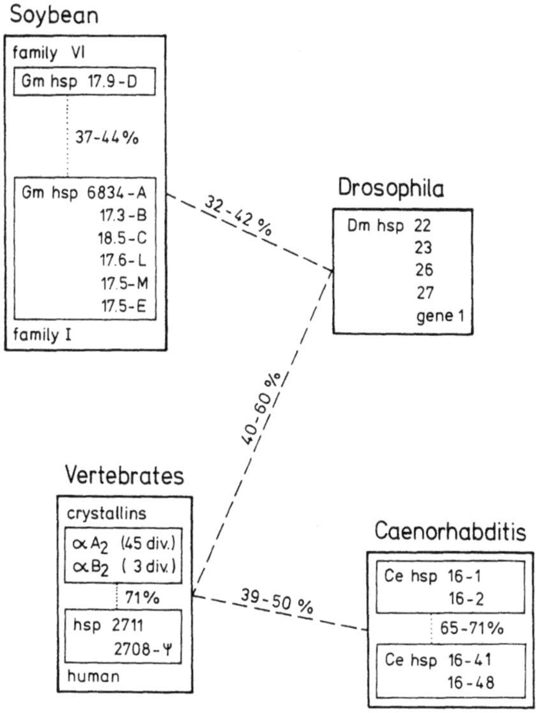

Fig. 2. Homologies among small hsps in and between different organisms. The homologies refer to regions encompassing the conserved domains in the C-terminal halves of the polypeptides, according to Raschke (1988)

(Yost and Lindquist, 1986). In contrast to those of other organisms, plant hsp70 genes seem to contain introns. Maize (Rochester *et al.*, 1986) and *Petunia* (J. Winter, personal communication) contain introns at the same position as the hsp70 cognate genes of *Drosophila* which are not expressed during heat shock.

B. Heat Shock Promoter and Upstream Sequences

The coordinate expression and transcriptional regulation of hs genes suggest a conservation of *cis*-active regulatory promoter and upstream sequences (Schöffl *et al.*, 1986, 1987). Dot matrix comparisons revealed local homologies in these regions between soybean hs genes of the two different gene families (Raschke *et al.*, 1988). Figure 3 shows an intergenic comparison with high resolution and stringency of the 5′-flanking regions of the genes listed in Table 1. Two patches of homologies appear in different areas of the plots. The proximal areas 5′ to the TATA boxes indicate homology of multiple HSE-like sequences. These palindromic sequence elements are clustered within about 150 bp upstream from the TATA box. A compilation of these sequences (Table 2) with 70% or more identity with the *Drosophila* consensus sequence 5′CT-GAA--TTC-AG results in an extended palindrome with the consensus 5′TCTAGAA--TTCTAGA for soybean. Seven out of 10 symmetrical nucleotides in synthetic HSEs are sufficient for heat-inducible transcription in animal cells (Pelham and Bienz, 1982). The hs genes of all organisms analyzed so far contain HSE-like sequences starting 1.5 helical turns upstream from the TATA box in most genes (for a review see Bienz, 1985). Many of these genes contain multiple copies of HSE and several of them overlap with each other by four nucleotides (for soybean see Schöffl *et al.*, 1986; Raschke *et al.*, 1988). This configuration suggests phasing of HSEs by one helix turn or multiples of it (see Table 2). The significance of the spacing of HSE repetitions for transcriptional activation is not clear but it may reflect the potential for fine tuning of the regulation by direct and cooperative

Table 2. Compilation of Soybean HSE-like Sequences

Drosophila Consensus-HSE: 5' --CT-GAA--TTC-AG-- 3'
(>> >>> <<< <<)

Soybean	Pos.		Homology with Drosophila Consensus-HSE
Gm hsp 17.3-B	-276	ATCCCGAAACTTCTAGTT	9 / 10
	-245	GTCCAGAATGTTTCTGAA	7 / 10
	-235	TTTCTGAAAGTTTCAGAA	7 / 10
	-225	TTTCAGAAAATTCTAGTT	8 / 10
	-173	AACAAGGACTTTCTCGAA	7 / 10
	-163	TTCTCGAAAGTACTATAT	8 / 10
Gm hsp 18.5-C	-196	CTGTAGAAAGCTCTAGAA	8 / 10
	-186	CTCTAGAACTTGGGATTT	7 / 10
	-146	AAACAGAATTTTCTGGAA	7 / 10
	-136	TTCTGGAAAACACAGGAT	7 / 10
Gm hsp 17.9-D	-280	TTCTGGACATTACTAGAA	8 / 10
	-270	TACTAGAAAGATCCGAAG	7 / 10
	-221	TACTGGAAGTTTCACAGC	8 / 10
	-190	CTCCAGAAACTTCCATTT	8 / 10
	-146	CTTCAGAAACTTCCATTT	7 / 10
	-126	TTCTCGAATTATCTATGT	8 / 10
Gm hsp 17.6-L	-275	ATCTAGAAGGTTGTAGAA	9 / 10
	-253	AGCTAGAACGTACGTATT	7 / 10
	-224	GTCCTGAAGTTTATCGAA	7 / 10
	-214	TTATCGAATCATCTAAAA	7 / 10
	-155	TTCTGGAACATACAAGAG	9 / 10
Gm hsp 17.5-M	-600	ATCTTCAAACTTCAAGTT	9 / 10
	-430	AAAAACAATATTCTAGAA	7 / 10
	-359	CACAACAATATTTCAGAA	7 / 10
	-243	CACTAGAACCTTCGTACA	8 / 10
	-223	GAGTGGAGAAGTCCAGAA	7 / 10
	-213	GTCCAGAAGTTTTTATAG	7 / 10
	-160	AACACGATTTTTCTGGAA	7 / 10
	-150	TTCTGGAACGTACACGAT	8 / 10
Gm hsp 17.5-E	-535	TCCTCTATGGTTTCAGTG	7 / 10
	-513	GTTTGAAATTTTTTAGAT	7 / 10
	-495	TTCTTTAACATTCTAAAA	8 / 10
	-144	TTCTGGAACATACAAGAT	9 / 10

Soybean Consensus-HSE: 5' -TCTAGAA--TTCTAGA- 3'
(>>>>>>> <<<<<<<)

interaction with the transcription factor HSTF. The efficient targeting of HSTF to HSE may require a DNA distortion provoked by overlapping repeats and/or by the central purine and pyrimidine clusters in palindromes (Nover, 1987).

A second cluster of homologous sequences starts upstream from the HSE-containing region (#1 in Fig. 3). These signals correspond to short AT-rich repeats, starting with runs of "simple sequences" $(A)_n$, $(T)_n$, and $(AT)_n$. The repeated structure of the AT-rich intergenic sequences is most

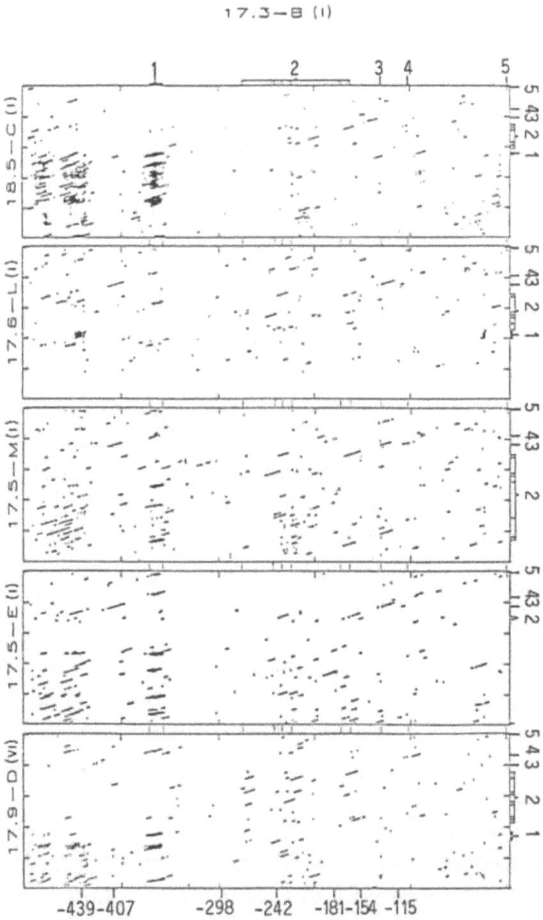

Fig. 3. Dot matrix analysis of the 5′ flanking sequences of soybean small hsp genes The analysis was performed for 500 nucleotides upstream from the translational start sites using a span of 15 with a match of 10 nucleotides (>67% homology). The numbers at the top and on the right-hand axes of panels refer to the positions of "simple sequences" (1), multiple HSE (2), TATA box (3), mRNA start (4), protein coding start (5). The numbers at the bottom panel refer to the endpoints of promoter deletions used for a functional analysis of Gmhsp17.3-B (see Fig. 4). For origin of DNA sequence data see Table 1

pronounced in the region upstream from Gmhsp 18.5-C, with two different consensus motifs, 5′TTTTTAA and 5′AAAAAT (Raschke *et al.*, 1988). Most other soybean hs genes are also proceded by blocks of AT-rich sequences. Emphasis is given to these simple sequences, which demarcate the upstream repeats (>75% A+T) and the proximal promoter regions (~55% A+T), because of their possible regulatory function in gene expression (see next chapter).

Dot matrix analyses also revealed the repetition of AT-rich motifs at positions downstream from the genes. Thus hs genes appear to be flanked by repetitive intergenic sequences. The full length of one repeat, upstream from Gmhsp17.9-D, is approximately 750 bp; those flanking the other genes were not completely sequenced. This organization of genomic sequences is in accordance with the short-period interspersion pattern (1.3 kb average unit length of single-copy DNA separated by 0.4 kb moderately repeated sequences) of the soybean genome (Gurley *et al.*, 1979). The flanking repeats may have structural and perhaps also regulatory importance, and they may have served as the targets for the processes leading to gene duplication.

C. *Heterologous Expression of hs Genes in Transgenic Plants*

The soybean lmm-hsp genes, Gmhsp17.3-B and Gmhsp17.5-E, were used to investigate hs gene expression in heterologous plant systems (Schöffl and Baumann, 1985; Gurley *et al.*, 1986). In these early studies, genes were introduced into sunflower using tumorigenic Ti-plasmid vectors. Thermoinduced synthesis of transcripts and the faithful initiation of the mRNAs demonstrated the functional integrity of plant hs promoters across phylogenetic barriers. The potential promoter sequences were delimited to approximately 1 kb (Gmhsp17.3-B) and 3.25 kb (Gmhsp17.5-E) upstream from the respective genes. A 5′-deletion to −1.175 kb resulted in a large increase in basal transcription, and a deletion to −95 bp curtailed hs-inducible RNA to 5% of the regular level of hs-induced transcription of Gmhsp17.5-E in primary tumors. These results rendered no definitive conclusions about the function of conserved promoter structures including the HSE-like sequences.

Sequences with homologies to the metal-responsive elements (MRE), the SV40 core enhancer sequence and other potential regulatory eukaryotic sequences, are not preferentially clustered in the plant hs gene promoters. MRE-like sequences upstream from Gmhsp17.5-E (Czarnecka *et al.*, 1985) may be important for the cadmium regulation of this gene (Gurley *et al.*, 1986), but a more critical analysis of this promoter will be required to discriminate between the two separate modes of induction via HSE and MRE respectively.

As a consequence of the known ambiguities of gene expression analyses in tumor tissues we changed the experimental strategy for our subsequent investigations of Gmhsp17.3-B. The gene was transferred into tobacco via leaf disk transformation with the binary T-DNA vector *Bin19*

(Schöffl *et al.*, 1986). Gene expression was studied in regenerated transgenic plants using Northern blot hybridization and S1-nuclease mapping of hs-induced transcripts. The analysis revealed equivalent levels of Gmhsp17.3-B mRNA in normal soybean and transgenic tobacco plants. The functional analysis of sequences required for transcriptional activation of this gene in tobacco was based on a series of 5′ deletions in the promoter region (Baumann *et al.*, 1987). For each deletion, seven or eight independently transformed plants were tested for the induction of mRNA synthesis by heat shock. Despite clonal variations of transcript levels in different plants, the average level of hs-induced mRNA declined in a nonlinear fashion with progressive deletions (Fig. 4). At least two different functions can be assigned to distinct parts of the hs promoter:

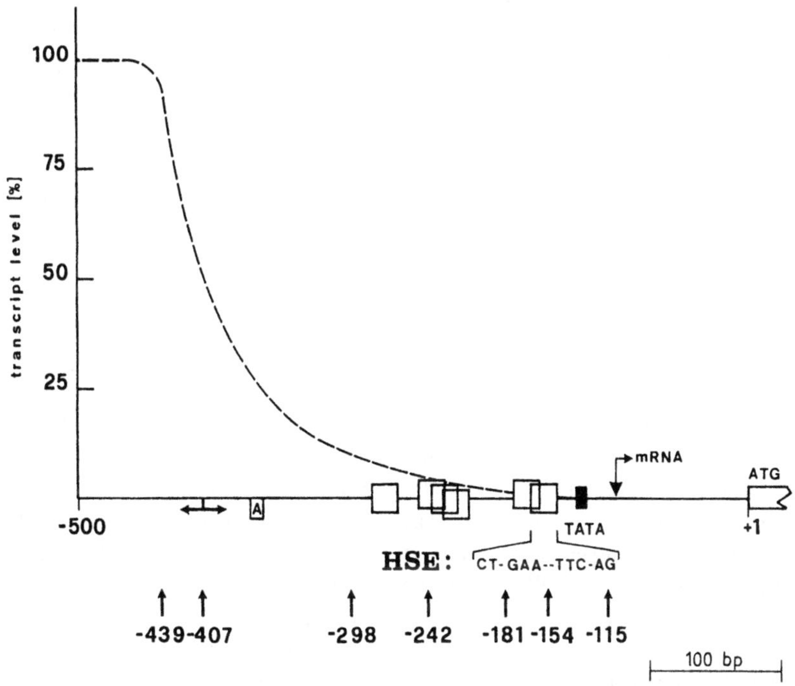

Fig. 4. Schematic diagram of the transcriptional activity of Gmhsp17.3-B upstream deletions in transgenic plants
The transcript levels obtained for the different deletions (vertical arrows) after heat shock in tobacco were determined by Northern hybridization using hs-induced mRNA in soybean as a standard (data from Baumann *et al.*, 1987). HSEs are marked by squares, A denotes a run of A ("simple sequences"), the bi-pointed horizontal arrow marks a hyphenated dyad

(i) Thermoinduced transcription requires the sequences between positions −181 and −154, a region containing the first two overlapping HSEs proximal to the TATA box. Additional HSEs between positions −298 and −181 seem to modulate moderately the level of transcripts.

(ii) The upstream sequences delimited by the deletions −439 and −298 lack HSEs; however, this region is required for maximal transcription. This stretch of DNA enhances the mRNA level by a factor of 10. The structural features within this region include an imperfect dyad symmetry encompassing 34 bp with its centre at position −407 and a run of adenine bases between the positions −357 and −371.

It is important to note that full promoter activity was achieved in plants transformed by the −439 deletion. Thus the essential *cis*-regulatory sequences are contained within 335 bp upstream from the start site of transcription (position −104). The bipartite structure of this plant hs gene promoter is quite unlike the known hs gene promoters in animals where transcriptional control seems to be solely based on HSE. The HSE-like elements are also essential for thermoinduced transcription in plants; enhancer-like sequences have not yet been discovered in the other hs genes. It is not known whether these characteristics of a hs gene promoter are typical for plant genes. The sparse data on the other hs genes in connection with inappropriate gene constructs and/or assay systems render it difficult to make a comparison. The conservation of the important structural features in most soybean hs gene promoter and upstream regions (see Fig. 3) and a growing body of information about enhancer based gene regulation in plants encourage us to predict the same principles for the other hs genes. In most other studies attention was given only to the presence of required HSE-like sequences, as for hs-induced transcription of Gmhsp17.5-E in sunflower tumors (Gurley *et al.*, 1986), Zmhsp70 in transgenic petunia (Rochester *et al.*, 1986), and a chimeric *npt*II gene controlled by the *Drosophila* hsp70 promoter in tobacco (Spena *et al.*, 1985; Spena and Schell, 1987). The combination of HSEs with upstream sequences from unrelated genes may generate plant promoters with novel functional properties. The fusion of an HSE-containing promoter fragment from soybean with the upstream region of a light-inducible gene from pea results in a light-dependent, heat inducibility of a linked reporter gene (Strittmatter and Chua, 1987). Bipartite promoter structures may also be an important feature of non-hs genes as indicated by the requirement of upstream sequences from constitutive genes for the anaerobic regulation of a maize *Adh*-1 promoter in tobacco (Ellis *et al.*, 1987).

In transgenic systems, except Gmhsp17.3-B gene in tobacco (Schöffl *et al.*, 1986; Baumann *et al.*, 1987), one could not achieve the same activity as the promoter in its native genetic background. This may be the result of the close proximity of promoter and enhancer sequences in Gmhsp17.3-B and the lack (or a larger distance from promoter) of enhancer-like sequences in the other constructs.

The upstream enhancer-like sequence shares no preferential homology with the SV40 enhancer core sequence or with any other known enhancer except for the dA:dT simple sequences that function as upstream enhancer elements in yeast (Struhl, 1985). Similar runs of simple sequences $(A)_{14}$, $(A)_{13}$, $(AT)_{15}$, and $(T)_{10}(A)_{12}$ occur within 350 bp upstream from the transcriptional start site of Gmhsp17.3-B, −18.5-C, −17.6-L and −17.9-D.

Soybean genes encoding the small subunit of ribulose bisphosphate carboxylase and leghemoglobin respectively also contain such sequences in their 5′-flanking regions or within introns (Grandbastien *et al.*, 1986; Brown *et al.*, 1984; Lee and Verma, 1984). The significance of simple sequences for an enhancer effect in gene regulation is unknown. However, it can be speculated that their possible function is related to the special stereochemical and torsional properties of the DNA helix and its potential to form cruciforms (Suggs and Wagner, 1986; Alexeev *et al.*, 1987; McClellan *et al.*, 1986; Koo *et al.*, 1986; Wright and Dixon, 1986). A new and exciting possibility could be the nuclear scaffold attachment of large chromosomal loops via "scaffold attachment regions" (SARs) that also contain A- and T-rich motifs; such sequences cohabit with some transcriptional enhancers in *Drosophila* (Gasser and Laemmli, 1986; 1987). The enhancer function may be viewed as a structural DNA effect keeping the nearby genes in close contact with the transcription machinery in the nuclear matrix (Jackson, 1986). Studies of the DNA : protein interaction in DNA regions with functional significance will be of great importance in elucidating the molecular mechanism of enhancer function.

III. General Conclusions

Small hsps (15—20 kDa) are the characteristic and most abundant stress proteins in plants following hyperthermic treatment. Their synthesis correlates with the development of thermotolerance, which may also be induced by other stresses (e.g. arsenite and cadmium). A number of the small hsps share homologous sequences ($\sim$90% for members of the same gene family, $<$50% for different families) in soybean; their genes are also related to hs genes in other plants. Common (shared) characteristics in the predicted secondary structures are particularly pronounced in the C-terminal halves of these proteins. Although the function of the small hsps is unknown, their possibly protective role may be connected with their ability to aggregate or to associate with important cellular macromolecules.

The synthesis of hsps is primarily regulated by a transcriptional control allowing rapid induction of high levels of mRNA and a preferential translation during hyperthermia.

Local homologies in the promoter and 5′-flanking regions of soybean hs genes can be assigned to two different types of conserved sequences: (i) short palindromic sequences with 70% or more identity with the *Drosophila* consensus HSE sequence, and (ii) "simple sequences" (runs of A, T or AT) which are present in most genes upstream from the HSE-containing regions. Functional analysis following progressive deletions from one gene suggests a bipartite promoter structure. Whereas the proximal HSE-containing sequences are required for thermoregulated transcription, the more distally located region including some of the simple sequences is necessary for maximal enhancement of transcription in transgenic tobacco plants. It appears that the HSEs interact with a hs transcription factor, promoting transcription in a way very similar to the HSTF: HSE interaction in animal cells. The enhancer-like function of the upstream region of a heat shock gene is without precedent. The binding of nuclear proteins to sequences in this region (F. Schöffl and K. Severin, unpublished results) is a first step towards an understanding of the mechanism of enhancer functions in heat shock and other environmentally controlled gene expression.

Acknowledgements

We thank Bob Kosier for his critical reading of the manuscript and Sieglinde Angermüller for her technical assistance in the preparation of graphs. The research was supported by grants from the Deutsche Forschungsgemeinschaft to F. S. We would also like to thank Drs. Jill Winter, Elisabeth Vierling, Joe Key, Günter Strittmatter and Nam-Hai Chua, for providing unpublished information and preprints.

References

Alexeev, D. G., Lipanov, A. A., Skuratovskii, I. Y., 1987: Poly(dA)/poly(dT) is a B-type double helix with a distinctively narrow minor groove. Nature (Lond.) **325,** 821—823.

Anathan, J., Goldberg, A. L., Voellmy, R., 1986: Abnormal proteins serve as eukaryotic stress signals and trigger the activity of heat shock genes. Science **232,** 522—524.

Ayme, A., Tissières, A., 1985: Locus 67B of *Drosophila melanogaster* contains seven, not four, closely related heat shock genes. EMBO J. **4,** 2949—2954.

Baumann, G., Raschke, E., Bevan, M., Schöffl, F., 1987: Functional analysis of sequences required for transcriptional activation of a soybean heat shock gene in transgenic tobacco plants. EMBO J., **6,** 1161—1166.

Berry, J., Björkman, O., 1980: Photosynthetic response and adaption to temperature in higher plants. Ann. Rev. Pl. Physiol. **31,** 491—543.

Bienz, M., 1985: Transient and developmental activation of heat shock genes. Trends Biochem. Sci. **10,** 157—161.

Blackman, R. K., Meselson, M., 1986: Interspecific nucleotide sequence comparisons used to identify regulatory and structural features of the *Drosophila* hsp82 gene. J. Mol. Biol. **188,** 499—516.

Brown, G. G., Lee, J. S., Brisson, N., Verma, D. P. S., 1984: The evolution of a plant globin gene family. J. Mol. Evol. **21,** 19—32.

Brugge, J., Erikson, E., Erikson, R. L., 1981: The specific interaction of the Rous sarcoma virus transforming protein, $pp60^{src}$, with two cellular proteins. Cell **25,** 363—372.

Burdon, R. H., 1986: Heat shock and the heat shock proteins. Biochem. J. **240,** 313—324.

Burke, J. J., Hatfield, J. L., Klein, R. R., Mullet, J. E., 1985: Accumulation of heat shock proteins in field grown cotton. Pl. Physiol. **78,** 394—398.

Catelli, M. G., Binart, N., Jung-Testas, I., Renoir, J. M., Baulieu, E. E., Feramisco,

J. R., Welch, W. J., 1985: The common 90 kd protein component of non-transformed "8S" steroid receptors is a heat shock protein. EMBO J. **4,** 3131—3135.

Cooper, P., Ho, T.-D., Hauptman, R. M., 1984: Tissue specificity of the heat shock response in maize. Pl. Physiol. **75,** 431—441.

Coldrion-Raichlen, P., Dietrich, P. S., Sinibaldi, R. M., 1986: The heat shock response of *Arabidopsis thaliana.* J. Cell Biol. **103,** 176a.

Czarnecka, E., Edelmann, L., Schöffl, F., Key, J. L., 1984: Comparative analysis of physical stress responses in soybean using cloned cDNAs. Pl. Molec. Biol. **3,** 45—58.

Czarnecka, E., Gurley, W. B., Nagao, R. T., Mosquera, L. A., Key, J. L., 1985: DNA sequence and transcript mapping of a soybean heat shock gene encoding a small heat shock protein. Proc. Natl. Acad. Sci. U. S. A. **82,** 3726—3730.

De Jong, W. W., 1982: Eye lens proteins and vertebrate phylogeny. In: Macromolecular sequences in systematic and evolutionary biology. pp. 75—114. Goodman, M. (ed.). Plenum Press, New York.

Dietrich, P. S., Bouchard, R. A., Sinibaldi, R. M., 1986: Isolation of maize 83, 70 and 18 kDa HS genes. J. Cell Biol. **103,** 311a.

Ellis, J. G., Llewellyn, D. J., Dennis, E. S., Peacock, W. J., 1987: Maize Adh-1 promoter sequences control anaerobic regulation: addition of upstream promoter elements from constitutive genes is necessary for expression in tobacco. EMBO J. **6,** 11—16.

Gasser, S. M., Laemmli, U. K., 1986: Cohabitation of scaffold binding regions with upstream enhancer elements of three developmentally regulated genes of *Drosophila melanogaster.* Cell **46,** 521—530.

Gasser, S. M., Laemmli, U. K., 1987: A glimpse at chromosomal order. Trends Genet. **3,** 16—22.

Glaczinski, H., Kloppstech, K., 1987: Heat induced alterations in the properties of thylakoid membranes. Europ. J. Cell Biol. **43,** 19.

Grandbastien, M. A., Berry-Lowe, S., Shirley, B. W., Meagher, R. B., 1986: Two soybean ribulose-1,5-bisphosphate carboxylase small subunit genes share extensive homology even in the distant flanking sequences. Pl. Molec. Biol. **7,** 451—465.

Guedon, G. F., Gilson, G. J. P., Ebel, J. P., Befort, N. M. T., Remy, P. M., 1986: Lack of correlation between extensive accumulation of bisnucleoside phosphates and the heat shock response in eukaryotic cells. J. Biol. Chem. **261,** 16459—16465.

Gurley, W. B., Hepburn, A. G., Key, J. L., 1979: Sequence organization of the soybean genome. Biochim. Biophys. Acta **561,** 167—183.

Gurley, W. B., Czarnecka, E., Nagao, R. T., Key, J. L., 1986: Upstream sequences required for efficient expression of a soybean heat shock gene. Molec. Cell. Biol. **6,** 559—565.

Hickey, E., Brandon, S. E., Potter, R., Stein, J., Weber, L. A., 1986: Sequence and organization of genes encoding the human hsp27 kDa heat shock protein. Nucl. Acids Res. **14,** 4127—4145.

Hoffman, E. P., Gerring, S. L., Corces, V. G., 1987: The ovarian, ecdysterone and heat shock responsive promoters of *Drosophila* hsp27 gene react very differently to perturbations of DNA sequence. Molec. Cell. Biol. **7,** 973—981.

Hultmark, D., Klemenz, R., Gehring, W. J., 1986: Translational and transcriptional control elements in the untranslated leader of the heat shock gene hsp22. Cell **44,** 429—438.

Ingolia, T. D., Craig, E. A., 1982: Four small *Drosophila* heat shock proteins are related to each other and to mammalian α-crystallin. Proc. Natl. Acad. Sci. U.S.A. **79,** 2360—2364.

Jackson, D. A., 1986: Organization beyond the gene. Trends Biochem. Sci. **11,** 249—252.

Jones, D., Russnak, R. H., Kay, R. J., Candido, E. P. M., 1986: Structure, expression and evolution of the heat shock gene locus in *Caenorhabditis elegans* that is flanked by repetitive elements. J. Biol. Chem. **261,** 12006—12015.

Kalish, F., Cannon, C., Brunke, K., 1986: Characterization of a putative hsp81 gene in *Brassica oleracea* which is both constitutive and inducible. J. Cell. Biol. **103,** 176a.

Karin, M., Haslinger, A., Holtgreve, H., Richards, R. I., Krauter, P., Westphal, H. M., Beats, M., 1984: Characterization of DNA sequences through which cadmium and glucocorticoid hormones induce human metallothionein IIa gene. Nature (Lond.) **308,** 513—519.

Kee, S. C., Nobel, P. S., 1986: Concomitant changes in high temperature tolerance and heat shock proteins in desert succulents. Pl. Physiol. **80,** 596—598.

Key, J. L., Czarnecka, E., Lin, C.-Y., Kimpel, J., Mothershed, C., Schöffl, F., 1983: A comparative analysis of the heat shock response in crop plants. In: Current topics in plant biochemistry and physiology. Vol. 2, pp. 107—118. Randall, D. D., Blevins, D. G., Larson, R. L., Rapp, B. J. (eds.). Columbia, Missouri. University of Missouri-Columbia.

Key, J. L., Gurley, W. B., Nagao, R. T., Czarnecka, E., Mansfield, M. A., 1985: Multigene families of soybean heat shock proteins. Vol. 83, pp. 81—100. Van Vloten-Doting, L., Groot, G., Hall, T. C. (eds.). Plenum Press, New York.

Kimpel, J. A., Key, J. L., 1985: Presence of heat shock mRNAs in field grown soybeans. Pl. Physiol. **79,** 672—678.

Kleinsek, D. A., Beattie, W. G., Tsai, M. J., O'Malley, B. W., 1986: Molecular cloning of a steroid-regulated 108 K heat shock protein gene from hen oviduct. Nucl. Acids Res. **14,** 10053—10069.

Klemenz, R., Hultmark, D., Gehring, W. J., 1985: Selective translation of heat shock mRNA in *Drosophila melanogaster* depends on sequence information in the leader. EMBO J. **4,** 2053—2060.

Kloppstech, K., Meyer, G., Schuster, G., Ohad, I., 1985: Synthesis, transport and localization of a nuclear coded 22 kDa heat shock protein in the chloroplast membranes of peas and *Chlamydomonas reinhardii.* EMBO J. **4,** 1901—1909.

Koo, H. S., Wu, H. M., Crothers, D. M., 1986: DNA bending at adenine/thymine tracts. Nature (Lond.) **320,** 501—506.

Koyasu, S., Nishida, E., Kadowaki, T., Matsuzaki, F., Iida, K., Harada, F., Kasuga, M., Sakai, H., Yahara, I., 1986: Two mammalian heat shock proteins, HSP90 and HSP100, are actin binding proteins. Proc. Natl. Acad. Sci. U.S.A. **83,** 8054—8058.

Lee, J. S., Verma, D. P. S., 1984: Structural and chromosomal arrangement of leghaemoglobin genes in kidney bean suggest divergence in soybean leghaemoglobin gene loci following tetraploidization. EMBO J. **3,** 2745—2752.

Lin, C.-Y., Roberts, J. R., Key, J. L., 1984: Acquisition of thermotolerance in soybean seedlings. Pl. Physiol. **74,** 152—160.

Lindquist, S., 1986: The heat shock response. Ann. Rev. Biochem. **55,** 1151—1191.

Lubben, T. H., Keegstra, K., 1986: Effective in vitro import of a cytosolic heat shock protein into pea chloroplasts. Proc. Natl. Acad. Sci. U.S.A. **83,** 5502—5506.

McClellan, J. A., Palecek, E., Lilley, M. J., 1986: $(AT)_n$ tracts embedded in random sequence DNA — formation of a structure which is chemically reactive and torsionally deformable. Nucl. Acids Res. **14,** 9291—9309.

McGarry, T. J., Lindquist, S., 1985: The preferential translation of *Drosophila* hsp70 mRNA requires sequences in the untranslated leader. Cell **42,** 903—911.

Mestril, R., Schiller, P., Amin, J., Klapper, H., Anathan J., Voellmy, R., 1986: Heat shock and ecdysterone activation of the *Drosophila melanogaster* hsp23 gene; a sequence element implied in developmental regulation. EMBO J. **5,** 1667—1673.

Munro, S., Pelham, H., 1985: What turns on heat shock genes? Nature (Lond.) **317,** 477—478.

Nagao, R. T., Czarnecka, E., Gurley, W. B., Schöffl, F., Key, J. L., 1985: Genes for low molecular weight heat shock proteins of soybean: sequence analysis of a multigene family. Molec. Cell. Biol. **5,** 3417—3428.

Nagao, R. T., Kimpel, J. A., Vierling, E., Key, J. L., 1986: The heat shock response: A comparative analysis. In: Oxford Surveys of Plant Molecular and Cell Biology. Vol. 3, pp. 384—438. Miflin, B. J. (ed.). Oxford University Press, Oxford, U.K.

Nover, L., Hellmund, D., Neumann, D., Scharf, K.-D., Serfling, E., 1984: The heat shock response of eukaryotic cells. Biol. Zbl. **103,** 357—435.

Nover, L., 1987: Expression of heat shock genes in homologous and heterologous systems. Enzyme Microb. Technol. **9,** 130—144.

Oppermann, H., Levinson, W., Bishop, J. M., 1981: A cellular protein that associates with the transforming protein of Rous sarcoma virus is also a heat shock protein. Proc. Natl. Acad. Sci. U.S.A. **78,** 1067—1071.

Ougham, H. J., Stoddart, J. L., 1986: Synthesis of heat shock protein and acquisition of thermotolerance in high-temperature tolerant and high-temperature susceptible lines of sorghum. Plant Sci. **44,** 163—167.

Pelham, H., 1985: Activation of heat shock genes in eukaryotes. Trends Genet. **1,** 31—35.

Pelham, H., 1986: Speculation on the function of the major heat shock and glucose regulated proteins. Cell **46,** 959—961,

Pelham, H. R. B., Bienz, M., 1982: A synthetic heat shock promoter element confers heat-inducibility on the herpes simplex virus thymidine kinase gene. EMBO J. **1,** 1473—1477.

Raschke, E., 1987: Molekulare Analyse verschiedener Gene für kleine Hitzeschockproteine der Sojabohne (*Glycine max* [L.] Merrill). Dissertation, Universität Bielefeld, F.R.G.

Raschke, E., Baumann, G., Schöffl, F., 1988: Nucleotide sequence analysis of soybean small heat shock protein genes belonging to two different multigene families. J. Mol. Biol. **199,** 549—557.

Roberts, J. K., Key, J. L., 1985: Characterization of the genes for the heat shock 70kD and 80kD proteins in soybean. First International Congress of Plant Molecular Biology, Savannah, GA, p. 137.

Rochester, D. E., Winter, J. A., Shah, D. M., 1986: The structure and expression of maize genes encoding the major heat shock protein, hsp70. EMBO J. **5,** 451—458.

Sachs, M., Ho, T. H., 1986: Alteration of gene expression during environmental stress in plants. Ann. Rev. Plant Physiol. **37,** 363—376.

Sargan, D. R., Tsai, M. J., O'Malley, B. W., 1986: Hsp108, a novel heat shock inducible protein of chicken. Biochem. **25,** 6252—6258.

Schlesinger, M. J., 1986: Heat shock proteins: the search for function. J. Cell Biol. **103,** 321—325.

Schlesinger, M. J., Ashburner, M., Tissières, A., 1982: Heat shock from bacteria to man. Cold Spring Harbor Laboratory, NY.

Schöffl, F., Baumann, G., 1985: Thermoinduced transcripts of a soybean heat shock gene after transfer into sunflower using a Ti plasmid vector. EMBO J. **4,** 1119—1124.

Schöffl, F., Raschke, E., Nagao, R. T., 1984: The DNA sequence analysis of soybean heat shock genes and identification of possible regulatory elements. EMBO J. **3,** 2491—2497.

Schöffl, F., Baumann, G., Raschke, E., Bevan, M., 1986: The expression of heat shock genes in higher plants. Phil. Trans. R. Soc. Lond. **B 314,** 453—468.

Schöffl, F., Rossol, I., Angermüller, S., 1987: Regulation of the transcription of heat shock genes in nuclei from soybean (*Glycine max*) seedlings. Plant Cell Envir. **10,** 113—119.

Singh, R. J., Hymowitz, T., 1985: The genomic relationship among six wild perennial species of the genus *Glycine* subgenus *Glycine willd.* Theor. Appl. Genet. **71,** 221—230.

Sinibaldi, R. M., Turpen, T., 1985: A heat shock protein is encoded within mitochondria of higher plants. J. Biol. Chem. **260,** 15382—15385.

Southgate, R., Ayme, A., Voellmy, R., 1983: Nucleotide sequence analysis of the *Drosophila* small heat shock gene cluster 67 B. J. Mol. Biol. **165,** 35—57.

Spena, A., Hain, R., Ziervogel, U., Saedler, H., Schell, J., 1985: Construction of a heat-inducible gene for plants. Demonstration of heat-inducible activity of the *Drosophila* hsp70 promoter in plants. EMBO J. **4,** 2739—2743.

Spena, A., Schell, J., 1987: The expression of a heat-inducible chimeric gene in transgenic tobacco plants. Mol. Gen. Genet. **206,** 436—440.

Strittmatter, G., Chua, N.-H., 1987: Artificial combination of two cis regulatory elements generates a unique pattern of expression in transgenic plants. Proc. Natl. Acad. Sci. U.S.A. **84,** 8986—8990.

Struhl, K., 1985: Naturally occurring poly(dA-dT) sequences are upstream promoter elements for constitutive transcription in yeast. Proc. Natl. Acad. Sci. U.S.A. **82,** 8419—8423.

Suggs, J. W., Wagner, R. W., 1986: Nuclease recognition of an alternating structure in $d(AT)_{14}$ plasmid insert. Nucl. Acids Res. **14,** 3703—3716.

Süss, K.-H., Yordanov, I. T., 1986: Biosynthetic cause of in vivo acquired thermotolerance of photosynthetic light reactions and metabolic responses of chloroplast to heat stress. Pl. Physiol. **81,** 192—199.

van den Heuvel, R., Hendriks, W., Quax, W., Bloemendahl, H., 1985: Complete structure of the hamster α A crystallin gene. J. Mol. Biol. **185,** 273—284.

Vierling, E., Mishkin, M. L., Schmidt, G. W., Key, J. L., 1986: Specific heat shock proteins are transported into chloroplasts. Proc. Natl. Acad. Sci. U.S.A. **83,** 361—365.

Vierling, E., Chen, Q., 1987a: Characterization and expression of chloroplast heat shock proteins in pea and petunia (in press).

Vierling, E., Roberts, J. K., Nagao, R. T., Key, J. L., 1987b: A chloroplast heat shock protein has homology to cytoplasmic heat shock proteins (in press).

Wright, J. M., Dixon, G. H., 1986: Induction by torsional stress of a cruciform conformation 5′ upstream of the gene for a high mobility group protein from trout and the specific binding to flanking sequences by the gene product. J. Cell Biol. **103,** 43a.

Wu, B. J., Morimoto, R. I., 1985: Transcription of the human hsp70 gene is induced by serum stimulation. Proc. Natl. Acad. Sci. U.S.A. **82,** 6070–6074.

Xiao, C.-M., Mascarenhas, J. P., 1985: High temperature-induced thermotolerance in pollen tubes of *Tradescantia* and heat shock proteins. Pl. Physiol. **78,** 887–890.

Yost, A. J., Lindquist, S., 1986: RNA splicing is interrupted by heat shock and is rescued by heat shock protein synthesis. Cell **45,** 185–193.

Zhang, H., Croes, A. F., Liskens, H. F., 1984: Qualitative changes in protein synthesis in germinating pollen in *Lilium longiflorum* after a heat shock. Pl. Cell Envir. **7,** 689–691.

Chapter 15

Protein Transport in Plant Cells

Peter Weisbeek and Sjef Smeekens

Department of Molecular Cell Biology and Institute of Molecular Biology, University of Utrecht, Padualaan 8, NL-3584 CH Utrecht, The Netherlands

With 6 Figures

Contents

I. Introduction

In plant cells several subcellular structures can be recognized. These include the nucleus, the endoplasmic reticulum, the Golgi system, endosomes (e. g. peroxisomes and glyoxysomes), mitochondria, chloroplasts and thylakoids. They are all surrounded by a single or a double lipid bilayer and constitute in general separate compartments within the cell. In addition the different membranes themselves and the outside of the cell can be viewed as other compartments. Each of these subcellular structures and their enclosing or internal membranes contains many different proteins, yet most of these compartments do not have the machinery to synthesize the proteins they need. Only the nucleus, the mitochondrion and the chloroplast contain DNA and are capable of transcribing this genetic information into RNA, and only the chloroplast and the mitochondrion are capable of translating this RNA into protein. All nuclear transcripts are transported out of the nucleus and exclusively translated in the cytosol.

Thus there are only three sites of protein synthesis in the plant cell. Of these the synthesis in the cytosol is the most important for the production of the many different proteins found in plant cells, since the nuclear DNA is thousands of times larger than the organelle DNA.

All proteins synthesized inside the mitochondria or chloroplast remain in these organelles and therefore all proteins found in the other cellular compartments are derived from the cytosol and are nuclear encoded. Cytosolic protein synthesis is the major source for transported and excreted proteins.

Mitochondria and chloroplasts are able to synthesize proteins, but the size of the DNA contained in these organelles and therefore their maximal protein-coding capacity is ten to one hundred times lower than what is needed for their functioning. For example chloroplast DNA contains no more than 50 genes for proteins (Shinozaki and Sugiura, 1986), whereas at least one thousand different proteins are required to sustain its photosynthetic and metabolic capacities. Of plant mitochondria little is known, but by extrapolation from yeast and animal mitochondria one may expect their DNA to code for five to ten different proteins (Grivell, 1983), whereas about 400 proteins are needed.

The proteins needed but not synthesized in these organelles are coded for by the nuclear DNA, synthesized in the cytosol, and transported into the organelles. The population of proteins present in chloroplasts and mitochondria is therefore of mixed genetic origin; one part is encoded by organellar DNA and the other much larger part by nuclear DNA.

The proteins made in the chloroplasts function in the stroma or in the thylakoid while imported proteins also have to be distributed over the different organellar compartments; an intra-organellar routing mechanism is therefore required. A schematic view of the sites of protein synthesis and the transport of proteins to and inside the various subcellular structures is given in Figure 1.

Proteins can not traverse membranes unless they are equipped with specific information for interaction with the membrane-associated translocation mechanisms.

In eukaryotic cells three different types of transport signals are known for the primary targeting of proteins. These are i) *signal peptides* which in general are used in the cotranslational (vectorial) transport of these proteins into or across the membranes of the endoplasmic reticulum, ii) *nuclear targeting signals* for, presumed posttranslational import into the nucleus, and iii) *transit peptides, presequences, and internal targeting signals* for posttranslational import into chloroplasts, mitochondria, and endosomes. Signal peptides, transit peptides and presequences are mostly present as amino-terminal extensions of the proteins that are cleaved off during or shortly after translocation; the precursor has a larger size than

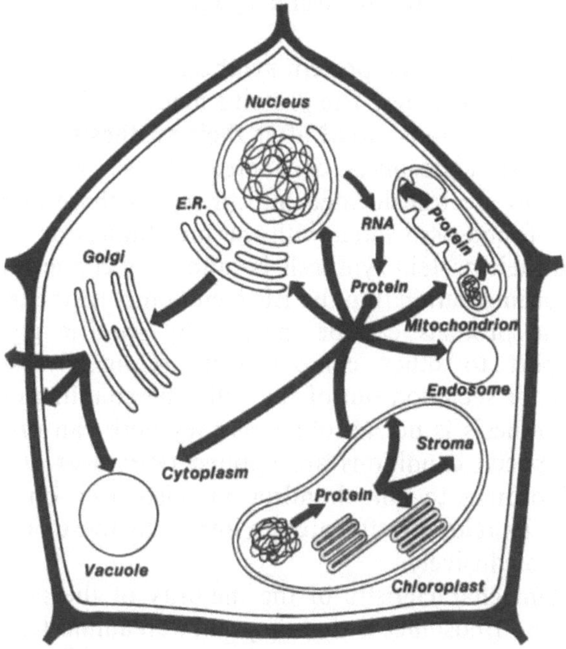

Fig. 1. The sites of protein synthesis and the routing of the proteins in plant cells

the functional mature protein. A number of proteins however has no cleavable peptide and the targeting information is not necessarily found at the amino terminus. All nuclear targeting signals analyzed so far have been found to be internal to the proteins to be transported and they are not removed after transport.

The first two types of signals will be discussed in sections II and III A. The import of proteins into chloroplasts, which is the best studied protein transport mechanism in plant cells, and their import into mitochondria and endosomes (peroxisomes and glyoxysomes) will be dealt with in more detail in sections III B, C and D. The routing of proteins towards the Golgi, into the envelope membranes, and out of the cell proceeds via the endoplasmic reticulum and can be determined by specific glycosylation within the ER and the Golgi system. This secondary protein transport process will not be dealt with in this review.

Protein translocation has mostly been studied in prokaryotic, animal, and yeast cells. Only with respect to the chloroplast is the plant situation explored to a certain depth. For that reason our present picture of how and where proteins move in plant cells is strongly influenced by the supposed analogy between different eukaryotic cells. This will be reflected in the next sections.

II. The Secretory Pathway

In eukaryotic cells cytosolic protein synthesis occurs either on free or on membrane-bound ribosomes. The proteins made on free ribosomes remain in the cytosol or are transported after their synthesis into the mitochondrion, chloroplast, endosomes or nucleus; those made on membrane-bound polysomes are translocated into or across these membranes. Such polysomes are found associated with the membranes of the endoplasmic reticulum and their protein synthesis is intimately connected with translocation across or integration into the ER membrane. Inside the endoplasmic reticulum these proteins can be marked by specific glycosylation for further transport to other compartments (Golgi, lysosomes, plasma membrane) or for secretion out of the cell. The coupling of translocation and protein-synthesis is not absolute, as they both can proceed independently when special conditions are applied (Perara *et al.*, 1986). *In vivo* binding of ribosomes to mitochondria has also been observed (Kellems *et al.*, 1974) but its relevance for protein import by these organelles has not been decided conclusively.

An important characteristic of the majority of the proteins made on membrane-bound ribosomes is that they carry an amino-terminal sequence that is cleaved off during the translocation process. This sequence contains the information for the binding of the polysome to the membrane and for the subsequent transfer of the growing polypeptide chain through the membrane. These sequences, signal peptides, have a variable length (between 15 and 30 amino acids), are strongly hydrophobic, and have in general one or a few positively charged amino acids at their amino terminus (Von Heijne, 1985). This pathway is e. g. used during embryogenesis for seed storage proteins and for lysosomal (vacuolar) enzymes.

All the known proteins made inside mitochondria and most of the proteins made in chloroplasts are integrated into the envelope membranes and thylakoid membrane respectively. Both these organelles have many properties in common with prokaryotic cells. Transport of their membrane proteins will therefore occur most likely in a prokaryotic fashion, by a signal peptide-directed targeting mechanism but without the obligate coupling of protein synthesis and translocation that is observed in eukaryotic cells. The presence of an such an amino-terminal extension has been documented for the chloroplast protein cytochrome f, and its structure and transport features resemble other eukaryotic signal sequences (Wiley *et al.*, 1984). Fusions of amino-terminal fragments of the cytochrome f precursor with beta-galactosidase (Rothstein *et al.*, 1985) are routed in *E. coli* to the cytoplasmic membrane; this involves the use of components of the bacterial secretory pathway. These results indicate that cytochrome f is synthesized and integrated into the thylakoid membrane very much like the membrane proteins of *E. coli*.

In the case of the thylakoid membrane proteins apocytochrome b-559 and P680 chlorophyll a apoprotein (Herrmann *et al.*, 1984; Morris and Herrmann, 1984) which are also encoded by the chloroplast DNA, no evidence for a cleavable amino terminus was found in *in vitro* translation experiments nor does the nucleotide sequence of their genes suggest a larger precursor.

III. Post-Translational Transport

Chloroplast and mitochondrial proteins for which the genetic information is located on the nuclear DNA and all proteins targeted for the nucleus, peroxisomes, and glyoxysomes are synthesized in the cytosol and are believed to be taken up by these organelles after their completion. Post-translational protein import is easily demonstrated *in vitro* where intact chloroplasts can import added precursor proteins. *In vivo*, however, it is difficult to determine whether the import is post- or cotranslational or maybe even both. In yeast it has been shown that mitochondrial precursors, after forced accumulation in the cytoplasm, can still be imported into the mitochondrion, indicating that *in vivo* import can occur posttranslationally (Ried and Schatz, 1982). This, however, does not exclude the possibility that the normal process is different.

For chloroplasts no precursors have yet been found in the cytoplasm. The precursor proteins for this organelle are synthesized preferentially on free polysomes and no association of ribosomes with chloroplasts has been found (Dobberstein *et al.*, 1977).

After the interaction of the precursor with the proper organelle it has to be routed to its ultimate location within this organelle. For the chloroplast this can be the outer or inner membrane, intermembrane space, stroma, thylakoid membrane, or lumen. The mitochondrion can be divided into outer and inner membranes, intermembrane space, and matrix. The endosomes have a matrix and a surrounding single membrane.

The major difference between the mitochondrion and the chloroplast is the additional targeting information necessary for a protein to reach the thylakoid system. The other transport steps are certainly distinct from each other but are thought to operate along the same lines.

The import of proteins into the chloroplast can be divided into the following steps: i) binding to the envelope membrane(s), ii) translocation into or across the envelope membranes, iii) eventual routing to the thylakoid system, and iv) proteolytic processing of the precursor.

A. The Nucleus

The membrane of the nucleus is unique among the membranes that are found in the eukaryotic cell in that it contains numerous large discontinuities in its phospholipid bilayer (pore complexes). Through these pores small polypeptides (up to 18,000 kDa) diffuse freely into the nucleus and accumulate there if they are true nuclear proteins. Larger proteins require a selective, active mechanism to pass the nuclear membrane. Nuclear proteins are synthesized in the cytoplasm as mature proteins, i. e. there is no proteolytic processing associated with this import process. Once accumulated in the nucleus these proteins are retained so strongly that they do not dissipate out of the nucleus even when the nuclear membrane is damaged severely (De Robertis, 1986; Dabauvalle and Franke, 1982). This property is thought to indicate that the binding of the imported protein to as yet unidentified nondiffusible nuclear components is a major aspect of this transport process.

It has been shown that nuclear proteins contain regions, internal or amino-terminal, that function as signals for import; these signals have a strong requirement for positive charges. They can be used efficiently to transport other non-nuclear proteins into the nucleus (Kalderon *et al.*, 1984); chemical crosslinking of a synthetic nuclear targeting signal to a non-nuclear protein at random positions resulted in the transport of these combinations to and inside the nucleus (Lanford *et al.*, 1986; Goldfarb *et al.*, 1986).

Nuclear import signals appear to act post-translationally and are not removed after completion of the nuclear accumulation. Such properties agree well with the fact that nuclear proteins become dispersed over the complete cell during mitosis and must reenter the nucleus at each telophase.

B. The Chloroplast

1. Import

Thus far, all nuclear-encoded chloroplast proteins analyzed in detail are synthesized as precursors with N-terminal transit peptides that are cleaved off during or immediately after import. A transit peptide can vary in size between 35 and 80 amino acids and it determines the binding of the precursor to the chloroplast envelope; removal of this peptide completely abolishes all binding to the envelope. Mature proteins, either isolated from the chloroplast or synthesized *in vitro,* do not bind to isolated chloroplasts or to purified envelope membranes. Linkage of a transit peptide to proteins that are not found normally in the chloroplast or that are not of plant origin results in recognition of and binding to the envelope membranes followed by uptake and processing. This has been shown both *in vivo* and *in vitro* (Van den Broeck *et al.*, 1985; Wasmann *et al.*, 1986).

The proteins that have been analyzed for their import were targeted for the stroma (e. g. small subunit and ferredoxin), the thylakoid membrane (chlorophyll a/b binding protein), or lumen (plastocyanin) (Wasmann *et al.*, 1986; Kohorn *et al.*, 1986; Smeekens *et al.*, 1986b). No envelope membrane or intermembrane space proteins have been tested yet but it has been demonstrated that the phosphate translocator, an inner membrane protein, and a 22,000 dalton protein of the outer membrane are synthesized in the cytoplasm as larger precursors (Flugge and Wessel, 1984).

Correct targeting in the cell means that a precursor has to discriminate between the various membranes present in the cell, e. g. chloroplast proteins should not import into the mitochondria and *vice versa*. There is in fact no information about possible misrouting of proteins *in vivo* but immunological electron microscopy analysis of thin sections of plant cells does not reveal the presence of chloroplast proteins in other compartments. This observation does not, however, exclude import followed by rapid degradation of such wrongly imported proteins. *In vitro* experiments have shown that intact chloroplast precursors and fusion products of the complete transit peptide with mature mitochondrial proteins are not imported by intact mitochondria (Smeekens, unpublished results). They did, however, show binding to these mitochondria. One case has been published in which an amino-terminal fragment of the transit peptide of the small subunit of ribulosebisphosphate carboxylase (RUBISCO) of *Chlamydomonas* was shown to cause the import of another protein into mitochondria. In this case no processing was observed (Hurt *et al.*, 1986).

The specific interaction of cytoplasmic precursor proteins with the chloroplast is thought to involve receptors on the surface of the chloroplast. The nature of such receptors is unclear. Proteolytic treatment of the chloroplast reduces subsequent binding of precursor to a large extent and removal of envelope proteins by treatment with detergents also has a negative effect on binding (Cline *et al.*, 1985; Bitsch and Kloppstech, 1986). These results indicate the involvement of membrane-associated proteins in the binding process. Work with mitochondrial presequences on the other hand, showed the competence of these peptides to interact directly with lipid bilayers (Roise *et al.*, 1986). Theoretical analyses of chloroplast precursors did not identify similar amphiphilic regions in their transit peptide. The binding of the precursor to the envelope will therefore probably involve a combination of protein-protein and protein-lipid interactions, either together or in sequence.

Another question is whether each precursor has its own receptor or whether all precursors can bind to the same receptor. Experiments in which synthetic peptides were used to study binding to mitochondria and competition among precursors indicate the presence of a limited set of different receptors (Yoshida *et al.*, 1985). The observation that etioplasts can bind and import the small subunit of RUBISCO and plastocyanin (Schindler and Soll, 1986; Hageman and Roper, unpublished results), two proteins that have no function in this plastid, and also that leukoplasts, the non-photosynthetic plastids of endosperm in developing seeds, are able to import and process the small subunit precursor (Boyle *et al.*, 1986) does suggest that the receptor(s) is (are) not tissue- and precursor-specific.

The binding of the precursor to intact chloroplasts or to purified envelope membranes is dependent on the presence of a transit peptide but it does not require energy in the form of ATP or a transmembrane electrochemical gradient (Pfisterer *et al.*, 1982). For the subsequent translocation over the outer membrane there is a strict requirement for ATP, and this ATP functions at the outside of the envelope membrane (Flugge and Hinz, 1986). It may be involved in a possible ATP-dependent phosphorylation/dephosphorylation cycle at the outer membrane. Proteins targeted for the outer membrane itself have not been analyzed in the chloroplast system but for mitochondria no energy is required for the insertion of proteins into the outer membrane. In this case the N-terminal sequence is not cleaved off during insertion although it does contain the information for targeting and membrane insertion (Hase *et al.*, 1984; Hurt *et al.*, 1985). Evidence about chloroplast outer membrane proteins indicates that proteolytic processing takes place during insertion into the membrane; the *in vitro* synthesized precursors are of larger molecular weight than the mature proteins (Flugge and Wessel, 1984).

The bound precursor is transported through the envelope membranes when ATP is present or added after binding. The mechanism of this protein transport is unknown, but it appears to be strongly coupled to the last step in the transport process, the proteolytic processing of the precursor to the mature protein. In chloroplasts all imported proteins analyzed so far are processed by the proteolytic removal of an amino-terminal part and no conditions have yet been found to separate this processing from translocation. Also the use of mutants in which the sequence containing the processing site was removed (next section) did not result in the import of intact precursor. *Lemna* chlorophyll a/b binding protein, however, can be imported into barley etioplasts such that part of the imported precursor still carries the transit peptide (Chitnis *et al.*, 1986).

In mitochondria chelating agents can block processing without interfering with the import of the precursor (Zwizinski and Neupert, 1983). This means that the processing protease (Miura *et al.*, 1982) is not an active part of the translocating machinery in this organelle.

An interesting question in this translocation process is whether the structure of the precursor is maintained during membrane passage or whether it becomes unfolded. This question has been addressed experimentally for the protein dihydrofolate reductase in the mitochondrial import system. When the structure of the precursor was stabilized by interaction with the specific ligand methotrexate prior to the import experiment, the import was blocked completely (Eilers and Schatz, 1986). This experiment suggests that unfolding of the precursor is important for its translocation over the envelope membranes.

The transit peptide is removed in a very precise manner from the precursor during or immediately after import. An enzyme has been identified and partly purified that can perform the cleavage *in vitro* with the same specificity as *in vivo*. This enzyme, a metalloprotease, is located in the stroma and is capable of processing stroma-targeted proteins to their mature size and the thylakoid-lumen protein plastocyanin to an intermediate form (Robinson and Ellis, 1984; Hageman *et al.*, 1986). The plastocyanin intermediate is subsequently transported over the thylakoid membrane into the lumen where the remainder of the transit peptide is removed (see next section). This second part of the plastocyanin transit peptide was found to be very similar to the signal peptide of a cyanobacterial plastocyanin (Van der Plas and Weisbeek, manuscript in preparation). In cyanobacteria only transport across the thylakoid membrane is necessary. It is therefore very tempting to speculate that the transport of nuclear-encoded proteins across the thylakoid membrane makes use of the same pathway that is used by the organelle-encoded proteins that are targeted for the lumen. In this model, targeting information for a new eukaryotic transport route has been fused to information for an already existing prokaryote-like transport system.

Stromal protease was unable to process the precursor for the thylakoid membrane-specific light-harvesting chlorophyll a/b protein (Robinson, personal communication), although the transit peptide of this precursor does resemble the transit peptides of precursors for stromal proteins (Smeekens *et al.*, 1986a). The interactions between the thylakoid membrane and this precursor for light-harvesting chlorophyll a/b protein has been probed *in vitro* (Cline, 1986). The precursor was found to integrate specifically and correctly into the thylakoid membrane. When combined with the observation that the precursor can be imported into etiochloroplasts without processing (Chitnis *et al.*, 1986), this raises the possibility that the transit peptide of this precursor is removed only when the structure of the precursor becomes changed by integration into the thylakoid membrane.

2. Fusion Proteins

The process of protein import into chloroplasts has been studied by *in vitro* synthesis of the precursor protein (via translation of total poly(A+) RNA, hybrid selection of the transcript for the protein of interest or *in vitro* transcription of isolated DNA sequences and subsequent translation) and incubation of the precursor with intact chloroplasts followed by fractionation of the chloroplasts to determine the intra-organellar location of the imported proteins. Such an experiment with the precursors for ferredoxin and plastocyanin, stromal and luminal proteins respectively, showed that both precursors bind efficiently and that both are internalized and processed. Ferredoxin is recovered only as the mature protein and it is exclusively located in the stroma; the mature form of plastocyanin is found only in the thylakoid fraction in a protease-protected manner. In addition to the mature plastocyanin, an intermediate-sized protein is found in the stroma attached to the stromal face of the thylakoid membrane (see results with fusion proteins).

On the basis of these experiments a model for the routing of plastocyanin to the lumen as shown in Figure 2 was developed. This routing involves two distinct steps. First, pre-plastocyanin is translocated over the envelope into the stroma where the amino-terminal part of the transit peptide is cleaved off. Next, this stromal intermediate recognizes the thylakoid membrane through the remaining part of the transit peptide and is then translocated across the thylakoid membrane into the lumen where it is processed to its mature size. Based on these and other observations a "two-domain transit peptide model" has been proposed. The plastocyanin transit peptide is composed of a chloroplast import domain, which mediates transport to the stroma, and a thylakoid transfer domain, involved in transport across the thylakoid membrane (Fig. 2). Transport

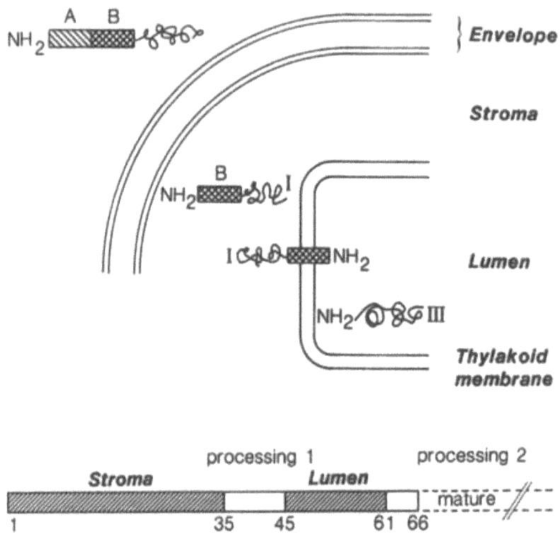

Fig. 2. The two-step transport of plastocyanin to the thylakoid lumen (upper part) and the two transport domains present in the transit peptide of the precursor (lower part)

of plastocyanin to the lumen involves the successive removal of the two domains in two different processing steps. A stromal transit peptidase processes pre-plastocyanin to the intermediate size. Processing to the mature size requires an additional protease, which subsequently was found to be present in the thylakoids (see next section).

The function of the transit peptide in chloroplast import has been probed by the construction of fusion proteins in which the transit peptide is linked to a reporter protein. This has been done for fusions of the transit peptide of the small subunit to the proteins neomycin phosphotransferase, LSU, heat shock etc. (Wasmann *et al.*, 1986; Kuntz *et al.*, 1986; Lubben and Keegstra, 1986), for the transit peptide of ferredoxin to the mature parts of plastocyanin and superoxide dismutase, and for the transit peptide of plastocyanin to the mature parts of ferredoxin and superoxide dismutase (Smeekens *et al.*, 1986a; Smeekens *et al.*, submitted).

In the *in vitro* experiments the fusion DNAs are expressed *in vitro* and the resulting radiochemically almost pure proteins are imported into intact chloroplasts. Fractionation of the chloroplasts is then used to determine the ultimate location of the hybrid protein tested. The *in vivo* experiments involve transformation of plant cells with the fusion DNA, regeneration of intact plants, and subsequent analysis of the chloroplasts.

The results for the stroma-directed transit peptides of the small subunit and ferredoxin proteins *in vivo* and *in vitro* lead to the same conclusion: all the fusion proteins tested so far are imported into the stroma and processed the same way as the original precursors. The efficiency of import, however, does vary considerably for each construction. The overall conclusion is that the complete transit peptide is sufficient for chloroplast-specific import but that the sequence around the proteolytic processing site, including several amino acids of the mature protein, does contribute considerably to the efficiency of the process.

The results with the transit peptide of the thylakoid-lumen directed plastocyanin fused to either ferredoxin or mitochondrial superoxide dismutase (SOD) are more complex (Fig. 3). In the SOD fusion the complete plastocyanin transit peptide was present; in the ferredoxin fusion the first amino acid (alanine) of the transit peptide was missing. The fused PCSOD precursor (Fig. 3 A) was imported efficiently and processed to a single protein of a molecular weight in between that of the precursor and the SOD mature protein. This molecule was found only in the stroma and was not associated with the thylakoid membrane. The fused PCFD precursor (Fig. 3 B) was equally well imported and processed but it gave

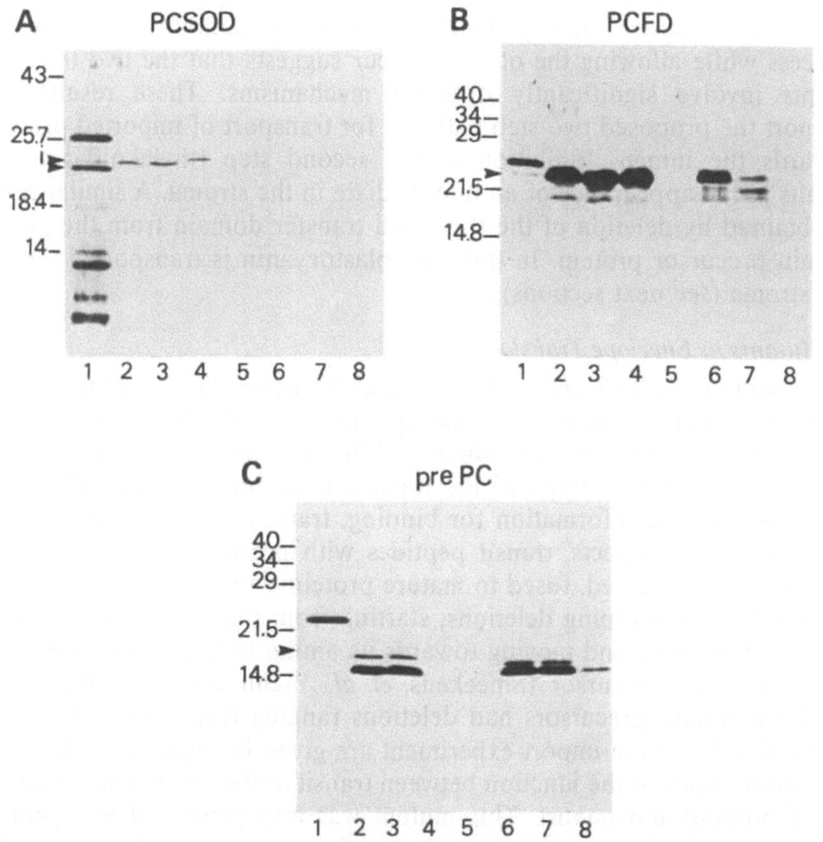

Fig. 3. Analysis on polyacrylamide gels of the import of fusion proteins by isolated chloroplasts. The translation mix (lane 1) was incubated with intact chloroplasts and the total mixture before (lane 2) and after protease treatment (lane 3) was lysed and fractionated in stroma (lane 4), envelope (lane 5), thylakoids (lane 6), protease-treated thylakoids (lane 7) and sonicated thylakoids (lane 8)

two bands. The major protein is intermediate in size between the precursor and the mature ferredoxin. It is found in part in the stroma and the rest is attached to the stromal face of the thylakoid membrane. The second, minor component has the mobility of the mature ferredoxin and is located in the thylakoid fraction in a protease-protected manner. It may well represent a small fraction of the precursor that is succesfully routed towards the lumen. In the case of the plastocyanin precursor (Fig. 3 C) the larger part of the protein is recovered as mature protein in the thylakoid lumen whereas an intermediate protein is found in the stroma or associated with the thylakoid membrane.

The results obtained with the two fusion proteins PCFD and PCSOD make it evident that transport across the thylakoid membrane is a much more sensitive process than translocation across the chloroplast envelope and that it critically depends on the nature of the passenger protein. In both cases the plastocyanin transit peptide could direct import of the passenger proteins into chloroplasts but was not capable of transporting the proteins across the thylakoid membrane. This ability to block one process while allowing the other to occur suggests that the two transport events involve significantly different mechanisms. These results also support the proposed two-step pathway for transport of imported proteins towards the lumen. Inhibition of the second step (thylakoid transfer) results in the appearance of an intermediate in the stroma. A similar result is obtained by deletion of the thylakoid transfer domain from the plastocyanin precursor protein. In this case plastocyanin is transported only to the stroma (see next sections).

3. Mutants in Envelope Translocation

The results described above demonstrate the essential role of the transit peptide in translocation. The next question is how the transit peptide performs this function and whether different parts of the peptide are involved in different steps of the translocation process; e. g. whether it contains separate information for binding, translocation, and processing. To probe these aspects, transit peptides with precisely determined deletions were constructed, fused to mature proteins, and tested for import.

A set of overlapping deletions, starting from the carboxy terminus of the transit peptide and moving towards its amino terminus were made for the ferredoxin precursor (Smeekens *et al.*, manuscript submitted); the resulting mutant precursors had deletions ranging from two to 36 amino acids. Results of the import experiment are given in Figure 4. Deletion of two amino acids at the junction between transit and mature protein did not affect binding and import. This mutant was also processed to a protein similar in size to mature ferredoxin; therefore it reacted indistinguishably from the wild-type ferredoxin precursor. In addition, the efficiency of mutant t2 binding and import was comparable to the wild-type precursor. Removal of the first seven or more amino acids (t7—t36) completely abolished binding and translocation. Similarly *in vivo* experiments with plants transformed with a fusion protein containing deletions in the transit

Ferredoxin				binding	import	processing
	-48 -40 -30 -20 -10 -1	1	5---98			
wt	MASTLSTLSVSASLLPKQQPMVASSLPTNMGQALFGLKAGSRGRVTAM	A	TYKVTL	+	+	+
t2	MASTLSTLSVSASLLPKQQPMVASSLPTNMGQALFGLKAGSRGRVT	A	TYKVTL	+	+	+
t7	MASTLSTLSVSASLLPKQQPMVASSLPTNMGQALFGLKAGS	P	TYKVTL	−	−	−
t10	MASTLSTLSVSASLLPKQQPMVASSLPTNMGQALFGLK	A	TYKVTL	−	−	−
t13	MASTLSTLSVSASLLPKQQPMVASSLPTNMGQALF	A	TYKVTL	−	−	−
t15	MASTLSTLSVSASLLPKQQPMVASSLPTNMGQA	S	TYKVTL	−	−	−
t16	MASTLSTLSVSASLLPKQQPMVASSLPTNMGQ	A	TYKVTL	−	−	−
t17	MASTLSTLSVSASLLPKQQPMVASSLPTNMG	P	TYKVTL	−	−	−
t18	MASTLSTLSVSASLLPKQQPMVASSLPTNM	A	TYKVTL	−	−	−
t20	MASTLSTLSVSASLLPKQQPMVASSLPT	T	TYKVTL	−	−	−
t36	MASTLSTLSVSA	S	TYKVTL	−	−	−

Fig. 4. Amino acid sequence of the transit peptides of mutant ferredoxin precursors and their properties in binding, import and processing when tested with intact chloroplasts

peptide of the small subunit showed that removal of the first C-terminal amino acid of the transit peptide had no effect on the import process but removal of the first sixteen C-terminal amino acids resulted in the inhibition of import (Kuntz *et al.*, 1986). The sequence between −2 and −7 appears therefore to be critical in the binding of the precursor to the envelope membrane.

At present transit peptide sequences are available for four distinct classes of chloroplast-specific precursor proteins: pSS and pCAB sequences from many different species (for a recent compilation see Karlin-Neumann and Tobin, 1986), preferredoxin (Smeekens *et al.*, 1985a), and preplastocyanin (Smeekens *et al.*, 1985b). The −2/−7 regions all contain the sequence 'GRV' or the functionally very similar sequence 'PRM' (helix destabilizer/positive charge/hydrophobic amino acid). The helix-destabilizing (G or P) and charged (R) amino acids in this region could be involved in its surface exposition. Internal deletions near the amino terminus of the transit peptide of the small subunit ranging from nine to 23 amino acids were found to reduce binding and import drastically although not completely (Wasmann *et al.*, 1986).

The main conclusions that can be drawn from these deletion analyses are that the C-terminus of the transit peptide contains essential information for binding of the precursor to the envelope but that the N-terminus is also necessary for these interactions.

4. Mutants in Thylakoid Translocation

A set of overlapping deletions, starting from the carboxy terminus of the transit peptide and moving towards its amino terminus, were also made for the plastocyanin precursor (Hageman and Weisbeek, manuscript in preparation). All deletions were made only in the thylakoid domain of the transit peptide. The amino acid sequences of the resulting mutant precursors with deletions ranging from two to 25 amino acids and the result of their import are given in Figures 5 and 6. The mutant precursors all bind to the envelope, are translocated into the chloroplast, and are processed to the intermediate form only.

	Plastocyanin transit		Import	Routing
	-66 -50 -40 -29 -20 -10 -1			
wt	MATVTSSAAVAIPSFAGLKASSTTRAATVKVAVATPRMSIKASLKDVGVVVAATAAAGILAGNAMA		+	lumen
t2	MATVTSSAAVAIPSFAGLKASSTTRAATVKVAVATPRMSIKASLKDVGVVVAATAAAGILAG	MA	±	thylakoid
t3	MATVTSSAAVAIPSFAGLKASSTTRAATVKVAVATPRMSIKASLKDVGVVVAATAAAGILA	MA	±	thylakoid
t7	MATVTSSAAVAIPSFAGLKASSTTRAATVKVAVATPRMSIKASLKDVGVVVAATAAA	MA	+	stroma
t11	MATVTSSAAVAIPSFAGLKASSTTRAATVKVAVATPRMSIKASLKDVGVVVAA	MA	+	stroma
t21	MATVTSSAAVAIPSFAGLKASSTTRAATVKVAVATPRMSIKAS	MA	+	stroma
t25	MATVTSSAAVAIPSFAGLKASSTTRAATVKVAVATPRMS	MA	±	stroma

Fig. 5. Amino acid sequence of the transit peptides of mutant plastocyanin precursors and their *in vitro* routing in chloroplasts

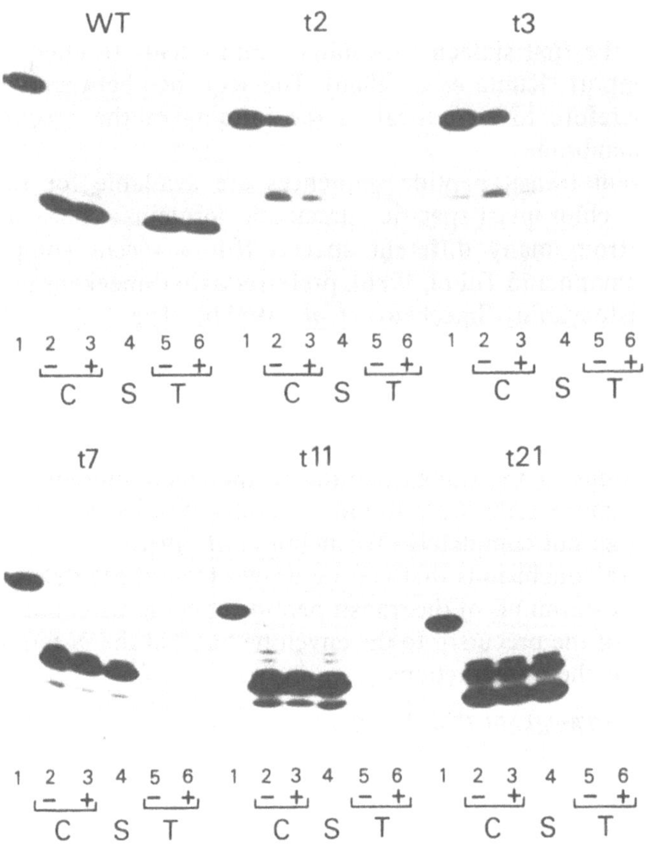

Fig. 6. Analysis on polyacrylamide gels of the import of mutant plastocyanin precursors by isolated chloroplasts. The translation mix (lane 1) was incubated with intact chloroplasts and the total mixture was lysed before (lane 2) and after protease-treatment (lane 3) and fractionated in stroma (lane 4) and thylakoids before (lane 5) and after protease treatment (lane 6)

Deletion of two or three amino acids near the junction between transit and mature protein slows down import considerably and the imported processed precursor binds to the thylakoid membranes without being translocated (they are protease-sensitive). Removal of seven or more amino acids restores the wild-type efficiency of import but no affinity for the thylakoid membrane is left; the imported protein is found only in the stromal fraction.

The conclusions from these analyses are that i) the thylakoid domain of the transit peptide affects the transport over the envelope, ii) the thylakoid transfer is more sensitive to disturbance of the regular transit sequence than the envelope transport and iii) binding to and translocation over the thylakoid membrane can be separated, whereas this has not been observed with the envelope transport mutants.

5. Processing

An important question is whether the stromal protease (Robinson and Ellis, 1984) can do the complete trimming of the precursor to the mature size or whether additional proteases are needed. The partially purified stromal processing protease was capable of removing the first part of the plastocyanin transit peptide, but it did not process the precursor to its mature size. A second and enzymatically different protease was subsequently identified in the thylakoid fraction (Hageman *et al.*, 1986) and this protease removes the last part of the PC transit peptide. This new protease is able to process the intermediate protein to its mature form, but it does not recognize the precursor as a substrate. Complete processing of plastocyanin precursor is therefore only accomplished by successive processing events; the first cleavage at the intermediate processing site mediated by the stromal protease and the second cleavage at the mature processing site mediated by the thylakoidal protease.

The involvement of two independent proteases in the processing of plastocyanin to its mature size emphasizes the special nature of the transport route taken by nuclear-encoded proteins that are functional in the lumen. It also confirms the division of the plastocyanin transit sequence into a chloroplast import domain and a thylakoid transfer domain; each domain is recognized by a different protease. The plastocyanin processing intermediate is not the only protein that is imported from the stroma into the thylakoid lumen. A transport mechanism also has to exist for chloroplast-synthesized lumen proteins. By analogy with the proposed evolutionary origin of the chloroplasts, this probably will be a prokaryote-like import mechanism. It will be very interesting to determine whether plastocyanin "hitch-hikes" with the chloroplast-synthesized proteins or passes through the thylakoid membrane by an alternative mechanism. If the first case, the newly identified protease may also be the signal peptidase of the transport mechanism for the proteins synthesized in the chloroplast.

C. *The Mitochondrion*

Our knowledge of mitochondrion-specific protein targeting is almost exclusively derived from studies of animal or yeast mitochondria. There are, however, good reasons to assume that this information is to a large extent also relevant for plants. The mitochondrion can be divided into four compartments: the outer membrane, the inner membrane, the matrix and the periplasmic space. Proteins targeted for the mitochondrion usually have an N-terminal extension called a presequence that determines the routing of the protein and that is cleaved off during import (except for outer membrane proteins). Such presequences usually are rich in positively charged amino acids, have serine and threonine in excess, and contain no acidic amino acids. In the presequences the following domains can be distinguished: a matrix-targeting domain, one or two cleavage sites, and stop-transfer sequences for the inner or outer membrane. The combination of different domains determines the routing of the protein within the mitochondrion (Hurt and Van Loon, 1986). In this view of import and sorting, the constitution of the presequence determines how far the precursor will go towards the matrix.

An alternative model (Hartl *et al.*, 1987) assumes that all precursors except the outer membrane proteins first move into the matrix and than eventually are redirected to the inner membrane or intermembrane space. This is based on homologies between parts of the presequence and the signal peptides of proteins synthesized in the mitochondria. In this attractive model (see also the previous section on the transport of the chloroplast lumen protein plastocyanin) it is assumed that the sorting information necessary for proteins synthesized inside the mitochondria is still present in the nuclear-encoded precursors but that it has been linked to mitochondrial import signals.

The presequence in general contains amino acid sequences that can insert spontaneously into phospholipid monolayers and that can form amphiphilic helices. It has been suggested that this feature plays a crucial role in transporting proteins into mitochondria (Roise *et al.*, 1986). The immobilization of the structure of a precursor by binding to a specific ligand (e. g. methotrexate to dihydrofolate reductase) reduces import to a large extent, suggesting that the precursor has to unfold during membrane translocation (Eilers and Schatz, 1986). The transport into or across the inner membrane requires a membrane potential together with ATP (Pfanner and Neupert, 1986). In this respect it differs from chloroplast import where only ATP is required.

It has been found for plant mitochondria that the precursor proteins also carry presequences with properties that appear similar to what has been found with other mitochondrial proteins (Boutry and Chua, 1985). This stresses the likelihood that plant mitochondria import their nuclear-encoded proteins very much the same way as mitochondria from other origins do.

D. Endosomes

Endosomes like peroxisomes and glyoxysomes are DNA-less organelles within an enclosing single membrane. They function in diverse oxidative and metabolic reactions. The genetic information for the many proteins that are found in the matrix or the membrane of these organelles is located in the nuclear DNA; the proteins are synthesized in the cytosol and are imported post-translationally (Fujiki *et al.*, 1984; Gietl and Hock, 1984).

Research with plants like watermelon, pumpkin, and cucumber as well as work with fungi and rats (see review by Borst, 1986) has shown that some of the imported proteins are synthesized as larger precursors but that many others are made as mature protein. In the pumpkin peroxisome two forms of the enzyme catalase were found, one of the size of the mature protein and the other with the size of the precursor (Yamaguchi *et al.*, 1984). The presence of both proteins inside the peroxisome indicates that processing is no prequisite for *in vivo* import, but it can also mean that the cleavable peptide (at either the N- or C-terminus) is not involved in transport at all. In the case of catalase the precursor is not enzymatically active and the processing may only serve to activate the protein. This would be in line with the many proteins that are imported as mature molecules (Borst, 1986; Nguyen *et al.*, 1985). A better insight into the details of this protein transport has to come from *in vitro* experiments with better-characterized proteins and with fusion proteins.

V. Conclusions

The major aspects of the mechanisms for protein transport and sorting in plant cells still have to be elucidated. It may be expected, however, that the strategies employed in other eukaryotic cells will also be found here. The first insights into how cytoplasmically-synthesized proteins reach the chloroplast have been obtained; the routing towards the chloroplast, the sorting within this organelle, and the mechanisms involved are all unique to these cells. Further research has to focus on the topology of the transit peptide to determine what amino acid sequence or other properties determine the target specificity for this organelle, whether the information for targeting and translocation can be separated, and how the transit peptide functions in the intra-organellar routing.

The analysis of transport mechanisms in the envelope membranes and the thylakoid membrane still has to be started. There are indications of a proteinaceous receptor on the envelope, but direct interactions between the transit peptide and the membrane is another interesting possibility that has not yet been worked out. The available information with regard to the thylakoid membrane is even more limited. Straightforward experiments have been done only for plastocyanin. It has to be tested whether the plastocyanin-derived model is valid for other luminal proteins too. So far thylakoid translocation appears to be a very sensitive process; knowledge of the real mechanism is necessary to learn how to overcome the inability to introduce foreign proteins into this compartment. The targeting and sorting of proteins of the inner membrane and of the thylakoid membrane itself, with the exception of the chlorophyll a/b protein, have not been studied yet. First experiments with chlorophyll a/b protein indicate that it behaves very differently from plastocyanin. It probably contains information for thylakoid interaction within the mature protein.

The results obtained thus far with stroma-targeted proteins show that we are able to use their transit peptides for import into the stroma of a variety of foreign proteins without knowledge of the translocation mechanism in which these peptides function. It enables us to study and interfere with many metabolic processes that occur in the chloroplast. Similar possibilities will soon be available for the other compartments of the plant cell.

VI. References

Bitsch, A., Kloppstech, K., 1986: Transport of proteins into chloroplasts: reconstitution of the binding capacity for nuclear-encoded precursor proteins after solubilization of envelopes with detergents. Eur. J. Biochem. **40,** 160—166.

Borst, P., 1986: How proteins get into microbodies (peroxisomes, glyoxysomes, glycosomes). Biochim. Biophys. Acta **866,** 179—203.

Boutry, M., Chua, N.-H., 1985: A nuclear gene encoding the beta subunit of the mitochondrial ATP synthase in *Nicotiana plumbaginafolia.* EMBO J. **4,** 2159—2165.

Boyle, S. A., Hemmingsen, S. M., Dennis, D. T., 1986: Uptake and processing of the precursor of the small subunit of ribulose 1,5-bisphosphate carboxylase by leucoplasts from the endosperm of developing castor oil seeds. Plant Physiol. **81,** 817—822.

Chitnis, P. R., Harel, E., Kohorn, B. D., Tobin, E. M., Thornber, J. P., 1986: Assembly of the precursor and processed light-harvesting chlorophyll a/b protein of *Lemna* into the light-harvesting complex II of barley etiochloroplasts. J. Cell Biol. **102,** 982—988.

Cline, K., Werner-Washburne, M., Lubben, T. H., Keegstra, K., 1985: Precursors to two nuclear-encoded chloroplast proteins bind to the outer envelope membrane before being imported into chloroplasts. J. Biol. Chem. **260,** 3691—3696.

Cline, K., 1986: Import of proteins in chloroplasts. Membrane integration of a thylakoid precursor protein reconstituted in chloroplast lysates. J. Biol. Chem. **261,** 14804—14810.

Dabauvalle, M. C., Franke, W. W., 1982: Karyophilic proteins: polypeptides synthesized *in vitro* accumulate in the nucleus upon microinjection into the cytoplasm of amphibian oocytes. Proc. Nat. Acad. Sciences U.S.A. **79,** 5302—5306.

De Robertis, E. M., 1986: Nucleocytoplasmic segregation of proteins and RNAs. Cell **32,** 1021—1025.

Dobberstein, B., Blobel, G., Chua, N.-H., 1977: *In vitro* synthesis and processing of a putative precursor for the small subunit of ribulose 1,5-bisphosphate carboxylase of *Chlamydomonas reinhardtii.* Proc. Nat. Acad. Sciences U.S.A. **74,** 1082—1085.

Eilers, M., Schatz, G., 1986: Binding of a specific ligand inhibits import of a purified precursor protein into mitochondria. Nature **322,** 228—232.

Flugge, U. I., Hinz, G., 1986: Energy dependence of protein translocation into chloroplasts. Eur. J. Biochem. **160,** 563—570.

Flugge, U. I., Wessel, D., 1984: Cell-free synthesis of putative precursors for envelope membrane polypeptides of spinach chloroplasts. FEBS **168,** 255—259.

Fujiki, Y., Rachibinski, R. A., Lazarow, P. B., 1984: Synthesis of a major integral membrane polypeptide of rat liver peroxisomes on free polysomes. Proc. Natl. Acad. Sci. U.S.A. **81,** 7127—7131.

Gietl, C., Hock, B., 1984: Import of *in-vitro*-synthesized glyoxysomal malate dehydrogenase into isolated watermelon glyoxysomes. Planta **162,** 261—267.

Goldfarb, D. S., Gariepy, J., Schoolnik, G., Kornberg, R. D., 1986: Synthetic peptides as nuclear localization signals. Nature **322,** 641—644.

Grivell, L., 1983: Mitochondrial DNA. Sci. Amer. March, 60—73.

Hageman, J., Robinson, C., Smeekens, S., Weisbeek, P., 1986: A thylakoid-located processing protease is required for complete maturation of the lumen protein plastocyanin. Nature **324,** 567—569.

Hartl, F.-U., Schmidt, B., Wachter, E., Weiss, H., Neupert, W., 1987: Transport into mitochondria and intramitochondrial sorting of the Fe/S protein of ubiquinol-cytochrome c reductase. Cell, 939—950.

Hase, T., Muller, U., Riezman, H., Schatz, G., 1984: A 70 kd protein of the yeast mitochondrial outer membrane is targeted and anchored via its extreme amino terminus. EMBO J. **3,** 3157—3164.

Herrmann, R. G., Alt, J., Schiller, B., Widger, W. R., Cramer, W. A., 1984: Nucleotide sequence of the gene for apocytochrome b-559 on the spinach plastid chromosome: implications for the structure of the membrane protein. FEBS **176,** 239—244.

Hurt, E., Muller, U., Schatz, G., 1985: The first twelve amino acids of a yeast mitochondrial outer membrane protein can direct a nuclearencoded cytochrome oxidase subunit to the mitochondrial inner membrane. EMBO J. **4,** 3509—3518.

Hurt, E. C., Van Loon, A. P. M., 1986: How proteins find mitochondria and intramitochondrial compartments. TIBS **11,** 204—207.

Hurt, E. C., Soltanifar, N., Goldschmidt-Clermont, M., Rochaix, J.-D., Schatz, G., 1986: The cleavable pre-sequence of an imported chloroplast protein directs attached polypeptides into yeast mitochondria. EMBO J. **5,** 1343—1350.

Kalderon, D., Roberts, B. L., Richardson, W. D., Smith, A. E., 1984: A short amino acid sequence able to specify nuclear location. Cell **39,** 499—509.

Karlin-Neumann, G. A., Tobin, E. M., 1986: Transit peptides of nuclearencoded chloroplast proteins share a common amino acid framework. EMBO J. **5,** 9—13.

Kellems, R., Allison, V., Butow, R., 1974: Cytoplasmic type 80S ribosomes associated with yeast mitochondria. IV. Attachment of ribosomes to the outer membrane of isolated mitochondria. J. Cell Biol. **65,** 1—14.

Kohorn, B. D., Harel, E., Chitnis, P. R., Thornber, J. P., Tobin, E. M., 1986: Funtional and mutational analysis of the light-harvesting chlorophyll a/b protein of thylakoid membranes. J. Cell Biol. **102,** 972—981.

Kuntz, M., Simons, A., Schell, J., Schreier, P. H., 1986: Targeting of protein to chloroplasts in transgenic tobacco by fusion to mutated transit peptide. Mol. Gen. Genet. **205,** 454—460.

Lanford, R. E., Kanda, P., Kennedy, R. C., 1986: Induction of nuclear transport with a synthetic peptide homologous to the SV40 T antigen transport signal. Cell **46,** 575—582.

Lubben, T., Keegstra, K., 1986: Efficient *in vitro* import of a cytosolic heatshock protein into pea chloroplasts. Proc. Natl. Acad. Sci. U.S.A. **83,** 5502—5506.

Miura, S., Mori, M., Amaya, Y., Tatibaba, M., 1982: A mitochondrial protease that cleaves the precursor of ornithine carbamoyltransferase. Purification and properties. Eur. J. Biochem. **122,** 641—647.

Morris, J., Herrmann, R. G., 1984: Nucleotide sequence of the gene for the P680 chlorophyll a apoprotein of the photosystem II reaction center of spinach. Nucl. Acid Res. **12,** 2837—2850.

Nguyen, T., Zelechowska, M., Foster, V., Bergmann, H., Verma, D. P. S., 1987: Primary structure of the soybean nodulin-35 gene encoding uricase II localized in the peroxisomes of uninfected cells of nodules. Proc. Natl. Acad. Sci. U.S.A. **82,** 5040—5044.

Perara, E., Rothman, R. E., Lingappa, V. R., 1986: Uncoupling of translocation from translation: implications for transport of proteins across membranes. Science **232,** 348—352.

Pfanner, N., Neupert, W., 1986: Transport of F1-ATPase subunit beta into mitochondria depends on both membrane potential and nucleoside triphosphates. FEBS **209,** 152—156.

Pfisterer, J., Lachmann, P., Kloppstech, K., 1982: Transport of proteins into chloroplasts. Binding of nuclear encoded proteins to the chloroplast envelope. Eur. J. Biochem. **126,** 143—148.

Ried, G. A., Schatz, G., 1982: Import of proteins into mitochondria. Extramitochondrial pools and post-translational import of mitochondrial protein precursors *in vivo.* J. Biol. Chem. **257,** 13062—13074.

Robinson, C., Ellis, R. J., 1984: Transport of proteins into chloroplasts. Partial purification of a chloroplast protease involved in the processing of imported precursor polypeptides. Eur. J. Biochem. **142,** 337—342.

Roise, D., Horvath, S. J., Tomich, J. M., Richards, J. H., Schatz, G., 1986: A chemically synthesized pre-sequence of an imported mitochondrial protein can form an amphiphilic helix and perturb natural and artificial phospholipid bilayers. EMBO J. **5,** 1327—1334.

Rothstein, S. J., Gatenby, A. A., Willey, D. L., Gray, J. C., 1985: Binding of pea cytochrome f to the inner membrane of *Escherichia coli* requires the bacterial secA gene product. Proc. Natl. Acad. Sciences **82,** 7955—7959.

Schindler, C., Soll, J., 1986: Protein transport in intact purified pea etioplasts. Arch. Biochem. Biophys. **247,** 211—220.

Shinozaki, K., Sugiura, M, 1986: The complete nucleotide sequence of the tobacco chloroplast genome: its gene organization and expression. EMBO J. **5,** 2043—2049.

Smeekens, S., Van Binsbergen, J., Weisbeek, P., 1985a: The plant ferredoxin precursor: nucleotide sequence of a full-length cDNA clone. Nucl. Acids Res. **13,** 3179—3194.

Smeekens, S., De Groot, M., Van Binsbergen, J., Weisbeek, P., 1985b: The sequence of the precursor of the chloroplast thylakoid lumen protein plastocyanin. Nature **317,** 456—458.

Smeekens, S., Bauerle, C., Hageman, J., Keegstra, K., Weisbeek, P., 1986a: The role of the transit peptide in the routing of precursors towards different chloroplast compartments. Cell **46,** 365—375.

Smeekens, S., Van Oosten, J., De Groot, M., Weisbeek, P., 1986b: Silene cDNA clones for a divergent chlorophyll-A/B-binding protein and a small subunit of ribulosebisphosphate carboxylase. Plant Mol. Biol. **7,** 433—441.

Van den Broeck, G., Timko, M. P., Kausch, A. P., Cashmore, A. R., Montagu, M. van, Herrera-Estrella, L., 1985: Targeting of foreign protein to chloroplasts by fusion to the transit peptide from the small subunit of ribulose 1,5-bisphosphate carboxylase. Nature **313,** 358—363.

Von Heijne, G., 1985: Signal sequences. The limit of variation. J. Mol. Biol. **184,** 99—105.

Wasmann, C., Reiss, B., Bartlett, S. G., Bohnert, H. J., 1986: The importance of the transit peptide and the transported protein for protein import into chloroplasts. Mol. Gen. Genet. **205,** 446—453.

Wiley, D. L., Auffret, A. D., Gray, J. C., 1984: Structure and topology of cytochrome f in pea chloroplast membranes. Cell **36,** 555—562.

Yamaguchi, J., Nishimura, M., Akazawa, T., 1984: Maturation of catalase precursor proceeds to a different extent in glyoxysomes and leaf peroxisomes of pumpkin cotyledons. Proc. Natl. Acad. Sci. U.S.A. **81,** 4809—4813.

Yoshida, Y., Hashimoto, T., Kimura, H., Sakakibara, S., Tagawa, K., 1985: Interaction with mitochondrial membranes of a synthetic peptide with a sequence common to extra peptides of mitochondrial precursor proteins. Biochem. Biophys. Res. Comm. **128,** 775—780.

Zwizinski, C., Neupert, W., 1983: Precursor proteins are transported into mitochondria in the absence of proteolytic cleavage of the additional sequences. J. Biol. Chem. **258,** 13340—13346.

Chapter 16

Genetic Engineering of Herbicide Resistance Genes

Dilip M. Shah, Charles S. Gasser, Guy della-Cioppa, and Ganesh M. Kishore

Plant Molecular Biology, Divison of Biological Sciences, Monsanto Company, 700 Chesterfield Village Parkway, St. Louis, MO 63198, U.S.A.

With 3 Figures

Contents

I. Introduction

The ability to integrate functional genes stably into a plant genome not only offers a powerful approach to address the fundamental questions of developmental gene expression but also provides valuable opportunities for crop improvement (for reviews, see Fraley *et al.*, 1986; Goodman *et al.*, 1987). Exciting progress has been made during the last four years in the identification and transfer of genes that confer resistance to plant viruses and insect pests. Gene transfer has also been used to engineer resistance to nonselective, environmentally safe herbicides. Over the past several years, the use of herbicides has become an established practice in world agriculture. By eliminating weeds that compete with crops for water and nutrients, herbicides increase the crop yield. New highly potent herbicides have been developed that inhibit plant growth by interfering with the biosynthesis of essential amino acids, rather than by inactivating a component of the photosynthetic apparatus (Table 1) (LaRossa and Falco, 1984). These structurally unrelated herbicides include: glyphosate which inhibits the synthesis of aromatic amino acids; the sulfonylurea and imidazolinone herbicides which block branched chain amino acid biosynthesis; and phosphinothricin which inhibits glutamine biosynthesis. Although potent and environmentally safe, these herbicides have broad-spectrum activity that discriminates poorly between weeds and crops. The genetic engineering of selective resistance to these herbicides in crop species will have substantial agronomic significance and has been the major focus of research in several labs. The biochemical basis for the action of these herbicides has been elucidated through physiological, biochemical, and genetic studies. These studies have identified the target enzymes of the herbicides and have made possible the cloning of genes encoding the herbicide-sensitive and -insensitive enzymes from microbes as well as higher plants. The enzymes detoxifying the herbicide have also been studied and the gene encoding one of these enzymes has been cloned. These genes have been shown to confer herbicide resistance when expressed in transgenic plants. In this chapter, we discuss the identification and characterization of these genes and review the different strategies that have been employed to engineer resistance to these herbicides in higher plants.

Table 1. Herbicides that inhibit the biosynthesis of essential amino acids in plants

Herbicide	Active Ingredient	Pathway Inhibited	Target Enzyme
Roundup®	Glyphosate	Shikimate	5-enolpyruvyl-shikimate-3-phosphate synthase
Oust®	Sulfometuron methyl	Branched chain	Acetolactate synthase
Glean®	Chlorsulfuron	Branched chain	Acetolactate synthase
Arsenal®	Imidazolinone	Branched chain	Acetolactate synthase
Basta®	Phosphinothricin	Glutamine	Glutamine synthetase
Herbiace®	Phosphinothricin	Glutamine	Glutamine synthetase
Amitrole®	Aminotriazole	Histidine	Imidazole glycerol phosphate dehydratase

II. Identification and Engineering of Herbicide Resistance Genes

A. Glyphosate Resistance

Glyphosate (N-[phosphonomethyl]glycine) is a broad-spectrum nonselective herbicide of widespread use in agriculture. It is nontoxic to animals, is rapidly degraded by soil microbes, and effectively controls 76 of the world's 78 worst weed species. The shikimate pathway enzyme 5-enolpyruvylshikimate-3-phosphate synthase (EPSPS), involved in aromatic amino acid biosynthesis has been identified to be the specific target of this herbicide in bacteria (Steinrucken and Amrhein, 1980). Subsequent studies have shown that glyphosate inhibits this enzyme also in higher plants (Mousdale and Coggins, 1984; Rubin, Gaines and Jensen, 1984; Nafziger *et al.*, 1984). Direct evidenc that EPSPS is the primary target of the herbicide stems from the observation that, in *E coli*, amplification of the cloned EPSPS gene results in increased resistance to glyphosate (Rogers *et al.*, 1983). Glyphosate-resistant mutants of *S. typhimurium* (Comai, Sen and Stalker, 1983; Stalker, Hiatt and Comai, 1985), *A. aerogenes* (Schultz, Sost and Amrhein, 1984; Sost, Schultz and Amrhein, 1984), and *E. coli* (Kishore *et al.*, 1986) have been isolated and shown to contain glyphosate-resistant forms of EPSPS. Glyphosate-resistant plant cell cultures containing elevated levels of EPSPS have been described (Nafziger *et al.*, 1984; Amrhein, Johanning and Smart, 1985; Steinrucken *et al.*, 1986; Smart *et al.*, 1985; Smith, Pratt and Thompson, 1985). Using tissue culture selection techniques, glyphosate-resistant cell lines have been obtained from haploid suspensions of tobacco and the expression of stable resistance has been demonstrated in tobacco calli and regenerated plants (Singer and McDaniel; 1985). The molecular basis for resistance in these tobacco plants, however, has not been determined. Despite some early controversy, it has now become clear, from the studies decribed below, that EPSPS is the primary target of the herbicide in plants.

First attempts to engineer glyphosate resistance in transgenic plants have taken advantage of the cloned bacterial gene encoding glyphosate-resistant EPSPS. A mutant gene encoding the resistant enzyme was isolated from *S. typhimurium* and was shown to contain a single base pair change resulting in a proline to serine amino acid substitution at position 101 of the protein (Comai, Sen and Stalker, 1983; Stalker, Hiatt and Comai, 1985). The chimeric genes were constructed in which the EPSPS coding sequence was driven by either the octopine synthase promoter or the mannopine synthase promoter (Comai *et al.*, 1985). The chimeric genes were introduced into tobacco cells using the root-inducing (Ri) plasmid vectors and plants were regenerated from the transformed cells. Chimeric mRNA of the expected size and bacterial EPSPS enzyme activity were detected in the leaves of the transformed plants. The transformed plants carrying the chimeric genes were two to three times more resistant to glyphosate than the control plants. Recently, using the binary *Agrobacterium tumefaciens* vector, the chimeric EPSPS gene has also been introduced into transgenic tomato plants (Fillatti *et al.*, 1987). The transgenic tomato plants expressing the gene were resistant to glyphosate sprayed at concentrations of 0.84 kg/ha. The transgenic tomato plants, however, were reduced in growth after spraying relative to unsprayed control plants. EPSPS and other shikimate pathway enzymes are known to be localized predominantly in chloroplasts (Mousdale and Coggins, 1985; Siehl, Singh and Conn, 1986). The bacterial gene used in this study, however, lacked the chloroplast transit peptide sequence and thus gave rise to a cytoplasmic form of the enzyme. It is surprising that a cytoplasmic form of the bacterial enzyme conferred herbicide resistance in transgenic plants. It is possible that the level of resistance might be significantly enhanced if the mutant enzyme had been delivered to chloroplasts.

In a second set of experiments, high-level expression of a plant gene encoding chloroplast EPSPS was used to engineer glyphosate-resistant plants (Shah *et al.*, 1986). A full-length cDNA clone for EPSPS has been isolated from a glyphosate-resistant suspension cell line of *Petunia hybrida*. The amino acid sequence predicted from the nucleotide sequence indicated that the enzyme is synthesized as a precursor polypeptide with an amino-terminal "transit peptide" sequence of 72 amino acids. The transit peptide is responsible for post-translational targeting of the precursor enzyme to the chloroplast (della-Cioppa *et al.*, 1986). After import into the chloroplast stroma, the transit peptide sequence is proteolytically removed to give a mature enzyme of relative molecular mass $(M_r) \sim 48{,}000$ daltons. Hybridization of the cDNA to Southern blots of DNA from glyphosate-resistant and -sensitive cells showed that a gene amplification event had occurred during the development of glyphosate resistance (Shah *et al.*, 1986). It is thought that this gene amplification is responsible for the overproduction of EPSPS in this cell line, as has been observed for other drug-resistant cell cultures (see Schimke, 1984 for a review). The amplified gene has been cloned and its exon-intron map has been determined (Gasser *et al.*, manuscript submitted). The gene spans 9.0 kb of DNA and is interrupted by seven introns. The sequence of the exons is identical to that of the cDNA.

A full-length cDNA clone encoding tomato EPSPS has also been isolated from a tomato pistil cDNA library (Gasser *et al.*, manuscript submitted). The deduced amino acid sequence for mature tomato EPSPS shows that the enzyme is identical in size to the petunia enzyme and has 93 % of the amino acid residues in common (Fig. 1). The tomato EPSPS transit peptide is 76 amino acids in length, four amino acids longer than that of the petunia enzyme. The degree of conservation in this region is much lower, with only 58 % of the residues identical in the two plants. This difference in rate of divergence between the two regions of the precursor enzyme indicates that the functional constraints on the transit peptide are much less stringent than those of mature polypeptide with enzymatic activity. A comparison of the plant EPSPS protein sequences to those of *E. coli* (Duncan *et al.*, 1984) and *Aspergillus* (Charles *et al.*, 1986) is shown in Fig. 1. The pattern of relatedness is striking. The plant EPSPS genes are more closely related to the procaryotic *E. coli* sequence than to the eucaryotic *Aspergillus* sequence. In plants, EPSPS appears to be present only in plastids. It has been hypothesized that plastids are derived from procaryotic endosymbionts (Margulis, 1970). If this is the case then it is possible that the cloned plant EPSPS gene, at some time during the evolution of higher plants, migrated from plastid to nucleus. Such a gene would be expected to share homology with the procaryotic genes.

In the experiments described by Shah *et al.*, (1986), *Agrobacterium tumefaciens* Ti-plasmid-mediated gene transfer was used to engineer plants with

```
Petunia    1 KPSEIVLQPI KEISGTVKLP GSKSLSNRIL LLAALSEGTT VVDNLLSSDD  50
 Tomato    1 ..H....X.. .D........ .......... ........R. ..........  50
 E.coli    1  MESLT.... ARVD..IN.. ...TV...A. .....AH.K. .LT...D...  49
 A.nidu  393 PSIEVHPGVA HSSNVICAP. ....I...A. V....GS..C RIKNLLHSDD 443

Petunia   51 IHYMLGALKT LGLHVEEDSA NQRAVVEGCG GLFPVGKESK EEIQLFLGNA 100
 Tomato   51 .......... ......D.NE ....I..... .Q.....K.E .......... 100
 E.coli   50 VRH..N..TA ..VSYTLSAD RT.CEII.N. .----PLHA. GALE......  95
 A.nidu  444 TEV..NA.ER ..----AATF SWEEEGEVLV VNGKG.NLQA SSSP.Y.... 490

Petunia  101 GTAMRPLTAA VTVAGG--NS RYVLDGVPRM RERPISDLVD GLKQLGAEVD 148
 Tomato  101 .......... ......--H. .......... .....G.... .......... 148
 E.coli   96 .......A.. --LCL.S--N DI..T.E... K....GH... A.RLG..KIT 141
 A.nidu  491 ...S.F..TV A.LANSSTVD SS..T.NN.. KQ...G.... A.TANVLPLN 540

Petunia  149 CFLGTKCPPV RIVSKGGLPG GKVKLSGSIS SQYLTALLMA APLA-LGDVE 197
 Tomato  149 .S...N.... .......... .......... .......... ....-..... 197
 E.coli  142 YLEQENY..L .--LQ..FT. .N.DVDG.V. ..F......T ....PE-.TV 188
 A.nidu  541 TSKGRASL.L K.AAS..FA. .NIN.AAKV. ..RVSS...C ..Y.KEPVTL 590

Petunia  198 IEIIDKLISV PYVEMTLKLM ERFGISVEHS SSWDRFFVRG GQKYKSPGKA 247
 Tomato  198 .......... .......... ....V..... .G....L.K. .......... 247
 E.coli  191 .R.KGD.V.K ..IDI..N.. KT..VEI.-N QHYQQ.V.K. ..S.Q...TY 237
 A.nidu  591 RVLGG.PI.Q ..ID..TAM. RS..ID.QK. TTEEHTYHIP QGR.VN.AEY 640

Petunia  248 FVEGDASSAS YFLAGAAVTG GTITVEGCGT NSLQGDVKFA -EVLEKMGAE 296
 Tomato  248 .......... .......... ..V....... S......... -......... 296
 E.coli  238 L......... ....A..IK. ..VK.T.I.R N.M...IR.. -D.......T 286
 A.nidu  641 VI.S...C.T .P..V..... T.C..PNI.S A.....AR.. V...RP..CT 690

Petunia  297 VTWTENSVTV KGPPRSSSGR KHLRAIDVNM NKMPDVAMTL AVVALYADGP 346
 Tomato  297 .......... .....N...M .......... .......... .....F.... 346
 E.coli  287 ICW------- -.DDYI.CTR GE.N...MD. .HI..A...I .TA..F.K.T 328
 A.nidu  691 .EQ..T.T.. T..SDGILRA TSK.GYGT.D RCV.RCFR.G SHRPMEKSQT 740

Petunia  347 T--AIRDVAS WRVKETERMI AICTELRKLG ATVEEGPDYC IITPPEKLN- 393
 Tomato  347 .--T...... .......... .......... ...V..S... .........- 393
 E.coli  329 .--RL.NIYN ......D.LF .MA.....V. .EV...H..I R........- 375
 A.nidu  741 .PPVSSGI.N Q....CN.IK .MKD..AKF. VICREHD.GL E.DGIDRS.L 790

Petunia  394 ---VTDIDTY DDHRMAMAFS LAACADVPVT INDPGCTRKT FPNYFDVLQQ 440
 Tomato  394 ---..E.... .......... .......... .KN....... ..D..E...K 440
 E.coli  376 ---FAEIA.. N......C.. .V.LS.T... .L..K..A.. ..D..EQ.AR 422
 A.nidu  791 RQPVGGVFC. ....V.FS.. VLSLVTPQP. LILEKECVGK TWPGWWDTLR 840

Petunia  441 YSKH- 445
 Tomato  441 ....- 445
 E.coli  423 I.QAA 427
 A.nidu  841 QLFKV 845
```

Fig. 1. Comparison of the amino acid sequences of the mature petunia and tomato EPSPS enzymes with those of *E. coli* (Duncan *et al.*, 1984) and *A. nidulans* (Charles *et al.*, 1986) enzymes. Only amino acids which differ from the petunia sequence are shown. Dots indicate amino acids that are identical to the petunia sequence, and dashes indicate insertions/deletions. The Aspergillus sequence is numbered with the first amino acid of the entire AROM complex as amino acid 1 (Charles *et al.*, 1986). Amino acids have been shown in single letter codes.

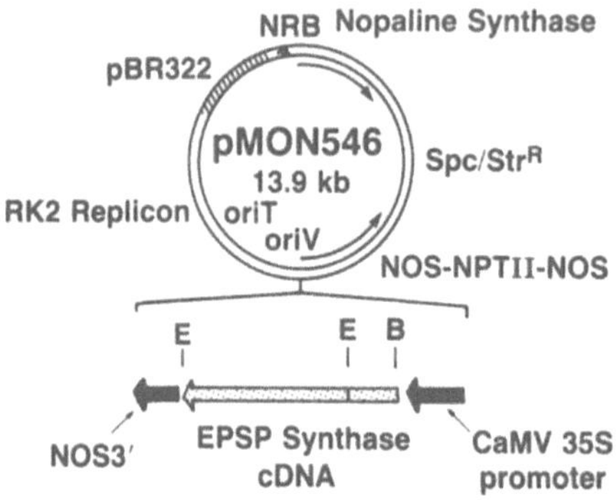

Fig. 2. The plant expression vector (pMON546) containing the chimeric petunia EPSPS gene for transformation into petunia cells. This vector contains a full-length EPSPS coding sequence flanked by the cauliflower mosaic virus 35S promoter and the polyadenylation signal from the nopaline synthase gene. The EPSPS coding sequence encodes the precursor enzyme consisting of 72 amino acids of the transit peptide and 444 amino acids of the mature enzyme. The vector contains the chimeric neomycin phosphotransferase gene as a selectable marker gene and nopaline synthase gene as a scorable marker gene. For details on the construction of this vector, reader is referred to Shah *et al.*, 1986, E = EcoRI; B = BglII; NRB = Nopaline right border

enhanced resistance to glyphosate. The wild-type petunia EPSPS cDNA was placed under control of the promoter for the 35S transcript of cauliflower mosaic virus (CaMV) and the resulting chimeric gene was transferred to a binary vector (Fig. 2). The CaMV 35S promoter directs the high-level expression of foreign genes in higher plants. Introduction of the chimeric EPSPS genes into petunia cells led to the growth of callus despite concentrations of glyphosate sufficient to inhibit completely proliferation of the wild-type callus. Transformed petunia plants were regenerated from similar transformation experiments. These plants were tolerant to a dose of formulated glyphosate equivalent to 0.9 kg/ha, approximately four times the quantity necessary to kill 100 % of the control plants transformed with a vector that did not contain the chimeric gene (Fig. 3). Analysis of tissue transformed with the chimeric gene showed that it contained ~20-fold more EPSPS activity than the wild-type controls. Thus, overproduction of wild-type chloroplast EPSPS confers glyphosate resistance in transgenic plants.

Since the endogenous EPSPS activity resides in chloroplasts, it is possible that the level of resistance might be significantly enhanced if the glyphosate-resistant bacterial enzyme could be delivered to chloroplasts. Della-Cioppa *et al.*, (1987) have recently demonstrated that a plant transit peptide sequence can be used to target a glyphosate-resistant bacterial enzyme to the chloroplasts of higher plants. They used a cloned gene from *E. coli* that encodes a highly glyphosate-resistant form of EPSPS (Kishore *et al.*, in preparation). A chimeric plant/bacterial gene containing the

Fig. 3. Glyphosate-resistant transgenic petunia plants. Transgenic petunia plants containing the chimeric petunia EPSPS gene (pMON546) were regenerated from the transformed calli selected for kanamycin resistance. These plants and the wild-type control plants were sprayed with Roundup® (formulated glyphosate with surfactant) at a dose equal to 0.8 lb/acre. The plants containing the chimeric gene (upper) survived glyphosate spraying and grew to maturity, while the control plants (lower) stopped growing and died. The photograph shows plants three weeks after spraying

chloroplast transit peptide sequence was constructed by joining portions of the EPSPS gene from the wild-type (K12 strain) and glyphosate-resistant mutant *E. coli* (designated SM-1) with the plant EPSPS cDNA at an Eco RI site. The chimeric gene thus encoded the 72-amino-acid transit peptide and the first 27 amino acids of the plant enzyme fused in frame to the remaining 400 amino acids from the *E. coli* enzyme. The gene was inserted into an *in vitro* transcription vector, pGEM-1. As a control, the full-length cDNA encoding petunia EPSPS synthase was inserted into plasmid pGEM-1. These constructs were transcribed *in vitro* to produce chimeric mRNA species. Translation of the plant/bacterial chimeric mRNA and wild-type petunia precursor mRNAs *in vitro* gave rise to polypeptides of $M_r \sim 53,000$ and $M_r \sim 55,000$ respectively.

The plant precursor EPSPS protein was previously shown to be enzymatically active (della-Cioppa *et al.*, 1986). The chimeric plant/bacterial precursor protein also retained its enzyme activity, and it was found to be 1000-fold more resistant to inhibition by glyphosate than the wild-type activity of the petunia precursor enzyme (della-Cioppa *et al.*, 1987). The translocation of the chimeric enzyme into chloroplasts *in vitro* was examined as described previously (della-Cioppa *et al.*, 1986). Both precursor enzymes were rapidly imported into chloroplasts and processed to their mature forms. Furthermore, the efficiency of chloroplast import of the chimeric precursor enzyme was nearly the same as that of the native plant enzyme. Like the mature plant enzyme, the mature bacterial enzyme was not imported into chloroplasts under any of the assay conditions used. Thus, the 72-amino acid transit peptide along with the first 27 amino acid residues of the mature enzyme contain all the determinants necessary to allow the heterologous bacterial enzyme to be imported into chloroplasts.

In order to determine if the glyphosate-resistant EPSPS could be imported by chloroplasts *in vivo*, the plant/bacterial chimeric gene was fused to the CaMV 35S promoter at the 5′ end and the polyadenylation signal of the nopaline synthase gene at the 3′ end and subsequently transferred to the binary vector shown in Fig. 2. The resulting vector (pMON542) containing the plant/bacterial chimeric gene was used to transform tobacco and petunia leaf discs. Leaf discs transformed with pMON542 produced a large quantity of kanamycin resistant callus from which plants were regenerated. Transformed petunia plants containing pMON542 and pMON546 (see Fig. 1) were sprayed with 0.45 kg/ha of glyphosate. The transformed plants containing pMON542 survived the glyphosate spraying and grew to maturity. In several experiments, the pMON542 plants expressing the glyphosate-resistant bacterial gene were found to be more resistant to glyphosate than the pMON546 plants expressing the wild-type petunia gene (Kishore *et al.*, in preparation). Glyphosate-resistant enzyme activity was demonstrated in the purified chloroplast fraction of the transgenic tobacco plants expressing the chimeric plant/bacterial EPSPS gene (della-Cioppa *et al.*, 1987). These results are significant from a genetic engineering standpoint in terms of demonstrating that a glyphosate-resistant bacterial enzyme when targeted to chloroplasts confers a high level of glyphosate resistance. Future attempts will focus on inserting genes encoding highly resistant plant EPSPS enzymes into transgenic crop plants.

B. Phosphinothricin Resistance

Phosphinothricin is a new herbicide being developed by the Hoechst Company. It is also an active ingredient in the herbicide bialaphos being developed by Meiji-Seika Ltd. Bialaphos is a tripeptide antibiotic produced by *Streptomyces hygroscopicus* and consists of phosphinothricin and two L-alanine residues. This molecule is cleaved by peptidases to release phosphinothricin. Phosphinothricin is a potent competitive inhibitor of glutamine synthetase (GS) from *E. coli* and higher plants (Bayer *et al.*, 1972; Leason *et al.*, 1982; Colanduoni and Villafranca, 1986; Manderscheid and Wild, 1986) and can be used as a nonselective herbicide. GS plays a central role in the assimilation of ammonia and in the regulation of nitrogen metabolism in plants (Miflin and Lea, 1977; Skokut *et al.*, 1978). The enzyme detoxifies ammonia released by nitrate reduction, amino acid degradation, and photorespiration. Inhibition of GS by phosphinothricin causes rapid accumulation of ammonia which is toxic to plant cells (Tachibana *et al.*, 1986). Initial attempts to confer resistance to this herbicide in plants focused on the isolation of a resistance gene from herbicide-resistant plant cells with the eventual goal of using this gene to establish specific herbicide resistance by gene transfer. Donn *et al.* (1984) selected alfalfa suspension cell lines that are 20- to 100-fold more resistant to the herbicide than the wild-type cells. GS activity was elevated 3- to 7-fold in the resistant cell lines. A cDNA clone encoding GS was isolated and used as a probe to show that the gene encoding GS was amplified 4- to 11-fold in those cell lines with increased GS mRNA levels. Increased enzyme synthesis was apparently sufficient to overcome the toxic effects of the inhibitor. To date, there is no evidence that high-level expression of the GS gene confers resistance to phosphinothricin at the whole plant level. Overexpression of GS may have undesirable effects on nitrogen metabolism. Another approach to engineering resistance would be to express herbicide-resistant GS in transgenic plants. Since plant GS complements the *glnA* defect in *E. coli* (DasSarma *et al.*, 1986), it may be possible to select for the resistant form of plant GS in *E. coli.* Plants contain multiple forms of GS. It is not clear whether the resistant form of a single GS would confer a commercially useful level of herbicide resistance.

A rather novel approach to confer complete resistance to phosphinothricin in transgenic plants has been recently tried (De Block *et al.*, manuscript submitted). The approach involves expressing an enzyme that detoxifies the herbicide. A gene that encodes the detoxifying enzyme phosphinothricin acetyl transferase in *S. hygroscopicus* has been cloned and characterized. The enzyme acetylates the free NH_2 group of phosphinothricin and protects the bacterium from the autotoxic effects of bialaphos. This gene was placed under the control of the cauliflower mosaic virus 35S promoter and introduced into tobacco cells. The calli transformed with the gene were resistant to high levels of the herbicide (500 mg/l). A number of plants were regenerated from the transformed calli and sprayed with 4 to 10 times the dose of herbicide required to effectively kill the control plants in 10 days. All transgenic plants assayed were completely resistant to the herbicide and survived additional applications of the herbicide. Treated plants produced normal flowers and set seed. Whereas the nontransformed plants treated with the herbicide accumulated toxic ammonia rapidly, the transgenic plants did not, indicating that the transgenic plants expressing the detoxifying enzyme were completely insensitive to the herbicide treatment. The expression of the detoxifying enzyme at a level as low as 0.001 % of the total extractable protein was sufficient to protect plants against the field dose application of the herbicide. The expression of this gene in transgenic tomato and potato plants also conferred complete resistance to the herbicide. Thus, the introduction of a detoxification pathway through gene transfer is a powerful approach for conferring selective herbicide resistance in crop plants.

C. Sulfonylurea and Imidazolinone Resistance

The sulfonylurea herbicides chlorsulfuron and sulfometuron methyl are, respectively, the active ingredients in the herbicides Glean® and Oust® sold by DuPont Company. These are highly potent herbicides with application rates of a few grams per hectare and low mammalian toxicity (Levitt *et al.*, 1981). The sulfometuron methyl is a broad-spectrum nonselective herbicide, whereas chlorsulfuron is a selective herbicide that can be used for weed control in crops such as wheat. This selectivity of chlorsulfuron is primarily due to the breakdown of the herbicide to inactive products in resistant species (Sweetser *et al.*, 1982). The biochemical site of action of sulfonylurea herbicides has been recently elucidated. LaRossa and Schloss (1984) first showed that sulfometuron methyl inhibited the growth of certain strains of *E. coli* and *S. typhimurium*. The sulfometuron methyl inhibition of bacterial growth could be reversed by the inclusion of branched-chain amino acids in the culture medium. It was subsequently determined that the enzyme acetolactate synthase (ALS) which is required for the synthesis of isoleucine, leucine, and valine was the target of sulfometuron methyl in *S. typhimurium*. Enterobacteria possess multiple isozymes for ALS, and it has been shown that ALS II and III but not ALS I are inhibited by sulfometuron methyl (LaRossa and Schloss, 1984; LaRossa and Smulski, 1984). ALS enzymes from yeast (Falco and Dumas, 1985) and a variety of plants (Ray, 1984; Chaleff and Mauvais, 1984; Haughn and Somerville, 1986) are also quite sensitive to inhibition by sulfonylureas.

The unequivocal proof that ALS is the primary target of sulfonylureas came from a large number of genetic studies carried out with microbes and plants. The mutants of *S. typhimurium* that are resistant to sulfometuron methyl have been isolated and mapped to the *ilvG* locus (LaRossa and Schloss, 1984). This gene codes for ALS isozyme II which is normally sensitive to inhibition by sulfonylureas. ALS II from the resistant strain of *S. typhimurium* was found to be far less sensitive to inhibiton by the herbicide than the enzyme from wild-type strains. Spontaneous mutations that confer resistance to the herbicide have also been obtained in the genes for ALS from *E. coli* and yeast (Yadav *et al.*, 1986). Nucleotide sequence analysis of the mutant genes revealed a single nucleotide change resulting in a valine-for-alanine substitution and a serine-for-proline substitution in the structurally homologous amino-terminal regions of the *E. coli* and yeast enzymes, respectively.

Analogous mutants have also been obtained in higher plants by selection for sulfonylurea-resistant mutants of haploid tobacco cells (Chaleff and Ray, 1984). The diploid tobacco plants which were regenerated from the mutant cells retained the herbicide-resistant phenotype even when grown under field conditions. It was established by genetic crosses that the herbicide-resistant phenotype was due to a single dominant or semi-dominant nuclear mutation at one of two unlinked loci which cosegregated with herbicide-resistant ALS activity. The sulfonylurea-resistant mutants of *Arabidopsis thaliana* have also been isolated by screening for growth of seedlings in the presence of the herbicide (Haughn and Somerville, 1986). The resistance was due to a single dominant nuclear mutation at the locus designated *csr*.

Like the glyphosate target enzyme EPSPS, ALS is the nuclear-encoded chloroplast-localized enzyme in higher plants (Chaleff and Ray, 1984; Jones, Young and Leto, 1985). The genes encoding the wild-type ALS have been isolated from tobacco and *Arabidopsis* and their nucleotide sequences have been determined (Mazur, Chui and Smith, manuscript submitted). The amino acid sequences deduced from the nucleotide sequences predict the presence of a presumptive chloroplast transit peptide at the amino-terminal ends of these two polypeptides. Although the tobacco gene was isolated from a sulfonylurea-resistant tobacco line, the gene encodes an enzyme that is sensitive to the herbicide. This is not surprising since two ALS loci that could mutate to confer a herbicide-resistant phenotype have been identified in tobacco (Chaleff and Ray, 1984). Southern blot analysis of tobacco DNA supports this observation. Genetic as well as genomic blot analyses indicate that the *Arabidopsis* genome contains a single gene for ALS. Using cloned wild-type genes as probes, it should be relatively straightforward to isolate the corresponding resistance genes from tobacco and *Arabidopsis*. The functional resistance genes can then be introduced into crop plants to determine if they confer resistance to sulfonylurea herbicides.

The imidazolinones are a new class of herbicides developed by the American Cyanamid Company. These herbicides kill monocots as well as dicots. In some crop species, selectivity is achieved by the differential metabolism of the herbicides to nonherbicidal forms (Orwick *et al.*,1983; Shaner *et al.*, 1983). The symptoms produced by this class of herbicides are similar to those produced by glyphosate in that the meristematic tissue dies first followed by the slow necrosis of the mature tissue. The mode of action of these herbicides is similar to that of sulfonylurea herbicides in that they interfere with the biosynthesis of the branched-chain amino acids valine, leucine, and isoleucine and also inhibit the enzyme ALS (Shaner *et al.*, 1984). Maize cell cultures that are tolerant of imidazolinones have been developed (Anderson and Georgeson, 1986). Several mutant maize cell lines having greater than 100-fold resistance have been isolated. Some of these cell lines have been characterized as having altered ALS. Plants were regenerated from one of these cell lines and the resistance was shown to be inherited as a single dominant nuclear gene. Plants homozygous for the resistance gene were 300-fold more resistant to a number of imidazolinone herbicides. Field studies have found no effect of the mutant gene on the growth and development of maize plants. It will be interesting to determine if the mutant maize plants show cross-resistance to sulfonylurea herbicides. The preliminary studies to determine the mode of inhibiton of ALS by these two structurally different herbicides suggest that the binding sites for these two herbicides may be different. It will also be of interest to isolate the mutant ALS gene from maize and decipher the amino acid substitution responsible for the resistance phenotype. Having the cloned gene, it should be feasible to introduce resistance to imidazolinone herbicides into different crop species through gene transfer.

III. Conclusions

The genetic engineering of selective resistance in crop species to a class of nonselective, environmentally safe herbicides promises to have enormous impact on world agriculture. The mechanism of action of several structurally unrelated herbicides that act by blocking essential amino acid biosynthesis has been unter intensive investigation. Similarities in the amino acid biosynthetic pathways of plant and microbes have led to the identification of specific enzymes that are primary targets of these herbicides and have made possible the cloning and modification of the microbial and plant genes for the target enzymes. In some instances, stable integration and expression of these genes in transformed plants has conferred herbicide resistance. In the limited number of studies that have been completed, no adverse effects on the growth and development of plants expressing these genes have been observed. Conferring an agronomically useful level of resistance in a variety of crop species remains the challenge of the future. Remarkable success has already been reported in engineering complete resistance to the herbicide phosphinothricin via expression in transgenic plants of a bacterial enzyme that inactivates the herbicide. The introduction and subsequent expression of a genetic pathway leading to the detoxification of the herbicide will be an important approach in our attempts to confer herbicide selectivity in crop species, because it should be independent of the plant species used. Such pathways, however, remain either unidentified or poorly characterized for most herbicides. This approach of engineering an herbicide detoxification pathway, although attractive, may not be feasible if several enzymes are involved in detoxification of the herbicide or if the enzymes have poor specificity for a given herbicide. Since the differential catabolysis of the herbicide is a major factor for herbicide selectivity, efforts in this area of research need to be intensified.

In addition to their agronomic significance, the genes for the herbicide target enzymes should serve as excellent tools for basic research. First, the genes for the enzymes EPSPS and ALS are excellent model systems for studying the import of non-photosynthetic proteins into the chloroplast. The identification of the domains of specificity for chloroplast localization of these enzymes will be of interest. Second, these enzymes are highly conserved in evolution and therefore the sequence analysis of genes encoding these enzymes from a wide range of species will provide insight into the evolution of these enzymes. Third, expression of the plant EPSPS and ALS genes in bacteria should permit the isolation of a large number of herbicide-resistant mutants for biochemical and structural analyses. The knowledge gained from these analyses will be crucial for the rational design of new environmentally safe herbicides. Finally, herbicide resistant alleles of these genes should have a broad potential as dominant selectable markers for genetic studies in plants.

Acknowledgements

We thank Dr. Robert Fraley for a careful review of the manuscript. We are grateful to our colleagues in the Plant Molecular Biology Group for making fruitful contributions to some of the work described in this chapter.

IV. References

Amrhein, N., Johanning, D., Smart, G. C., 1985: A glyphosate-tolerant plant tissue culture. In: Primary and Secondary Metabolism of Plant Cell Cultures. Neumann, K. H. (ed.), pp. 356—361, Berlin, Springer-Verlag.

Anderson, P. C., Georgeson, M., 1986: Selection and characterization of imidazolinone tolerant mutants of maize. In: The Biochemical Basis of Herbicide Action. Twenty-seventh Harden Conference Programme and Abstracts. Wye College, Ashford, United Kingdom.

Bayer, E., Gugel, K. H., Hagele, K., Hagemaier, H., Jessipow, S., Konig, W. A., Zahner, Z., 1972: Stoffwechselprodukte von Mikroorganismen, 98. Mitteilung — Phosphinothricin and phosphinothricyl-alanyl-alanin. Helv. Chim. Acta **55**, 224—239.

Chaleff, R. S., Mauvais, C. J., 1984: Acetolactate synthase is the site of action of two sulfonylurea herbicides in higher plants. Science **224**, 1443—1445.

Chaleff, R. S., Ray, T. B., 1984: Herbicide-resistant mutants from tobacco cell cultures. Science **223**, 1148—1151.

Charles, I. G., Keyte, J. W., Brammar, W. J., Smith, M., Hawkins, A. R., 1986: The isolation and nucleotide sequence of the complex AROM locus of *Aspergillus nidulans*. Nucleic Acids Research **14**, 2201—2213.

Colanduoni, J. A., Villafranca, J. J., 1986: Inhibiton of *Escherichia coli* glutamine synthetase by phosphinothricin. Bioorg. Chem. **14**, 163—169.

Comai, L., Sen, L. C., Stalker, D. M., 1983: An altered *aroA* gene product confers resistance to the herbicide glyphosate. Science **221**, 370—371.

Comai, L., Facciotti, D., Hiatt, W. R., Thompson, G., Rose, R. E., Stalker, D. M., 1985: Expression in plants of a mutant *aroA* gene from *Salmonella typhimurium* confers tolerance to glyphosate. Nature **317**, 741—744.

DasSarma, S., Tischer, E., Goodman, H. M., 1986: Plant glutamine synthetase complements a *glnA* mutation in *Escherichia coli*. Science **232**, 1242—1244.

della-Cioppa, G., Bauer, S. C., Klein, B. K., Shah, D. M., Fraley, R. T., Kishore, G. M., 1986: Translocation of the precursor of 5-enolpyruvylshikimate-3-phosphate synthase into chloroplasts of higher plants *in vitro*. Proc. Natl. Acad. Sci. U.S.A. **83**, 6873—6877.

della-Cioppa, G., Bauer, S. C., Taylor, M. L., Rochester, D. E., Klein, B. K., Shah, D. M., Fraley, R. T., Kishore, G. M., 1987: Targeting a herbicide-resistant enzyme from *Escherichia coli* to chloroplasts of higher plants. Bio/technology **5**, 579—584.

Donn, G., Tischer, E., Smith, J. A., Goodman, H. M., 1984: Herbicide-resistant alfalfa cells: an example of gene amplification in plants. J. Mol. Appl. Genet. **2,** 621—635.

Duncan, K., Lewendon, A., Coggins, J. R., 1984: The complete amino acid sequence of *Escherichia coli* 5-enolpyruvylshikimate 3-phosphate synthase. FEBS Lett. **170,** 59—63.

Falco, S. C., Dumas, K. S., 1985: Genetic analysis of mutants of *Saccharomyces cerevisiae* resistant to the herbicide sulfometuron methyl. Genetics **109,** 21—35.

Fillatti, J. J., Kiser, J., Rose, R., Comai, L., 1987: Efficient transfer of a glyphosate tolerance gene into tomato using a binary *Agrobacterium tumefaciens* vector. Bio/technology, in press.

Fraley, R. T., Rogers, S. G., Horsch, R. B., 1986: Genetic transformation in higher plants. CRC Crit. Rev. in Plant Sci. **4,** 1—46.

Goodman, R. M., Hauptli, H., Crossway, A., Knauf, V. C., 1987: Gene transfer in crop improvement. Science **236,** 48—54.

Haughn, G. W., Somerville, C., 1986: Sulfonylurea resistant mutants of *Arabidopsis thaliana.* Mol. Gen. Genet. **204,** 430—434.

Jones, A. V., Young, R. M., Leto, K., 1985: Subcellular localization and properties of acetolactate synthase, target site of the sulfonylurea herbicides. Plant Physiol. **77,** S293.

Kishore, G. M., Brundage, L., Kolk, K., Padgette, S. R., Rochester, D., Huynh, K., della-Cioppa, G., 1986: Isolation, purification and characterization of a glyphosate tolerant mutant *E. coli* EPSP synthase. Fed. Proc. **45,** 1506.

LaRossa, R. A., Falco, S. C., 1984: Amino acid biosynthetic enzymes as targets of herbicide action. Trends in Biotech. **2,** 158—161.

LaRossa, R. A., Schloss, J. V., 1984: The sulfonylurea herbicide sulfometuron methyl is an extremely potent and selective inhibitor of acetolactate synthase in *Salmonella typhimurium.* J. Biol. Chem. **259,** 8753—8757.

LaRossa, R. A., Smulski, D. R., 1984: ilvB-encoded acetolactate synthase is resistant to the herbicide sulfometuron methyl. J. Bacteriol. **160,** 391—394.

Leason, M., Cunliffe, D., Parkin, D., Lea, P. J., Miflin, B. J., 1982: Inhibition of pea glutamine synthetase by methionine-sulfoxamine, phosphinothricin and other glutamate analogues. Phytochem. **21,** 855—857.

Levitt, G., Ploeg, H. L., Weigel, R. C., Fitzgerald, D. J., 1981: 2-chloro-N-[4-methoxy-6-methyl-1,3,5-triazin-2-yl)amino-carbonyl] benzenesulfonamide, a new herbicide. J. Agric. Food. Chem. **29,** 416—424.

Manderscheid, R., Wild, A., 1986: Studies on the mechanism of inhibition by phosphinothricin of glutamine synthetase isolated from *Triticum aestivum L.* J. Plant Physiol. **123,** 135—142.

Margulis, L., 1970: Origin of Eukaryotic Cells. Yale University Press, New Haven, Connecticut.

Miflin, B. J., Lea, P. J., 1977: Amino acid metabolism. Ann. Rev. Plant Physiol. **28,** 299—329.

Mousdale, D. M., Coggins, J. R., 1984: Purification and properties of 5-enolpyruvylshikimate-3-phosphate synthase from seedlings of *Pisum sativum L.* Planta **160,** 78—83.

Mousdale, D. M., Coggins, J. R., 1985: Subcellular localization of the common shikimate pathway enzymes in *Pusum sativum L.* Planta **163,** 241—249.

Nafziger, E. D., Widholm, J. M., Steinrucken, H. C., Kilmer, J. L., 1984: Selection and characterization of a carrot cell line tolerant to glyphosate. Plant Physiol. **76,** 571—574.

Orwick, P. L., Marc, P. A., Umeda, K., Shaner, D. L., Los, M., Ciarlante, D. R., 1983: AC 252,214 — A new broad spectrum herbicide for soybeans: greenhouse studies. Proc. South Weed. Sci. Soc. **36,** 90.

Ray, T. B., 1984: Site of action of chlorsulfuron: inhibition of valine and isoleucine biosynthesis in plants. Plant Physiol. **75,** 827—831.

Rogers, S. G., Brand, L. A., Holder, S. B., Sharps, E. S., Brackin, M. J., 1983: Amplification of the *aroA* gene from *Escherichia coli* results in tolerance to the herbicide glyphosate. Appl. Environ. Microbiol. **46,** 37—43.

Rubin, J. L., Gaines, C., Jensen, R. A., 1984: Glyphosate inhibiton of 5-enolpyruvylshikimate-3-phosphate synthase from suspension cultured cells of *Nicotiana silvestris.* Plant Physiol. **75,** 839—846.

Schimke, R. T., 1984: Gene amplification in cultured animal cells. Cell **37,** 705—713.

Schulz, A., Sost, D., Amrhein, N., 1984: Insensitivity of 5-enolpyruvylshikimic acid 3-phosphate synthase confers resistance to this herbicide in a strain of *Aerobacter aerogenes.* Arch. Microbiol. **137,** 121—123.

Shah, D. M., Horsch, R. B., Klee, H. J., Kishore, G. M., Winter, J. A., Tumer, N. E., Hironaka, C. M., Sanders, P. R., Gasser, C. S., Aykent, S., Siegel, N. R., Rogers, S. G., Fraley, R. T., 1986: Engineering herbicide tolerance in transgenic plants. Science **233,** 478—481.

Shaner, D. L., Robson, P., Simcox, P. D., Ciarlante, D. R., 1983: Absorption, translocation and metabolism of AC 252,214 in soybeans, cocklebur and velvetleaf. Proc. South Weed. Sci. Soc. **36,** 92.

Shaner, D. L., Anderson, P. C., Stidham, M. A., 1984: Imidazolinones-potent inhibitors of acetohydroxyacid synthase. Plant Physiol. **76,** 545—546.

Siehl, D. L., Singh, B. K., Conn, E. E., 1986: Tissue distribution and subcellular localization of prephenate aminotransferase in leaves of *Sorghum bicolor.* Plant Physiol. **81,** 711—713.

Singer, S. R., McDaniel, C. N., 1985: Selection of glyphosate-tolerant calli and the expression of this tolerance in regenerated plants. Plant Physiol.**78,** 411—416.

Skokut, T. A., Wolk, C. P., Thomas, J., Meeks, J. C., Shaffer, P. W., 1987: Initial organic products of assimilation of [^{13}N] ammonium and [^{13}N] nitrate by tobacco cells cultured on different sources of nitrogen. Plant Physiol. **62,** 299—304.

Smart, C., Johanning, D., Muller, G., Amrhein, N., 1985: Selective overproduction of 5-enolpyruvylshikimic acid 3-phosphate synthase in a plant cell culture which tolerates high doses of the herbicide glyphosate. J. Biol. Chem. **260,** 16338—16346.

Smith, C. M., Pratt, D., Thompson, G. A., 1985: Increased 5-enolpyruvylshikimic acid 3-phosphate synthase activity in a glyphosate-tolerant variant strain of tomato cells. Plant Cell Rep. **5,** 298—301.

Sost, D., Schulz, A., Amrhein, N., 1984: Characterization of a glyphosate-insensitive 5-enolpyruvylshikimic acid-3-phosphate synthase. FEBS Lett. 238—242.

Stalker, D. M., Hiatt, W. R., Comai, L., 1985: A single amino acid substitution of the enzyme 5-enolpyruvylshikimate-3-phosphate synthase confers resistance to the herbicide glyphosate. J. Biol. Chem. **260,** 4724—4728.

Steinrucken, H. C., Amrhein, N., 1980: The herbicide glyphosate is a potent inhibitor of 5-enolpyruvylshikimic acid-3-phosphate synthase. Biochem. Biophys. Res. Commun. **94,** 1207—1212.

Steinrucken, H. C., Schulz, A., Amrhein, N., Porter, C. A., Fraley, R. T., 1986: Overproduction of 5-enolpyruvylshikimate-3-phosphate synthase in glyphosate-tolerant *petunia hybrida* cell line. Arch. Biochem. Biophys. **244,** 169—173.

Sweetser, P. B., Schow, G. S., Hutchison, J. M., 1982: Metabolism of chlorsulfuron by plants: biological basis for selectivity of a new herbicide for cereals. Pestic. Biochem. Physiol. **17,** 18—23.

Tachibana, K., Watanabe, T., Sekizuwa, Y., Takematsu, T., 1986: Action mechanism of bialaphos. 2. Accumulation of ammonia in plants treated with bialaphos. J. Pest. Sci. **11,** 33—37.

Yadav, N., McDevitt, R. E., Benard, S., Falco, S. C., 1986: Single amino acid substitutions in the enzyme acetolactate synthase confer resistance to the herbicide sulfometuron methyl. Proc. Natl. Acad. Sci. **83,** 4418—4422.

Chapter 17

Virus Cross-Protection in Transgenic Plants

Roger N. Beachy

Department of Biology, Washington University, St. Louis, MO 63130, U.S.A

With 2 Figures

Contents

I. Introduction

A. Classical Cross-Protection: Applications and Limitations

Plant virus diseases have traditionally been controlled by sanitary agronomic practices, use of virus-free propagates, controlling the insect vectors that spread the pathogen, and incorporating genes for disease resistance. These practices can, in most instances substantially reduce crop losses normally attributed to virus infections. However, changing agronomic practices and changes in virus strains often lead to significant disease losses, and the search for genetic resistance begins anew. In the majority of cases, sources of disease resistance are unavailable to the plant breeder, or if available, require many plant generations to incorporate the resistance trait into the desired cultivar. Furthermore, most of the resistance genes are effective against only a limited number of strains of the virus.

In instances where genetic resistance is unavailable a cross-protection approach has been taken. First described by McKinney (1929), cross protection refers to the observation that tobacco plants infected with a mild strain of tobacco mosaic virus (TMV) were less susceptible when subsequently inoculated (challenged) with a more severe TMV strain. The large body of work conducted in the study of cross-protection during the last 55 years has been described in outstanding detail by others (Hamilton, 1980; Ponz and Bruening, 1986). Therefore only an abbreviated description of cross-protection will be presented here.

Cross-protection has been adapted to protect glasshouse-grown tomatoes from TMV (Rast, 1972), papaya plants from papaya ringspot virus (Yeh and Gonsalves, 1984), and citrus from citrus tristeza virus (Costa and Müller, 1980). Indeed, significant protection against yield loss has been reported in some cases. There are, however, several drawbacks and potentially serious problems with adapting a cross-protection approach to broad scale use in agriculture. First, it can be difficult to isolate a cross-protecting strain of virus. Theoretically, the virus should be attenuated for symptom production, but able to replicate without attenuation. Workers may search for many years to isolate an appropriate strain. Second, introducing even a mild virus strain into a host increases the possibility of a synergistic interaction with a second non-related virus. Third, infections by mild virus strains often lead to low but significant yield losses. Fourth, widespread virus infections increase the likelihood that a more severe virus strain can arise, leading to greater disease, rather than protection. Despite these potential problems, cross-protection is being applied in some agricultural settings to control important viral diseases.

The molecular or cellular basis for cross-protection is not identified, and it is therefore largely defined operationally. (1) It can cause a reduction in the number of sites where infection is established following challenge inoculation, as in the case of TMV on *Nicotiana sylvestris* (Sherwood and Fulton, 1982). (2) Cross-protected plants do not show signs of infection by the challenge infection, or develop symptoms much later (after inoculation) than do non-protected plants. The goal in the use of cross-protection is to delay symptom development until after fruit is formed so that yield losses are minimized. (3) In some cases, cross-protection can be overcome by challenging the plant with viral RNA as the inoculum, rather than intact virions (Sherwood and Fulton, 1982; Dodds *et al.*, 1985). (4) Cross-protection is effective against "closely related" strains of viruses, but not more distantly related viruses or virus strains. Cross-protection between strains is not however, always reciprocal (Fulton, 1978).

B. Proposed Mechanisms of Cross-Protection

A number of studies have been carried out to identify the mechanism or mechanisms responsible for cross-protection, but have not provided conclusive evidence for any particular model. Several types of experiments implicate a role for RNA : RNA interactions in cross-protection. First, Niblett *et al.* (1978) reported cross-protection amongst four viroids. Because viroids are not encapsidated in proteins, and apparently do not encode any proteins, it was logical to conclude that RNA : RNA interactions are involved in viroid cross-protection. In the second study, Zaitlin (1976) demonstrated cross-protection using a TMV strain that encoded coat protein incapable of encapsidating viral RNA, and led to the postulate that RNA : RNA interactions may be involved in cross-protection. Palukaitis and Zaitlin (1984) further developed this model and suggested that cross-protection resulted from intermolecular hybridization between nascent antisense strands on challenger replicative intermediate molecules with the RNA of the protecting virus. Indeed, such a model was logical based upon other results which indicated that antisense RNA can control the expression of genes in *E. coli* (Coleman *et al.*, 1984; Green *et al.*, 1986).

de Zoeten and Fulton (1975) proposed that cross-protection occurs when RNA of the challenge virus is re-encapsidated in the coat protein molecules produced by the protecting strain. However, it was later reported that cross-protection can, in some cases, be overcome when the challenge inoculum is viral nucleic acid rather than virions. This is the case with TMV (Sherwood and Fulton, 1982) and cucumber mosaic virus (CMV) (Dodds *et al.*, 1985). The rationale developed that protection occurs prior to the full release of viral RNA from the virion, and led to the suggestion that capsid protein of the protecting strain re-encapsidated the RNA of the challenger during the process of uncoating, or stripping, of the genome. This model has been supported by the results of *in vitro* experiments of Wilson (1984), who reported that when TMV particles were treated for short periods of time at elevated pH (e.g., pH 8.0 and above) they served as good templates for *in vitro* translation reactions. Similar effects were observed with particles of alfalfa mosaic virus and CMV. Such treatments do not release viral RNA from the virus, but apparently cause a structural change that permits ribosomes to bind to the viral RNA. Wilson and Watkins (1986) also reported that the addition of high concentrations of TMV capsid protein to the translation reactions blocked expression. This led the authors to suggest that capsid protein may re-encapsidate partially uncoated virus particles to block translation, and may be a mechanism of cross-protection.

Sherwood and Fulton (1982) presented a somewhat different model when they showed that the nature of the capsid protein of the challenge strain was important in the protection mechanism. In their study they encapsidated challenger TMV-RNA with the capsid protein from brome mosaic virus and observed no protection. However, when TMV-RNA was reencapsidated in TMV-capsid protein, protection was observed. Similar results were obtained when the challenger was encapsidated in sunn hemp mosaic virus capsid protein (Zinnen and Fulton, 1986). These types of data led Sherwood and Fulton (1982) to propose that the specificity of cross-protection is reflective of the capsid protein of the challenger, and somehow blocks the uncoating reaction. Furthermore, they speculated that a lipid-containing structure may be involved in the uncoating process. Their hypothesis was somewhat clouded by the observation that, in contrast to protection late in infection, inoculation with challenge RNA in early stages of infection by the protecting strain were less successful than challenge during late stages of infection. This led to the suggestion that factors other than blocking uncoating of the challenger were at work in cross-protection against TMV in *N. sylvestris.*

The results of studies to define the mechanism(s) responsible for cross-protection have been important to formulate approaches that could be taken to effect a similar protection in genetically transformed plants. However, these studies did not provide a single model to be tested.

II. Genetic Transformation to Produce Virus Resistant Plants

In a review article in 1980, R. I. Hamilton predicted that the development of techniques to genetically transform plants might make it possible to produce plants that were similar to those that were cross-protected against virus infection. The caveat in the prediction was that the molecule(s) responsible for cross-protection was not determined. With the development of techniques for plant transformation (reviewed by many authors including Chilton, 1983; Caplan *et al.*, 1983; Fraley *et al.*, 1986) came the opportunity to test different approaches to confer protection to transgenic plants. For several reasons the TMV-tobacco system was a logical choice for the initial experiments: (1) There was exhaustive literature on the biology, molecular biology, and biophysics of the virus. Such a background would make it possible for studies of the molecular biology of protection. (2) The full sequence of TMV-RNA had been published earlier by Goelet *et al.* (1982), and would facilitate cDNA cloning of selected sequences of TMV. (3) Cross-protection was first demonstrated against TMV in 1929 and had been applied to protect glasshouse crops of tomatoes. This precedent would provide a basis of comparison for cross-protection with protection produced in transgenic plants. (4) The transformation and regeneration of plants was early demonstrated to be relatively easy in tobacco. Although the first report of successful protection of transgenic plants was in tobacco against TMV infection, other recent reports also documented that similar approach provides protection against alfalfa mosaic virus (AlMV).

A. Expression of Viral Coat Protein Coding Sequences in Transgenic Plants

As described above, it had been suggested that different products of virus infection were responsible for cross-protection, i. e., nucleic acid sequences of viral (+) or (−) polarity, or viral capsid protein. To test each possibility, experiments were designed to cause the expression of a variety of viral sequences. The first published reports of the expression of viral cDNAs in transgenic plants (Beachy *et al.*, 1985; Bevan *et al.*, 1985) described the expression of sequences of TMV-RNA that encoded the capsid protein (CP). In the report by Bevan *et al.*, a cloned cDNA was obtained which represented the CP gene of the OM strain of TMV (Meshi *et al.*, 1982). The cDNA contained the entire CP coding sequence flanked on the 5′ end by 4 nucleotides of TMV-RNA, and on the 3′ end by most of the non-translated region (205 out of 207 nucleotides) of the genome. This includes the highly structured tRNA-like region (in brome mosaic virus such a tRNA-like region is the site to which replicase binds to initiate synthase of complementary strand RNA; Miller *et al.*, 1986). This cDNA was ligated to the promoter from cauliflower mosaic virus that causes the expression of the 35S RNA viral transcript (35S promoter) (Guilley *et al.*, 1982), and the 3′ regulatory sequences from the nopaline synthetase (NOS) gene of the Ti-plasmid. The chimeric gene was transferred to the intermediate plasmid, pBIN 6, (Bevan, 1984) and used in plant transformation reactions with the *A. tumefaciens* strain LBA 4404 (Hoekema *et al.*, 1983). Tobacco plants regenerated from transformed leaf pieces expressed the chimeric gene and accumulated at least 6 different RNA molecules that were less than, and greater than, the predicted size. Plants accumulated coat protein to the level of about 170 μg/gm fresh weight, which represents approximately 0.001% (w/w) of soluble leaf protein. However, when R1 progeny of the transgenic plants were inoculated with TMV they were found equally sensitive to infection as non-transgenic plants (Bevan and Harrison, 1986).

Similar approaches were taken by Beachy and colleagues to express the CP gene of the U_1 strain of TMV in transgenic plants. In this instance the cloned cDNA produced was identical at the 5′ end with that used by Bevan and colleagues. The entire 3′ non-translated region (207 nucleotides) was retained, and an additional nine nucleotides were contributed by the oligonucleotides used to prime the synthesis of the first strand cDNA (Powell Abel *et al.*, 1986). The cloned cDNA was inserted between the 35S promoter of CaMV and the NOS 3′ polyadenylation signal on pMON 316 (Sanders *et al.*, 1987). pMON 316 is an intermediate plasmid used to deliver genes into a disarmed Ti-plasmid homologous integration. An *A. tumefaciens* strain carrying the disarmed plasmid pTi B6S3-SE (Fraley *et al.*, 1985) was used to introduce the chimeric CP gene into tobacco leaf discs from which transgenic plants were regenerated as described by Horsch *et al.* (1985).

The chimeric mRNAs that accumulated in the transgenic plants described by Powell Abel (1986) were approximately 0.95 kb and 2.0 kb. The smaller molecule was of the expected size, while the larger molecule probably represents an extension of the molecule at the 3′ end, and polyadenylation downstream of the NOS 3′ polyadenylation signal in the intermediate plasmid. TMV-CP accumulated to the level of approximately 0.02 to 0.1% (w/w) of the soluble protein of transgenic leaf tissue, or 20—100 X higher levels than those reported by Bevan *et al.* (1985). There are several possible reasons for the differences in the levels of accumulated CP gene in the transgenic plants produced by the two groups. (1) There may be different rates of transcription since each lab used a different DNA fragment to provide the 35S promoter. Bevan *et al.* used a fragment of 1050 nucleotides including approximately 800 nucleotides upstream of the promoter site (Guilley *et al.*, 1982) and 132 nucleotides 3′ of the cap site. The 35S promoter fragment used in pMON 316 contains 293 n. t. 5′ of the CAP site and 32 n. t. 3′ of the CAP (Rogers *et al.*, 1985 b). (2) Differences in the untranslated sequences 5′ of the AUG initiation codon of the CP coding sequences (shown in Fig. 1) may control the rate of translation of the transcript. Of particular interest is the fact that the Bevan construct contains an out of frame AUG codon 5′ of the CP coding sequence (indicated in Fig. 1). Rogers *et al.* (1985 a) reported that additional AUG codons in this untranslated leader sequence can lower the level of protein synthesis from mRNAs in plant cells, as it can in other eukaryotic systems. However, there are other significant differences in the sequence of the two 5′ untranslated leaders that may also contribute to difference in translation of the mRNAs. (3) Differences in 3′ non-translated sequences may contribute to the differences in stability or translatability of the transcripts in the two experimental systems.

Recently three research groups reported the expression of the CP coding sequence of alfalfa mosaic virus in transgenic plants (Tumer *et al.*, 1987; Loesch-Fries *et al.*, 1987; Van Dun *et al.*, 1987). Tumer *et al.* and Van Dun *et al.* each used the CaMV 35S promoter while Loesch-Fries *et al.*, used the CaMV 19S promoter. In each case the CP cDNA was flanked at the 3′ end by the NOS 3′ regulatory sequence. Tumer *et al.* and Loesch-Fries *et al.* detected many different RNA molecules that were larger as well as smaller than the expected size, when total RNA was analyzed by Northern blot hybridization reactions. However, Tumer *et al.* found that only the RNA of the predicted 1.2 kb and a larger RNA of 1.9 kb were polyadenylated. Identical results were obtained with RNA isolated from transgenic tomato plants (Tumer *et al.*, 1987). On the contrary, Van Dun *et al.* (1987) reported finding only one RNA species, that being the expected size. The reason for the differences in results between the various labs has not been addressed.

In each of the instances in which AlMV-CP mRNA was expressed, viral CP also accumulated. Tumer *et al.* reported that AlMV-CP constituted 0.1% to 0.4% of the protein extracted from leaves of transgenic plants, while Loesch-Fries *et al.* (1987) reported that CP accumulated between 0.004% and 0.08% of extracted leaf protein. The differences in protein accumulation probably reflect differences in gene expression levels of the 35S and 19S promoter, as has been previously reported (Lawton *et al.*, 1987). Van Dun estimated the accumulation of AlMV CP to be between 0.01% and 0.05% of protein in juice squeezed from leaves of transgenic plants. These levels may actually be similar to those reported by Tumer *et al.* (1987) with the differences reflecting methods of protein extraction.

Expression of tobacco rattle virus CP coding sequence has also been reported using a similar experimental approach (Van Dun *et al.*, 1987). A single mRNA of expected size was detected, and CP accumulated to between 0.01% and 0.05% (w/w) of protein in leaf sap.

```
    ↓
PA: G A C A C G C U G A A A U C A C C A G U C U C U C U A C A G A U C U G U
    | | | | | | | | | | | | | | | | | | | | | | | | | | | | |   | | | |   |
B:  G A C A C G C U G A A A U C A C C A G U C U C U C U A C A A A U C U A U

PA: U G U A A G G A G U . . . . . . . . . . . . . . . . . . . . . . . . . .

B:  A A A U C U A U C U C U C U A U A U C U U A U A G G G U A U A A U G U G U G

PA: . . . . . . . . . . . . . . . . . . . . . . . . . . . . . . . . . . . .

B:  A G U A G U U C C C A G A U A A G G G A A U U A G G G U U U C G C U C A U G

PA: . . . . . . . . . . . . . . . . . . . . . . . . C U A G A A A U A U G...(Coat Protein
                                                      | | | | | | | | |    Sequence)
B:  U G U U G A G C A U A U A A G A A A C C C(C G)C G A A A U A U G
```

Fig. 1. Comparison of the 5′ untranslated nucleotides of the TMV coat protein coding sequences expressed in transgenic plants. The upper sequence (PA) is that of the construct reported by Powell Abel *et al.* (1986) and the lower sequence (B) is the sequence constructed by Bevan *et al.* (1985). Sequences in parenthesis are presumed based upon the construction described by Bevan *et al.* (1985). "AUG" initiation codons are indicated by ⌐¬. ↓ indicates the cap site

Collaboration between N. Tumer and colleagues (Monsanto Company) and N.-H. Chua and colleagues (Rockefeller University) resulted in the isolation and expression in transgenic plants of CP coding sequences of cucumber mosaic virus (CMV) and potato virus X (PVX) (Hemenway *et al.*, 1988; Cuozzo *et al.*, 1988). The levels of PVX and CMV coat protein that accumulated in these plants were approximately equivalent to those described above for other viral coat proteins.

B. Resistance to Virus Infection in Transgenic Plants

Powell Abel *et al.* (1986) reported that transgenic tobacco plants expressing the TMV CP were resistant to infection by TMV. Plants expressing the TMV CP sequences (CP+) either escaped infection, or developed disease symptoms significantly later than plants not expressing the gene. Furthermore they also found that each of the different CP expressing transgenic lines tested were similarly resistant. This important result demonstrated that resistance was due to expression of the gene rather than to changes in the plant caused by tissue culture or plant regeneration. By contrast the transgenic TMV CP+ plants described by Bevan *et al.* (1985) were not resistant to infection by TMV (Bevan and Harrison, 1986). The authors suggested that the level of gene expression in these plants was too low to produce resistance.

The experiments of Powell Abel *et al.* (1986) were carried out in a tobacco cultivar that is a systemic host for TMV (*Nicotiana tabacum* cv. Xanthi). Resistance was therefore manifested as a delay or escape of systemic disease development. Resistance might have resulted from lack of

Table 1. Protection Against AlMV in Transgenic Plants Expressing AlMV CP

Reference	Plant	Promotor	Accumulation CP (%, w/w)	Relative Protection [Inoc][1]	Chlorotic Lesions +CP	Chlorotic Lesions −CP	[Inoc]	Systemic Symptoms 14 d.p.i.[2] +CP	Systemic Symptoms 14 d.p.i.[2] −CP
Tumer *et al.* (1987)	Tobacco	35 S	0.1%—0.8%	~ 50 μg/ml	0	6—55/leaf	20 μg/ml	20%	100%
	Tomato	35 S	0.1%—0.8%	~ 50 μg/ml	0	14/plant	—[3]	—	—
Van Dun *et al.* (1987)	Tobacco	35 S	0.01%—0.05%	~ 2.5 μg/ml	0	110—745/ 1/2 leaf	—	—	—
Loesch-Fries *et al.* (1987)	Tobacco	19 S	0.004%	2—15 μg/ml	34 ± 20	27 ± 17	—	—	—
	Tobacco	19 S	0.08%	2—15 μg/ml	3 ± 5	22 ± 9	2—15 μg/ml	20%	100%

[1] Except in the report by Van Dun *et al.* (1987) inocula concentration of AlMV were estimated.

[2] d.p.i. = days post inoculation.

[3] — indicates experiment not done.

infection, or lack of short-distance spread, long-distance spread, or from combinations of factors. Plants that did not develop systemic symptoms did not accumulate TMV, indicating that they were resistant to infection and/or virus spread, rather than suppressing symptom development *per se* (Nelson *et al.*, 1987). Using a strain of TMV that caused chlorotic lesions on inoculated leaves, Nelson *et al.* found that transgenic CP(+) plants produced 80—90% fewer lesions than CP(−) plants, supporting the idea that CP(+) plants somehow resist infection. This was confirmed by studies with transgenic Xanthi *nc* tobacco plants, a local lesion host for TMV. The size of the necrotic lesions that developed, however, were identical in CP(+) and CP(−) plants, suggesting that once infection was initiated in transgenic plants, the infection cycle and eventual formation of the local lesion was not affected in CP(+) plants.

In agreement with the experiments with TMV, transgenic plants that were CP(+) for expression of an AlMV capsid protein gene were resistant to infection by AlMV (Tumer *et al.*, 1987; Van Dun *et al.*, 1987; Loesch-Fries, 1987). Each of the reports on AlMV CP expression conclusively demonstrated that CP(+) plants had fewer chlorotic/necrotic lesions on inoculated leaves than did CP(−) plants, indicating reduced susceptibility to infection. Tumer *et al.* and Loesch-Fries *et al.* also reported a delay in development of systemic disease symptoms in transgenic plants. The degree of protection appears to be correlated with the level of accumulation of CP. As shown in Table 1, the degree of protection is apparently greater in the reports by Tumer *et al.* (1987) and Van Dun *et al.* (1987) than in the report from Loesch-Fries *et al.* (1987); the latter group reporting lower amounts of CP accumulation in transgenic plants. However, it is possible that the specific infectivity and concentrations of the inocula and the conditions in which the plants were grown, also played a role in the degree of protection recorded by the different laboratory groups.

There are similarities in the protection in the transgenic plants that express the CP coding sequence of TMV or AlMV. First, protection is manifested by a reduced number of sites where infection occurs on inoculated leaves. Second, inoculation with viral RNA largely overcame the protection that was recorded against TMV (Nelson *et al.*, 1987) or AlMV (Van Dun *et al.*, 1987; Loesch-Fries *et al.*, 1987). Third, protection is partially or fully overcome by increasing the concentration of the viral inoculum. Powell Abel *et al.* (1986) reported that increasing the concentration of TMV substantially reduced the length of delay in systemic spread of the virus in tobacco plants. An example of results of one such experiment is presented in Figure 2. Tumer *et al.* (1987) also found that increasing the concentration of AlMV in the inoculum increase the rate of disease development in transgenic plants.

By expanding these types of experiments it is possible to establish criteria to determine the relative degree of resistance (or susceptibility) of the transgenic plants compared to that of the parent cultivar. Results of two types of experiments are presented in Figure 2. In Figure 2A CP(+) plants were inoculated with increasing concentrations of TMV and disease

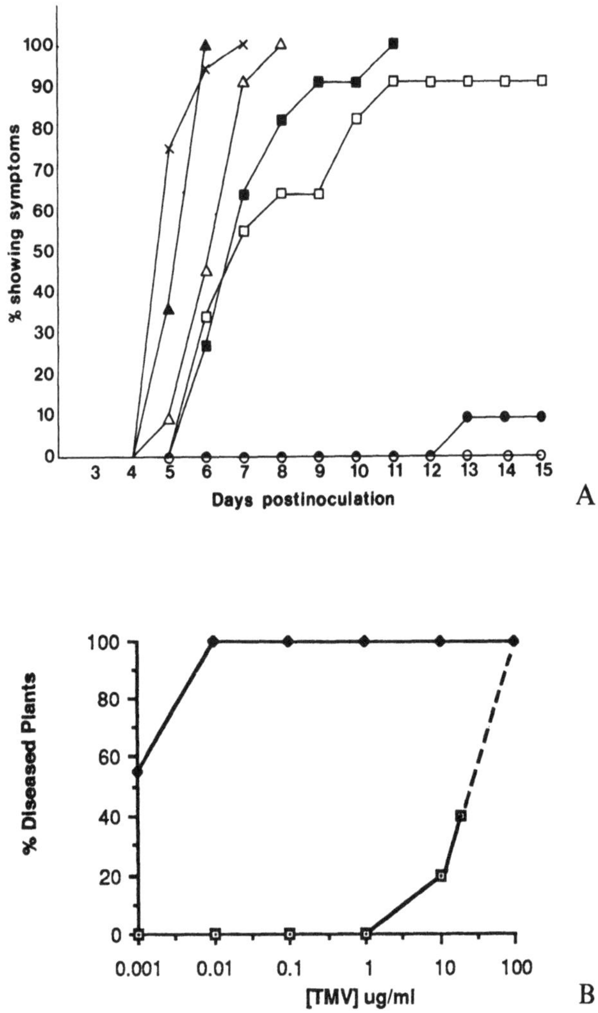

Fig. 2. Relative resistance or susceptibility of transgenic and non-transgenic tobacco plants to TMV. Seedlings (3—4 leaf stage) were inoculated with increasing concentrations of TMV and observed for systemic symptoms of disease (i. e., vein clearing, followed by chlorosis). A. Disease development in plants inoculated with increasing concentrations of U_1-TMV. x-x, control plants (transgenic, but not expressing the TMV CP gene) inoculated with 0.1 μg TMV/ml. Similar disease pattern was observed when plants were inoculated with 0.1 μg TMV/ml. Similar disease pattern was observed when plants were inoculated with 0.01 and 0.001 μg TMV/ml. The remaining seedlings were transgenic and expressed the TMV CP gene, inoculated as follows: ○ - ○, 0.001 μg TMV/ml; ●-●, 0.01 μg TMV/ml; □ - □, 0.1 μg TMV/ml; ■ - ■, 1.0 μg TMV/ml; △ - △, 5.0 μg TMV/ml; ▲ - ▲, 20 μg TMV/ml. (C. M. Deom and R. N. Beachy, previously unpublished. B. Disease development by 5 days post inoculation with increasing concentrations of TMV on plants not expressing the TMV CP gene, ■ - ■, or on those expressing the CP gene □ - □ (E. Anderson, previously unpublished)

development was recorded over a two week period. All the control plants inoculated with the U_1 strain at 0.001 µg/ml developed systemic symptoms within 6 days post infection (d. p. i.) (not shown). Transgenic CP(+) plants exhibited a delay in disease symptom development unless inoculated with 20 µg of TMV/ml. In a similar type of experiment, disease symptoms were recorded at 5 d.p.i. on plants inoculated with increasing concentrations of TMV. As shown in Figure 2B, CP(+) plants required approximately 10^4 higher concentrations of TMV to produce systemic symptoms in the same period of time as the controls. This level of resistance is therefore referred to as being 10^4 greater than control plants. It is proposed that this type of assay be used to assess protection whenever appropriate.

C. Elucidating the Mechanism(s) of Engineering Protection

Efforts to identify the molecular and cellular mechanism(s) responsible for resistance are in the early stages and will undoubtedly yield important new information in the near future. Indications from recent experiments, however, support the hypothesis that protection results because of interference with an event early in the infection process. Delay in systemic disease development may result from the same or a different mechanism.

A key feature in elucidating the mechanism(s) of protection lies in identifying the molecule(s) responsible for the protection. In each of the cases of protection described above, protection might have resulted due to accumulation of coat protein coding sequences (mRNA) or coat protein *per se.* Since each of the mRNAs used includes the 3′ end of the viral RNA and contain the viral sequences to which viral replicase binds, the mRNA may interfere with replication of viral RNA. Alternately it may bind to antisense viral RNA during replication. To address these issues P. A. Powell *et al.* (manuscript in preparation) generated a series of gene constructs that would produce mRNA sequences but not coat protein, or mRNA that lacked the replicase binding site, but would encode the synthesis of coat protein. The results of these experiments conclusively implicate the CP as the important moiety for protection rather than the mRNA.

As discussed earlier, cross-protection is largely overcome when viral RNA (rather than virions) is used as inoculum (Sherwood and Fulton, 1982; Dodds *et al.*, 1985). One of the first experiments with the CP(+) transgenic plants was to determine if the protection was effective against viral RNA. In each instance thus far reported, resistance was largely, but not fully overcome when viral RNA was used (TMV, Nelson *et al.*, 1987; AlMV, Van Dun *et al.*, 1987; and Loesch-Fries *et al.*, 1987). Since release of viral RNA from its capsid protein is an early step in infection, these results were interpreted to indicate that protection is primarily but not solely effected at a step in infection that proceeds or is concurrent with virus uncoating. However, the low degree of protection against viral RNA indicates that other steps in infection may also be inhibited or slowed in these transgenic plants.

Because of the difficulty of studying early events in whole leaves, protoplasts are often used for such studies. Protoplasts have also been used to study cross-protection (Ostuke and Takebe, 1976; Barker and Harrison, 1978). In a preliminary report Loesch-Fries *et al.* (1987a) demonstrated that fewer transgenic protoplasts (prepared from AlMV CP(+) plants) became infected following infection with AlMV than protoplasts from control plants. Resistance in protoplasts was not effective when cells were inoculated with AlMV RNAs 1+2+3. In a more detailed study, Register and Beachy (submitted) found that TMV CP(+) transgenic protoplasts were resistant to TMV introduced either via electroporation or by a PEG treatment, thereby demonstrating that resistance is independent of the method of virus introduction. These results lend support to the hypothesis that protection is an intracellular, rather than an extracellular reaction as suggested by de Zoeten and Gaard (1984). As demonstrated in whole plants (Nelson *et al.*, 1987) resistance in transgenic protoplasts was largely overcome by inoculation with TMV-RNA. Transgenic CP(+) protoplasts that became infected following inoculation either with TMV or TMV-RNA produced the same amount of virus per protoplast as did the control protoplasts. This indicates that transgenic CP(+) protoplasts are fully competent for replication of TMV and that protoplasts either became fully infected or escaped infection. Whereas resistance to systemic disease development was overcome by inoculating plants with high (20 to 100 μg/ml) concentrations of TMV (Fig. 2), it was not possible to overcome protection in protoplasts with TMV concentrations as high as 1 mg/ml (Register and Beachy, submitted). Not more than 9% of the transgenic protoplasts become infected following inoculation with TMV, compared to 55% infection of the control protoplasts. These results indicate that most but not all of the protoplasts are highly resistant to infection. It will be important to determine the basis for the susceptibility of this subset of cells.

The results of the experiments described above indicated that infection of transgenic CP(+) leaves and isolated protoplasts is blocked at a stage of infection prior to release of RNA from the virus. Although little is known about the steps in virus uncoating, the experiments of Wilson and colleagues (described above) lead to the suggestion that an early step may involve a subtle change in virus structure, such as that mimicked by treating TMV at pH 8.0. To see if transgenic cells were resistant or susceptible to partially disrupted virus, Register and Beachy (submitted) used virus treated at pH 8 to inoculate protoplasts or plants isolated from CP(+) Xanthi or Xanthi *nc* tobacco plants. In these experiments inoculation with virus particles treated at pH 8.0 or higher largely overcame the protection, much as does infection with viral RNA. This report argues against the hypothesis that protection results from re-encapsidation of viral RNA or partially disrupted virus. Rather it sugests that protection results from blockage at an early stage of the uncoating process. This early event occurs inside, rather than outside the cell. Since the level of gene expression is correlated with protection, we suggest that a critical threshold of CP accumulation is required to effect protection, perhaps in a concentration dependent manner. This leads to the proposition that protection is the result of competition between CP and virus for a "site of virus uncoating". Although other mechanisms can be presented, this model can form the focus for further experiments.

Another manifestation of protection is the delay of the spread of virus from inoculated leaves to upper leaves (Powell Abel *et al.*, 1986; Tumer *et al.*, 1987; Loesch-Fries *et al.*, 1987). Such a delay may be due to reduced spread of virus from the inoculated cells to surrounding cells, into, or out of the plant's vascular system (through which long distance spread of virus occurs; reviewed by Atabekov and Dorokhov, 1984), or establishing infection centers after virus moves into upper leaves. Little or no experimental data is yet available which addresses the delay in virus movement in CP(+) transgenic plants.

D. *Expression of Other Viral Sequences in Transgenic Plants*

There are only few reports of expression of viral coding sequences (other than those encoding capsid proteins) in transgenic plants. Deom *et al.* (1987) produced transgenic tobacco plants that expressed the TMV movement protein (MP; 30 kDa protein, TMV nucleotides 4903–5706). However, the expression of the coding sequence and accumulation of the MP did not protect plants from infection nor delay disease development. Beachy *et al.* (1987) reported production of plants that expressed TMV nucleotides 3335 to 6395. Although the RNA was detected, neither the MP or CP cistrons, each of which are found on the transcribed RNA, were translated. Plants expressing this sequence, either in the sense or antisense orientation, were as susceptible to infection and disease development as the controls.

Garcia *et al.* (1987) reported expression of a cDNA encoding the middle M-RNA of cowpea mosaic virus in transgenic cowpea cells. The M-RNA is essential for virus replication since it encodes a component of the viral replicase. Although a transcript of the M-RNA was identified in transgenic tissue no protein product was found. No attempt was made to determine if these transgenic cells were protected against virus infection.

As discussed earlier, antisense RNA has been proposed as a method for conferring protection against virus infection. However, other than the report by Beachy *et al.* (1987) there is little published experimental data to support or dismiss the hypothesis, although it is likely that considerable experimentation is in progress. It can be anticipated that results of such experiments will be forthcoming, and it will be important to compare the protection to plants that express antisense RNA or accumulate viral CP.

Expression of satellite RNAs in transgenic plants. During the past 25 years satellite (SAT) RNAs were discovered to be associated with a number of virus infections (see review by Murant and Mayo, 1982). Satellites depend on a helper virus to provide gene products needed for replication and spread. Satellites can either attenuate or exacerbate disease symptoms depending upon the host : virus system. For example, a satellite of cucumber mosaic virus (CMV) referred to as CARNA 5 can attenuate symptoms of CMV infection in tomato plants. However, when present with a different CMV strain, it can lead to a disease known as tomato lethal necrosis (Waterworth *et al.*, 1979). Recently it was reported that expression of SAT RNA of CMV (Harrison *et al.*, 1987) in transgenic tobacco plants decreased CMV replication in these plants and largely suppressed diseases symptoms therein. Furthermore, CARNA 5 similarly suppressed symptoms following infection with a closely related virus, tomato aspermy virus, although there was not a decrease in virus replication.

In a similar series of experiments Gerlach *et al.* (1987) reported that expression of the SAT RNA of tobacco ringspot virus in transgenic plants provided attenuation of the disease caused by infection by the helper virus. As in the case with expression of the CMV satellite, replication of the helper virus was likewise reduced. An interesting aspect of this work was that expression of antisense SAT RNA also provided attenuation, although somewhat later after infection than did sense SAT RNA.

The mechanism of disease attenuation by SAT RNAs is not well understood. In some instances SAT RNA may be replicated preferentially, resulting in reduced replication of the helper virus, but in other cases there are not such effects. It is also not clear which SAT RNA sequences are required for replication and/or disease attenuation. However, the reports described above indicate that virus SAT RNAs can provide a method for reducing disease in plants by mechanisms that are different from the protection provided by CP expression. There are several problems with using SAT RNAs. First, it is not immediately apparent how one would produce SAT RNAs for viruses for which satellites are not identified. Second, SAT RNAs can, when associated with other strains of the helper, result in increased, rather than decreased, disease severity. These features may make it difficult to develop a SAT RNA approach to control a wide range of different viruses.

III. Field Testing of Virus Protection in Transgenic Plants

Transgenic tomato plants (*Lycopersicon esculentum*) that express the coat protein coding sequence of the U_1 strain of TMV were produced by Nelson *et al.* (submitted) using the construct previously described by Powell Abel *et al.* (1986). Extensive greenhouse experiments demonstated that these plants possessed a high level of protection against several strains of TMV, as well as strains of tomato mosaic virus. After receiving approval from appropriate regulatory agencies transgenic plants were transplanted to the field, inoculated with U_1-TMV, and observed for symptom development, and virus replication. The results of the analysis confirmed that the CP(+) transgenic plants were highly resistant to TMV infection and disease development. Furthermore, whereas control plants suffered a 25—30% loss in fruit yield due to TMV infection, transgenic plants showed no yield loss (Nelson *et al.*, submitted). The results of these early field experiments, therefore, clearly demonstrated that CP gene expression very effectively controls infection of tomato by TMV under field situations.

IV. Conclusions

The expression of viral CP genes can confer protection to infection to the virus from which the CP coding sequence was obtained. Thus far protection has been conferred against TMV and AlMV, and unpublished evidence (from N. Tumer, Monsanto Company, and N.-H. Chua, Rockefeller University, and their colleagues) indicates that protection against CMV and PVX has also been derived by a similar approach. Research is ongoing in a number of laboratories around the world to extend the experimental approach to control many different viruses. Recent field experiments have demonstrated that transgenic tomato plants are protected against infection by TMV, and add support to the suggestion that such plants will be important in plant agriculture.

The basic cellular and molecular mechanisms responsible for the engineered protection are not well understood although the results of experiments indicate that an early step in infection is affected. It has been suggested that CP expression in transgenic plants interferes with virus uncoating (Register and Beachy, submitted); Sherwood and Fulton (1982) hypothesized that (classical) cross-protection also blocks an early event in infection. A second manifestation of protection is the delay in systemic spread of virus in transgenic CP(+) plants: in many cases, systemic disease symptoms fail to develop even though inoculated leaves are infected. Whether these two resistance phenotypes reflect a single or multiple mechanisms of protection remains to be determined.

Acknowledgements

The author is grateful to the many colleagues at Washington University and Monsanto Company (St. Louis, Missouri) for the pioneering work that has led to many of the results presented in this paper. To have added all of their names on page 1 would have trivialized the efforts of each. To have failed to mention them, however, would be unconscionable. So, I thank *at least* the following who contributed at Washington University: Edwin Anderson, W. Gregg Clark, Cristina Conesa, Barun De, C. Michael Deom, Philip Dubé, Curtis Holt, Maya Kaniewska, Richard S. Nelson, Melvin J. Oliver, Patricia Powell Abel, James C. Register III, and David Stark. At Monsanto Company: Xavier Delannay, Cynthia Fink, Robert T. Fraley, Nancy Hoffmann, Robert B. Horsch, Steven G. Rogers, Patricia R. Sanders, and Nilgen Tumer. Finally, the author is grateful to Ms. Nancy Burkhart for preparing the manuscript. Much of the work presented here was supported by a grant from the Monsanto Company.

References

Atabekov, J. G., Korokhov, Y. L., 1984: Plant virus-specific transport function and resistance of plants to viruses. Adv. in Virus Research. **29,** 313—364.

Barker, H., Harrison, B. D., 1978: Double infection, interference and superinfection in protoplasts exposed to two strains of raspberry ringspot virus. J. Gen. Virol. **40,** 647—658.

Beachy, R. N., Abel, P., Oliver, M. J., De, B., Fraley, R. T., Rogers, S. G., Horsch, R. B., 1985: Potential for applying genetic transformation to studies of viral pathogenesis and cross-protection. In: Biotechnology in plant science. Relevance to agriculture in the eighties. pp. 265—275. Zaitlin, M., Day, P., Hollaender, A., Wilson, C. M. (eds.). Academic Press, NY.

Beachy, R. N., Stark, D. M., Deom, C. M., Oliver, M. J., Fraley, R. T., 1987: Expression of sequences of tobacco mosaic virus in transgenic plants and their role in disease resistance. In: Tailoring genes for crop improvement. pp. 169—180. Bruening, G., Harada, J., Kosuge, T., Hollaender, A. (eds.). Plenum Publishing Co., NY.

Bevan, M., 1984: Binary *Agrobacterium* vectors for plant transformation. Nucleic Acids Res. **12,** 8711—8721.

Bevan, M. W., Harrison, B. D., 1986: Genetic engineering of plants for tobacco mosaic virus resistance using the mechanisms of cross-protection. In: Molecular strategies for crop protection. pp. 215—220. Arntzen, C. J., Ryan, C. (eds.). Alan R. Liss, Inc., NY.

Bevan, M. W., Mason, S. E., Goelet, P., 1985: Expression of tobacco mosaic virus coat protein by a cauliflower mosaic virus promoter in plants transformed by *Agrobacterium*. EMBO J. **4,** 1921—1926.

Caplan, A., Herrera-Estrella, Inzeé, D., Van Haute, E., Van Montague, M., Schell, J., Zambryski, P., 1983: Introduction of genetic material into plant cells. Science **222,** 815—821.

Chilton, M.-D., 1983: A vector for introducing new genes into plants. Scientific American **284,** 51—59.

Coleman, J., Green, P. J., Inouye. M., 1984: The use of RNAs complementary to specific mRNAs to regulate the expression of individual bacterial genes. Cell **37,** 429—436.

Costa, A. S., Müller, G. W., 1980: Tristeza control by cross-protection: A U.S.-Brazil cooperative success. Plant Dis. **64,** 538—541.

Cuozzo, M., O'Connell, K. M., Kaniewski, W., Fang, R.-X., Chua, N.-H., Tumer, N. E., 1988: Viral protection in transgenic plants expressing the cucumber mosaic virus coat protein or its antisense RNA. Bio/Technology, in press.

de Zoeten, G. A., Fulton, R. W., 1975: Understanding generates possibilities. Phytopathology **65,** 221—222.

de Zoeten, G. A., Gaard, G., 1984: The presence of viral antigen in the apoplast of systemically virus-infected plants. Virus Research **1,** 713—725.

Deom, C. M., Oliver, M. J., Beachy, R. N., 1987: The 30-kilodalton gene product of tobacco mosaic virus potentiates virus movement. Science **237,** 389—394.

Dodds, J. A., Lee, S. Q., Tiffany, M., 1985: Cross-protection between strains of cucumber mosaic virus: Effect of host and type of inoculum on accumulation of virions and double-stranded RNA of the challenge strain: Virology **144,** 301—309.

Fraley, R. T., Rogers, S. G., Horsch, R. B., Eichholtz, D. A., Flick, J. S., Fink, C. L., Hoffmann, N. L., Sanders, P. R., 1985: Bio/Technology **3,** 629—635.

Fraley, R. T., Rogers, S. G., Horsch, R. B., 1986: Genetic transformation in higher plants. CRC Critical Reviews in Plant Sciences. **4 (1),** 1—45.

Fulton, R. W., 1978: Superinfection by strains of tobacco streak virus. Virology **85,** 1—8.

Garcia, J. A., Hille, J., Vos, P., Goldbach, R., 1987: Transformation of cowpea *Vigna unguiculata* with a full-length DNA copy of cowpea mosaic virus M-RNA. Plant Science **48,** 89—98.

Gerlach, W. L., Llewellyn, D., Haseloff, J., 1987: Construction of a plant disease resistance gene from the satellite RNA of tobacco ringspot virus. Nature **328,** 802—805.

Goelet, P., Lomonossoff, G. P., Butter, P. J. G., Akam, M. E., Gait, M. J., Karn, J., 1982: Nucleotide sequence of tobacco mosaic virus. Proc. Natl. Acad. Sci. U.S.A. **79,** 5818—5822.

Green, P. J., Pines, Inouye, M., 1986: The role of antisense RNA in gene regulation. Ann. Rev. Biochem. **55,** 569—597.

Guilley, H., Dudley, R. K., Jonard, G., Balazs, E., Richards, K. E., 1982: Transcription of cauliflower mosaic virus DNA: Detection of promoter sequences and characterization of the transcripts. Cell **30,** 763—773.

Hamilton, R. I., 1980: Defenses triggered by previous invaders: Viruses. In: Plant Disease: An Advanced Treatise. Vol. 5, pp. 279—303. Horsfall, J. G., Cowling, E. B. (eds.). Academic Press, NY.

Harrison, B. D., Mayo, M. A., Baulcombe, D. C., 1987: Virus resistance in transgenic plants that express cucumber mosaic virus satellite RNA. Nature **328,** 799—802.

Hemenway, C., Fang, R.-X., Kaniewski, W., Chua, N.-H., Tumer, N. E., 1988: Analysis of the mechanism of protection in transgenic plants expressing potato virus X protein or its antisense RNA. EMBO J., in press.

Hoekema, A., Hirsch, P. R., Hooykaas, P. J. J., Schilperoort, R. A., 1983: A binary plant vector strategy based on separation of *vir* and T-region of the *Agrobacterium tumefaciens* Ti-plasmid. Nature **303,** 179—180.

Horsch, R. B., Fry, J. E., Hoffmann, N. L., Eichholtz, D., Rogers, S. G., Fraley, R. T., 1985: Science **227,** 1229—1231.

Lawton, M. A., Tierney, M. A., Nakamura, I., Anderson, E., Komeda, Y., Dubé, P., Hoffmann, N., Fraley, R. T., Beachy, 1987: Expression of a soybean β-conglycinin gene under the control of the califlower mosaic virus 35S and 19S promoters in transformed petunia tissues. Plant Molec. Biol. **9,** 315—324.

Loesch-Fries, L. S., Halk, E., Merlo, D., Jarvis, N., Nelson, S., Krahn, K., Burhop, L., 1987a: Expression of alfalfa mosaic virus coat protein gene and anti-sense cDNA in transform tobacco tissue. In: Molecular strategies for crop protection pp. 221—234. Arntzen, C. J., Ryan, C. (eds.). A. R. Liss, Inc., NY.

Loesch-Fries, L. S., Merlo, D., Zinnen, T., Burhop, L., Hill, K., Drahn, K., Jarvis, N., Nelson, S., Halk, E., 1987b: Expression of alfalfa mosaic virus RNA 4 in transgenic plants confers resistance. EMBO J. **6,** 1845—1852.

McKinney, H. H., 1929: Mosaic diseases in the Canary Islands, West Africa, and Gibraltar, J. Agric. Res. **39,** 557—578.

Meshi, T., Takamatsu, N., Ohno, T., Okada, Y., 1982: Molecular cloning of the complementary DNA copies of the common and cowpea strains of tobacco mosaic virus. Virology **118,** 64—75.

Miller, A. W., Bujarski, J. J., Dreher, T. H., Hall, T. C., 1986: Minus-strand initiation by brome mosaic virus replicase within the 3′ tRNA-like structure of native and modified RNA templates. J. Mol. Biol. **187**, 537—546.

Murant, A. F., Mayo, M. A., 1982: Satellites of plant viruses. Ann. Rev. Phytopathol. **20**, 49—70.

Nelson, R. S., McCormick, S. M., Delannay, X., Dubé, P., Layton, J., Anderson, E. J., Kaniewska, M., Proksch, R. K., Horsch, R. B., Rogers, S. G., Fraley, R. T., Beachy, R. N., 1988: Virus tolerance, plant growth, and field performance of transgenic tomato plants expressing coat protein from tobacco mosaic virus. Bio/Technology **6**, 403—409.

Niblett, C. L., Dickson, E., Fernow, K. H., Horsch, R. K., Zaitlin, M., 1978: Cross-protection amongst four viroids. Virology **91**, 198—203.

Otsuki, Y., Takebe, I., 1976: Double infection of isolated tobacco leaf protoplasts by two strains of tobacco mosaic. In: Biochemistry and cytology of plant-parasite interaction. pp. 214—222. Tomiyama, K., Daly, J. M., Uranti, F., Oku, H., Ouchi, S. (eds.). Elsevier, Amsterdam.

Palukaitis, P., Zaitlin, M., 1984: A model to explain the "cross-protection" phenomenon shown by plant viruses and viroids. In: Plantmicrobe interactions: Molecular and genetic perspectives. pp. 420—430. Kosuge, T., Nester, E. W. (eds.). Macmillan Publishing Co., NY.

Ponz, F., Bruening, G., 1986: Mechanisms of resistance to plant viruses. Ann. Rev. Phytopathol. **24**, 355—381.

Powell Abel, P., Nelson, R. S., De, B., Hoffmann, N., Rogers, S. G., Rogers, Fraley, R. T., Beachy, R. N., 1986: Delay of disease development in transgenic plants that express the tobacco virus coat protein gene. Science **232**, 738—743.

Rast, A. Th. B., 1972: MII-16, an artificial symptomless mutant of tobacco mosaic virus for seedling inoculation of tomato crops. Neth. J. Plant Pathol. **78**, 110—112.

Register, J. C. III, Beachy, R. N.: Tobacco mosaic virus infection of transgenic plants is blocked at an early event in infection (submitted).

Rogers, S. G., Fraley, R. T., Horsch, R. B., Levine, A. D., Flick, J. S., Brand, L. A., Fink, C. L., Mozer, T., O'Connell, K., Sanders, P. R., 1985a: Plant Molecular Biology Reporter **3**, 111—116.

Rogers, S. G., O'Connell, K., Horsch, R. B., Fraley, R. T., 1985b: Investigation of factors involved in foreign protein expression in transformed plants. In: Biotechnology in plant science: Relevance to agriculture in the eighties. pp. 219—226. Zaitlin, M., Day, P., Hollaender, A. (eds.). Academic Press, NY.

Sanders, R. P., Winter, J. A., Barnason, A. R., Rogers, S. G., Fraley, R. T., 1987: Comparison of cauliflower mosaic virus 35S and nopaline synthase promoters in transgenic plants. Nucleic Acids Research **15**, 1543—1558.

Sherwood, J. L., Fulton, R. W., 1982: The specific involvement of coat protein in tobacco mosaic virus cross protection. Virology **119**, 150—158.

Tumer, N. E., O'Connell, K. M., Nelson, R. S., Sanders, P. R., Beachy, R. N., Fraley, R. T., Shah, D. M., 1987: Expression of alfalfa mosaic virus coat protein gene confers cross-protection in transgenic tobacco and tomato plants. EMBO J. **6**, 1181—1189.

Van Dun, C. M. P., Bol, J. F., Van Vloten-Doting, L., 1987: Expression of alfalfa mosaic virus and tobacco rattle virus coat protein genes in transgenic tobacco plants. Virology **159**, 299—305.

Waterworth, H. E., Kaper, J. M., Tousignant, M. E., 1979: CARNA 5, the small cucumber mosaic virus-dependent replicating RNA, regulates disease expression. Science **204,** 845—847.

Wilson, T. M. A., 1984: Cotranslational disassembly of tobacco mosaic virus *in vitro.* Virology **137,** 255—265.

Wilson, T. M. A., Watkins, P. A. C., 1986: Influence of exogenous viral coat protein on the cotranslational disassembly of tobacco mosaic virus (TMV) particles *in vitro.* Virology **149,** 132—135.

Yeh, S.-D., Gansalves, D., 1984: Evaluation of induced mutants of papaya ringspot virus for control by cross-protection. Phytopathology **74,** 1086—1091.

Zaitlin, M., 1976: Viral cross-protection: More understanding is needed. Phytopathology **66,** 382—383.

Zinnen, T. M., Fulton, R. W., 1986: Cross-protection between sunn-hemp mosaic and tobacco mosaic viruses. J. Gen. Virol. **67,** 1679—1687.

Subject Index

MIX
Papier aus verantwortungsvollen Quellen
Paper from responsible sources
FSC® C105338

If you have any concerns about our products,
you can contact us on
ProductSafety@springernature.com

In case Publisher is established outside the EU,
the EU authorized representative is:
Springer Nature Customer Service Center GmbH
Europaplatz 3, 69115 Heidelberg, Germany

Printed by Libri Plureos GmbH
in Hamburg, Germany